生物学英语与论文写作

于湘晖　吴永革　马俊锋　编

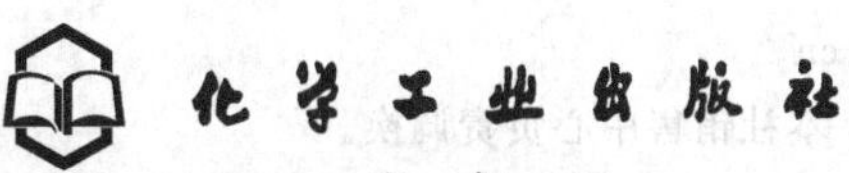

·北京·

图书在版编目（CIP）数据

生物学英语与论文写作/于湘晖，吴永革，马俊锋
编．—2版．—北京：化学工业出版社，2018.8（2023.8重印）
ISBN 978-7-122-32333-0

Ⅰ.①生… Ⅱ.①于… ②吴… ③马… Ⅲ.①生物学-
英语-论文-写作 Ⅳ.①Q1

中国版本图书馆CIP数据核字（2018）第123705号

责任编辑：傅四周 郎红旗　　装帧设计：王晓宇
责任校对：边 涛

出版发行：化学工业出版社（北京市东城区青年湖南街13号 邮政编码100011）
印　装：北京科印技术咨询服务有限公司数码印刷分部
787mm×1092mm 1/16 印张20 字数540千字　2023年8月北京第2版第3次印刷

购书咨询：010-64518888　　售后服务：010-64518899
网　址：http://www.cip.com.cn
凡购买本书，如有缺损质量问题，本社销售中心负责调换。

定　价：59.80元

前言

随着生命科学的迅猛发展和国际学术交流、学生国际化培养等的日益加强，生物学专业英语的教和学越来越重要。许多高校要求专业课程采用英文原版教材，用双语或英语讲授，这都极大促进了专业英语的教学工作。为了学生们能够熟练地阅读英文文献、书籍，将研究成果整理成英文科研论文在期刊杂志上发表；为了学生们能够进入国际高水平的大学研修，编写高水平的、内容前沿的生物学专业英语教材就显得十分重要。

吉林大学从1980年以来开设“生物学专业英语”课程至今，先后编写并出版了两部教材——《生物化学与分子生物学英语》（1993年）和《生物学英语与写作》（2008年）。近几年生物学的发展日新月异，专业英语教材的内容需要不断更新补充，才能适应新的知识体系。因此，本书编者在对近几年的实际教学经验总结基础上，对2008年版教材进行修订和再版。本教材对英文科技论文写作进行了系统讲述，第二版书名相应更名为《生物学英语与论文写作》。

全书共分为两部分内容，第一部分是课文，内容涵盖生物化学、酶工程、分子生物学、动物学、微生物学、遗传学、细胞生物学、基因组学、生物信息学、系统发育生物学和基因编辑等学科领域，力求扩大专业覆盖面和单词量。在素材的选择方面既注意内容的广泛性，又注意技术的新颖性，特别是较前一版增加了在生物学研究领域的新研究方向和技术，使学生既学习了英文语言又能了解专业知识的前沿，例如二代基因测序和CRISPR/Cas9基因编辑等。为了便于学习，课文后附有单词表（注明音标）和难句分析。第二部分是英语学术论文写作，全面介绍英文科技文章的写作知识，这也是本书的一个特色。本书不仅可以作为生命科学各学科本科生的专业英语教材，并且可供研究生专业阅读和写作学习，也可以作为青年教师和生物学专业相关人员的专业英语学习材料。

参加本书编写的于湘晖教授、吴永革教授和马俊锋副教授是吉林大学生命科学学院双语课程《生物化学》教学第一线的老师，都有国外的学习或工作经历，具有丰富的教学经验。李青山教授虽然已经退休，但长期从事生物学专业英语的教学，经验十分丰富，为本书把好质量关。编写英语写作部分的刘永新教授多年来担任英文版专业杂志的总编辑工作，英文文字能力强，文风严谨。

由于时间、水平有限，书中可能存在疏漏和不足，敬请广大读者批评指正，以便进一步完善和提高。

在本书的设计、编写和出版过程中，化学工业出版社编辑给予了认真、细致和耐心的指导，在此表示衷心的感谢。

编者

2018年5月于长春

第一版前言

随着大学英语教学质量的不断提高和四、六级考试不断普及，我国大学本科学生的基础英语水平越来越好，这为进一步提高英语的运用能力奠定了坚实的基础。近年来，各学校在加强基础英语教学的同时，十分重视基础英语之后的专业英语的教学，这对于学生提高英语语言运用能力，适应专业教材和文献的阅读、写作和翻译是十分重要的。为了适应教育和科学技术国际化的需要，许多学校要求专业课采用原版教材，用英语或双语讲授，这就更促进了专业英语的教学工作。

吉林大学从1980年以来，开设“生物学专业英语”课，编写了讲义；又于1993年编写出版了《生物化学与分子生物学英语》教材。这次的编写是在多年教学经验基础上的总结。本书分为两部分，第一部分是课文，内容包括生物化学、分子生物学、细胞生物学、微生物学、免疫学、生理学和生态学等，力求扩大专业覆盖面，扩大单词量。为了便于学习，课文后有单词表（并注明音标）和难句分析。第二部分是科技英语写作，全面介绍英文科技文章的写作知识。这部分也是本书的一个特点。在素材的选择方面既注意内容的广泛性又注意新颖性，特别是选择那些代表目前生物学新发展的材料，使学生既学习了语言又能得到专业知识的前沿内容，例如基因组学、蛋白质组学、基因治疗等。因此，本书不仅可以作为生物学各学科本科生的专业英语教材，而且适合于研究生的专业阅读和写作的需要，也可以作为青年教师和从事生物学相关人员的专业英语学习材料。

参加本书编写的于湘晖教授、吴永革教授是我院专业英语教学第一线的老师，他们都有国外的工作经历，具有丰富的教学经验。编写英语写作的刘永新教授多年来从事英文版专业杂志的总编辑工作，英文文字能力强，文风严谨。他们为本书的问世做了大量的工作，保证了本书的质量。但是，由于时间、水平有限，书中可能存有疏漏和不足，敬请广大读者批评指正，以便进一步完善和提高。

在本书的设计、编写和出版过程中，化学工业出版社编辑给予了认真、细致和耐心的指导，在此表示衷心的感谢。

吉林大学生命科学学院
李青山
2008年1月于长春

Contents

Section Ⅰ Lessons

The Structure and Function of Protein

Proteins are the most abundant macromolecules in living cells and constitute 50 percent or more of their dry weight. They are found in all cells and all parts of cells. Proteins also occur in great variety; hundreds of different kinds may be found in a single cell. Moreover, proteins have many different biological roles since they are the molecular instruments through which genetic information is expressed. It is therefore appropriate to begin the study of biological macromolecules with the proteins, whose name means "first" or "foremost".

The key to the structure of the thousands of different proteins is the group of relatively simple building-block molecules from which proteins are built. All proteins whether from the most ancient lines of bacteria or from the highest forms of life, are constructed from the same basic set of 20 amino acids, covalently linked in characteristic sequences. Because each of these amino acids has a distinctive side chain which lends it chemical individuality, this group of 20 building-block molecules may be regarded as the alphabet of protein structure.

In this paper we shall also examine peptides, short chains of two or more amino acids joined by covalent bonds. What is most remarkable is that cells can join the 20 amino acids in many different combinations and sequences, yielding peptides and proteins having strikingly different properties and activities. From these building blocks different organisms can make such widely diverse products as enzymes, hormones, the lens protein of the eye, feathers, spider webs, tortoise shell, nutritive milk proteins, enkephalins (the body's own opiates), antibiotics, mushroom poisons, and many other substances having specific biological activity.

Amino acids have common structural features

When proteins are boiled with strong acid or base, their amino acid building blocks are released from the covalent linkages that join them into chains. The free amino acids so formed are relatively small molecules, and their structures are all known. The first amino acid to be discovered was asparagines, in 1806. The last of the 20 to be found, threonine, was not identified until 1938. All the amino acids have trivial or common names, sometimes derived from the source from which they were first isolated. Asparagines was first found in asparagus, as one might guess; glutamic acid was found in wheat gluten; and glycine (Greek, glykos, "sweet") was so named because of its sweet taste.

All of the 20 amino acids found in proteins have as common denominators a carboxyl group and an amino group bonded to the same carbon atom. They differ from each other in their side chains, or R groups, which vary in structure, size, electric charge, and solubility in water. The 20 amino acids of proteins are often referred to as the standard, primary, or normal amino acids, to distinguish them from other kinds of amino acids present in living organisms but not in proteins. The standard amino acids have been assigned three-letter abbreviations and one-letter symbols, which are used as shorthand to indicate to composition and sequence of amino acids in polypeptide chains.

General structure of an amino acid. This structure is common to all but one of the α-amino acid (proline, a cyclic amino acid, is the exception) . The R group or side chain attached to the α carbon is different in each amino acid. $H_3\overset{\oplus}{N}—\underset{\alpha}{\overset{R}{\overset{|}{C}}}H—COOH$

Nearly all amino acids have an asymmetric carbon atom

We note that all the standard amino acids except one have an asymmetric carbon atom, the α carbon, to which are bonded four different substituent groups, i. e. , a carboxyl group, an amino group, a R group, and a hydrogen atom. The asymmetric α carbon atom is thus a chiral center. As we have seen, compounds with a chiral center occur in two different isomeric forms, which are identical in all chemical and physical properties except one, the direction in which they can cause the rotation of plane-polarized light in a polarimeter. With the single exception of glycine, which has no asymmetric carbon atom, all of the 20 amino acids obtained from the hydrolysis of proteins under sufficiently mild conditions are optically active. i. e. , they can rotate the plane-polarized light in one direction or the other. Because of the tetrahedral arrangement of the valence bonds around the α carbon atom of amino acids the four different substituent groups can occupy two different arrangements in space, which are nonsuperimposable, mirror images of each other. These two forms are called optical isomers, enantiomers, or stereoisomers. A solution of one stereoisomer of a given amino acid will rotate plane-polarized light to the left (counterclockwise) and is called the levorotatory isomer [designated (−)]; the other stereoisomer will rotate plane-polarized light to the same extent but to the right (clockwise) and is called the dextrorotatory isomer [designated (+)] . An equimolar mixture of the (+) and (−) forms will not rotate plane-polarized light. Because all the amino acids (except glycine) when carefully isolated from proteins do rotate plane-polarized light, they evidently occur in only one of their stereoisomeric forms in protein molecules.

Optical activity of a stereoisomer is expressed quantitatively by its specific rotation, determined from measurements of the degree of rotation of a solution of the pure stereoisomer at a given concentration in a tube of a given length in a polarimeter:

$$[\alpha]_D^{25}=\frac{\text{observed rotation, deg}}{\text{length of tube, dm}\times\text{concentration, g/mL}}$$

the abbreviation dm stands for decimeters (0. 1m) .

The temperature and the wavelength of the light employed (usually the D line of sodium, 598nm) must be specified. For the specific rotation of several amino acids, some are levorotatory and others dextrorotatory.

Periodic structures: the alpha helix, beta pleated sheet, and collagen helix

Can a polypeptide chain fold into a regularly repeating structure? To answer this question, Pauling and Corey evaluated a variety of potential polypeptide conformations by building precise molecular models of them. They adhered closely to the experimentally observed bond angles and distances for amino acids and small peptides. In 1951, they proposed two periodic polypeptide structures, called α helix and β pleated sheet.

The α helix is a rod-like structure. The tightly coiled polypeptide main chain forms the inner part of the rod, and the side chains extend outward in a helical array. The α helix is stabilized by hydrogen bonds between the NH and CO groups of the main chain. The CO group of each amino acid is hydrogen bonded to the NH group of the amino acid that is situated four residues ahead in the linear sequence. Thus, all the main-chain CO and NH groups are hydrogen bonded. Each residue is related to the next one by a translation of 1.5 Å[1] along the helix axis and a rotation of 100°, which gives 3.6 amino acid residues per turn of helix. Thus, amino acids spaced three and four apart in the linear sequence are spatially quite close to one another in an α helix. In contrast, amino acids two apart in the linear sequence are situated on opposite sides of the helix and so are unlikely to make contact. The pitch of the α helix is 5.4 Å, the product of the translation (1.5 Å) and the number of residues per turn (3.6). The screw-sense of α helix can be right-handed (clockwise) or left-handed (counterclockwise); the α helices found in proteins are right-handed.

The α helix content of proteins of known three-dimensional structure is highly variable. In some, such as myoglobin and hemoglobin, the α helix is the major structural motif. Other proteins, such as the digestive enzyme chymotrypsin, are virtually devoid of α helix. The single-stranded α helix discussed above is usually a rather short rod, typically less than 40 Å in length. A variation of the α helical theme is used to construct much longer rods, extending to 1000 Å or more. Two or more α helices can entwine around each other to from a cable. Such α helical coiled coils are found in several proteins: keratin in hair, myosin and tropomyosin in muscle, epidermin in skin, and fibrin in blood clots. The helical cables in these proteins serve a mechanical role in forming stiff bundles of fibers.

The structure of the α helix was deduced by Pauling and Corey six years before it was actually to be seen in the X-ray reconstruction of the structure of myoglobin. The elucidation of the structure of the α helix is a landmark in molecular biology because it demonstrated that the conformation of a polypeptide chain can be predicted if the properties of its components are rigorously and precisely known.

In the same year, Pauling and Corey discovered another periodic structural motif, which they named the β pleated sheet (β because it was the second structure they elucidated, the α helix having been the first). The β pleated sheet differs markedly from the α helix in that it is a sheet rather than a rod. The polypeptide chain in the β pleated sheet is almost fully extended rather than being tightly coiled as in the α helix. The axial distance between adjacent amino acids is 3.5 Å in contrast with 1.5 Å for the α helix. Another difference is that the β pleated sheet is stabilized by hydrogen bonds between NH and CO groups in different polypeptide strands, whereas in the α helix the hydrogen bonds are between NH and CO groups in the same polypeptide chain. Adjacent strands in a β pleated sheet can run in the same direction (parallel β sheet) or in opposite directions (antiparallel β sheet). For example, silk fibroin consists almost entirely of stacks of antiparallel β sheets. Such β sheet regions are a recurring structural motif in many proteins. Structural units comprising from two to five parallel or antiparallel β strands are especially common.

---

[1] 1Å=0.1nm。

The collagen helix, a third periodic structure, will be discussed in detail. This specialized structure is responsible for the high tensile strength of collagen, the major component of skin, bone, and tendon.

Polypeptide chains can reverse direction by making β turn

Most proteins have compact, globular shapes due to frequent reversals of the direction of their polypeptide chains. Analyses of the three-dimensional structures of numerous proteins have revealed that many of these chain reversals are accomplished by a common structural element called the β turn. The essence of this hairpin turn is that the CO group of residue n of a polypeptide is hydrogen bonded to the NH group of residue $(n+3)$. Thus, a polypeptide chain can abruptly reverse its direction.

Levels of structure in protein architecture

In discussing the architecture of proteins, it is convenient to refer to four levels of structure (Fig. 1. 1). Primary structure is simply the sequence of amino acids and location of disulfide bridges, if there are any. The primary structure is thus a complete description of the covalent connections of a protein. Secondary structure refers to the steric relationship of amino acid residues that are close to one another in the linear sequence. Some of these steric relationships are of a regular kind, giving rise to a periodic structure. The α helix, the β pleated sheet, and the collagen helix are examples of secondary structure. Tertiary structure refers to the steric relationship of amino acid residues that are far apart in the linear sequence. It should be noted that the dividing line between secondary and tertiary structure is arbitrary. Proteins that contain more than one polypeptide chain display an additional level of structural organization, namely quaternary structure, which refers to the way in which the

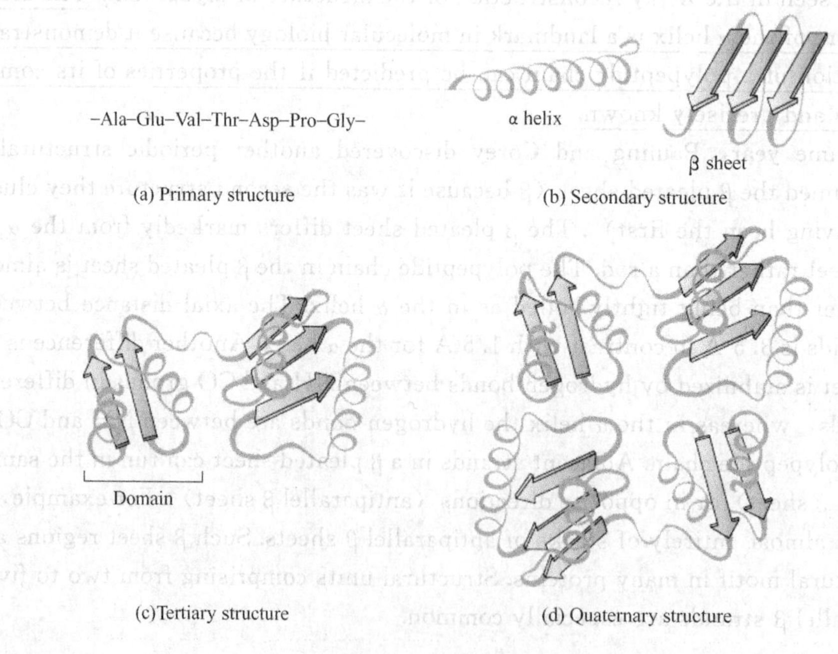

Fig. 1. 1 Protein structure
From primary to quaternary structure

chains are packed together. Each polypeptide chain in such a protein is called a subunit. Another useful term is domain, which refers to a compact, globular unit of protein structure. Many proteins fold into domains having masses that range from 10 to 20 kD. The domains of large proteins are usually connected by relatively flexible regions of polypeptide chain.

Amino acid sequence specifies three-dimensional structure

Insight into the relation between the amino acid sequence of a protein and its conformation came from the work of Christian Anfinsen on ribonuclease, an enzyme that hydrolyzes RNA. Ribonuclease is a single polypeptide chain consisting of 124 amino acid residues. It contains four disulfide bonds, which can be irreversibly oxidized by performic acid to give cysteic acid residues. Alternatively, these disulfide bonds can be cleaved reversibly by reducing them with a reagent such as β-mercaptoethanol, which forms mixed disulfides with cysteine side chains. In the presence of a large excess of β-mercaptoethanol, the mixed disulfides also are reduced, so that the final product is a protein in which the disulfides (cystines) are fully converted into sulfhydryls (cysteines). However, it was found that ribonuclease at 37℃ and pH7 cannot be readily reduced by β-mercaptoethanol unless the protein is partially unfolded by denaturing agents such as urea or guanidine hydrochloride. Although the mechanism of action of these denaturing agents is not fully understood, it is evident that they disrupt noncovalent interactions. Polypeptide chains devoid of cross-links usually assume a random-coil conformation in 8 mol/L urea or 6 mol/L guanidine HCl, as evidenced by physical properties such as viscosity and optical rotary spectra. When ribonuclease was treated with β-mercaptoethanol in 8 mol/L urea, the product was a fully reduced, randomly coiled polypeptide chain devoid of enzymatic activity. In other words, ribonuclease was denatured by this treatment.

Anfinsen then made the critical observation that the denatured ribonuclease, freed of urea and β-mercaptoethanol by dialysis, slowly regained enzymatic activity. He immediately perceived the significance of this chance finding: the sulfhydryls of the denatured enzyme became oxidized by air and the enzyme spontaneously refolded into a catalytically active form. Detailed studies then showed that nearly all of the original enzymatic activity was regained if the sulfhydryls were oxidized under suitable conditions. All of the measured physical and chemical properties of the refolded enzyme were virtually identical with those of the native enzyme. These experiments showed that the information needed to specify the complex three-dimensional structure of ribonuclease is contained in its amino acid sequence. Subsequent studies of other proteins have established the generality of this principle, which is a central one in molecular biology: sequence specifies conformation.

A quite different result was obtained when reduced ribonuclease was reoxidized while it was still in 8 mol/L urea, this preparation was then dialyzed to remove the urea. Ribonuclease reoxidized in this way had only 1% of the enzymatic activity of the native protein. Why was the outcome of the experiment different from the one in which reduced ribonuclease was reoxidized in a solution free of urea? The reason is that wrong disulfide pairings were formed when the random-coil form of the reduced molecule was reoxidized. There are 105 different ways of pairing eight cysteines to form four disulfides; only one of these combinations is enzymatically active. The 104 wrong pairings have been picturesquely termed "scrambled" ribonuclease spontaneously converted into fully active, native ribonuclease when trace amounts of β-mercaptoethanol were added to the aqueous solution of the reoxidized protein. The added β-mercaptoethanol catalyzed the rearrangement of disulfide pairings until the native structure was regained, which took about the hours. This process was driven entirely by the decrease in free energy as the "scrambled" conformations were converted into the stable, native conformation of the enzyme. Thus, the native form of ribonuclesae appears to be the thermodynamically most stable structure.

Functions of proteins

Proteins are undoubtedly the most functionally diverse of biomolecules. In general, globular proteins function by recognizing other molecules to which they specifically bind. Such precise binding is possible because the protein molecule has a site that is complementary to a site on the molecule recognized. However, this binding is not fixed and rigid but, rather, exists in a dynamic equilibrium with recognition, binding and release occurring continuously. At any moment the proportion of bound molecules depends upon (a) the relative concentrations of the protein and the molecule to which it binds, and (b) the strength of association between them. The latter depends on how well the complementary sites fit together and the types of interactions involved, for example, hydrophobic, ionic, hydrogen bonding, which occurs between the sites. Once binding does occur, there is a conformational change in the protein-bound molecule, or in the complex of protein and protein-bound molecule. This change is the signal that initiates the biochemical activity associated with the protein and forms the basis for the remarkable range of biological roles exhibited by proteins.

The formation of proteins is under the direct control of DNA. The growth and differentiation of cells, organs and organisms result from the orderly expression of information contained in the DNA molecules. However a chicken and egg situation exists, since the formation of proteins, and indeed the replication of DNA, requires the activity of pre-existing proteins.

Much of biochemistry is concerned with the remarkable protein catalysts called enzymes. Many reactions that normally proceed at barely measurable rates are typically accelerated by a factor of 10^8-10^{11} by the presence of the appropriate enzyme. In comparison with chemical catalysts, enzymes are also amazingly specific; a given enzyme catalyses only a single transformation or group of similar reactions. Their catalytic power, their specificity and the fact that their activity can be regulated, mean that enzymes ensure that metabolism proceeds in an orderly fashion.

Specific transport proteins are a feature of living systems. The well-known blood protein, haemoglobin, transports O_2 in the blood of vertebrates. Examples of transport proteins in the serum are albumin, which can transport fatty acids; lipoproteins, which carry cholesterol and other lipids; and transferring, which transports iron. Invertebrates have copper-containing proteins called haemocyanins, which have O_2-carrying roles parallel to those of the vertebrate haemoglobins.

Other transport proteins have a different function. They are situated in biological membranes and allow materials to be transported across the membrane. For example, the Na^+, K^+-ATPase is a protein that pumps Na^+, out of cells and K^+ into cells, at the expense of metabolic energy.

Proteins play a key role in the co-ordination of metabolism. For example, neurons respond to specific signals via protein receptors on their surfaces. Indeed, many of these "signals" are chemical ones, consisting of peptides or small proteins. Co-ordination in multicellular types is often mediated by hormonal signals; and in animals the hormone receptors and, indeed, some hormones themselves are proteins.

The movement of organisms is achieved by a dynamic function of protein molecules. Some bacteria are motile using extended appendages called flagella. Eukaryotic cells use cilia and flagella in locomotion, but multicellular animals move using skeletal muscle. All of these locomotory activities depend upon the co-ordinated movements of sets of fibrous proteins.

Protein molecules are responsible for the mechanisms by which organisms protect themselves against parasites and toxins. Scavenging white blood corpuscles, called leukocytes, recognize invading microorganisms by means of protein receptor molecules on their surfaces, and then engulf them. Antibodies are serum proteins that can combine with antigens such as bacterial toxins, leading to their neutralization. Other proteins such as fibrinogen, circulate in the blood and are able to from fibrous mats to seal wounds.

Mechanical support is given by several types of fibrous proteins both inside and outside cells. Tubulin forms extended microtubules within the cytoplasm, which help determine the shape of the cell. Other proteins are found extracellularly and help organize the matrix that surrounds the cell. Collagen is a widely distributed extracellular protein, which imparts a high tensile strength to tissues such as cartilage, bone and the skin.

Haemoglobinopathies

Survival of vertebrates is not possible without haemoglobin. However, many humans survive with partially defective haemoglobins. One such condition is sickle cell anaemia where because of mutation, and amino acid on the surface of the protein molecule is altered producing a haemoglobin that precipitates in the decoy state and therefore does not transport O_2 effectively. This condition leads to deformation of the red cells ("sickling") which become trapped in the capillaries and haemolysis occurs, resulting in anaemia.

Sickle cell disease is fairly common, especially amongst the North American black population, but it is rather unusual as a "haemoglobinopathy" or haemoglobin disease. The amino acid change results in there being a "sticky patch" on the β-polypeptide chain of deoxyhaemoglobin, leading to the aggregation and precipitation described above. The mutation arose by chance at some time in the past. Much more likely events to occur (and many hundreds of haemoglobin mutations are now known) are ones in which the haem pocket is modified so that haem does not bind or function properly, or ones in which the α or the β chains are not constructed properly.

The various parts of the haemoglobin, like all quaternary proteins, fit and stay together because they are complementary in shape, charge, hydrophobicity, etc. In particular, the Fe^{2+}-containing haem group is a highly hydrophobic molecule and requires to be placed in a hydrophobic pocket in the molecule, where it is held and carries out its function of binding oxygen reversibly. Mutations that result in the amino acid residues lining the haem pocket being replaced by ones that are less hydrophobic or more bulky may result in a failure to bind haem or failure to bind oxygen properly (i. e. not at all or irreversibly). Many such mutations are known and characterized. In the majority of cases only one type of subunit is affected. Thus, although the remaining unmutated subunits can potentially bind haem and oxygen normally, they often do not do so. Having only two oxygen-binding centers in the molecule, instead of four, does not allow for the usual subunit interactions which influence the binding and release of O_2, instead of behaving in the required way generating a sigmoidal binding curve, the oxygen-binding curve may be much more like that of myoglobin. Consequently, oxygen is not transported successfully.

Many patients with haemoglobinopathies are heterozygotic for the haemoglobinopathy in question; they have both a defective and a normal gene, so that effectively 50% of the haemoglobin they synthesize is normal. There may be a high rate of destruction of the abnormal haemoglobin which further lessens the problem. Also several of the genes for the polypeptide chains of haemoglobin are present in multiple copies. Consequently, only one of the genes may have mutated, while the others, even in homozygotes, still produce normal polypeptides.

In some individuals the results of the mutation may be slight and not noticed until sensitive blood screening is carried out. In others, it may be sufficiently severe as to cause debilitating anemia and other conditions. Many individuals probably do not survive because they are homozygous for the condition. However, this depends partly on how severe the defect is.

Haemoglobin variants are commonly detected by electrophoresis of a solution of the protein. When amino acids are changed as a result of a mutation, there may be a modification of the charge on the molecule, which may then display a higher or lower mobility than that of normal haemoglobin. Such screening may be done cheaply. Obviously, to determine which amino acid is altered requires a more extensive study, including peptide mapping and partial sequencing. Haemoglobin variants are usually named from the town/hospital where the case was first detected (e. g. Hb "Memphis"), although this gives the uninitiated little useful information. Hb Memphis is actually a rather unusual variant in which there are mutations in both the α and the β chains. It might more helpfully be described as:

$$\alpha_2^{23\text{Glu-Gln}}\beta_2^{6\text{Glu-Val}}$$

Obviously, mutations do not necessarily have to be single amino acid substitutions, and do not have only to affect the haem pocket. Many mutations on the surface of the molecule are known, which have almost no effect on the properties of the molecule (sickle cell haemoglobin is the exception to this rule).

As well as single amino acid changes, there may be double changes, changes in both α and β chains, deletions resulting in a failure to make chains, mutations that change a stop codon so that a much larger than normal polypeptide is produced, and so on. It is probably true to say that almost all variations have been encountered. The present-day distribution of defective haemoglobins has arisen from the accumulation of harmless mutations, early death of individuals with harmful mutations, and survival of some individuals because although they have a harmful mutation, this confers a selective survival advantage such as increased resistance to malaria, as is the case with sickle cell disease.

As a result of a great deal of experimental work (protein sequencing and, later DNA sequencing), an enormous amount is known about the haemoglobinopathies called thalassaemias. Almost all the possibilities that potentially could occur, do so. These include: deletion of one or more α chain genes per haploid genome; deletion of the β chain genes (unbalanced synthesis of chains may result in the production of homotetrameric molecules such as $Hb\alpha_4$ in β-thalassaemia, which are unstable and precipitate or oxidize very rapidly); chain-termination mutation (e. g. Hb "Seal Rock"); absent, reduced or inactive mRNA; gene fusion; and increased globin chain degradation.

Haemoglobin variants may now be detected in the fetus by molecular biology techniques and parents may be counselled about abortion. Although sickle cell disease may confer resistance to malaria, and consequently a selective survival advantage, there is usually little that can be done in any of the haemoglobinopathies in terms of medical treatment, other than to cope with crises and pain. Because there is anaemia, blood transfusions may be used, but in the longer term, repeated blood transfusion is not helpful.

Glossary

abortion [ə'bɔːʃən] n. 流产，堕胎，失败，夭折，中止，早产
albumin [æl'bjumin] n. 清蛋白，白蛋白
amino acid n. 氨基酸，胺
amino group 氨基
appendage [ə'pendidʒ] n. 附属物，附肢，附属丝
aqueous ['eikwiəs] n. 水的，眼房水的
asparagine [əs'pærədʒiːn] n. 天（门）冬素，天冬酰胺酸
asymmetric [æsi'metrik] adj. 不均匀的，不对称的
axis ['æksis] n. 轴
carboxyl group 羧基
chiral ['tʃirəl] adj. （化）手（征）性的
cholesterol [kə'lestərəul,- rɔl] n. 胆固醇
chymotrypsin [ˌkaimə'tripsin] n. （生化❶）胰凝乳蛋白酶，糜蛋白酶
cilia ['siliə] n. 睫，纤毛
codon ['kəudən] n. （遗❷）密码子
coil [kɔil] v. 盘绕，卷
collagen ['kɔləˌdʒən] n. 胶原质，胶原
corpuscle ['kɔːpʌs(ə)l] n. 血细胞
covalent [kəu'veilənt] adj. （化❸）共有原子价的，共价的
cysteic acid 半胱氨酸
cystine ['sistiːn,- tin] n. （生化）胱氨酸，双硫丙氨酸
decimeter ['desiˌmiːtə(r)] n. 分米
denominator [di'nɔmineitə] n. （数）分母，命名者
dextrorotatory isomer 右旋异构体
dialysis [dai'ælisis] n. （化）透析，分离
disulfide bridge 二硫键
domain [dəu'mein] n. 结构域
dynamic [dai'næmik] adj. 动力的，动力学的，动态的
enantiomer [i'næntiəumə] n. （化）对映（结构）体
enkephalin [enkefæliːn] n. 脑啡肽
enzyme ['enzaim] n. （生化）酶

❶"生化"表示"生物化学"，后同。
❷"遗"表示"遗传学"，后同。
❸"化"表示"化学"，后同。

epidermin [epidermin] n. 表皮素（一种构成表皮主要成分的纤维蛋白）
fiber ['faibə] n. 纤维
fibrin ['faibrin] n. （生化）（血）纤维蛋白，（血）纤维
flagella [flə'dʒelə] n. 鞭节，鞭毛
glutamic acid [glu:'tæmik æsid] n. 谷氨酸
gluten ['glu:tən] n. 谷蛋白，黏菌膜，黏胶质
guanidine hydrochloride n. 盐酸胍
haemocyanin [ˌhi:məu'saiənin] n. 血清蛋白，血细胞，血蓝蛋白
haemoglobinopathy ['hi:məuˌgləubi'nɔpəθi] n. （医❶）血红蛋白病
haemoglobin [ˌhi:məu'gləubin] n. 血色素，血红蛋白
hairpin ['hɛəpin] n. 发夹
hydrogen bond 氢键
hydrogen ['haidrəudʒən] n. 氢
hydrophobicity n. 疏水性
hydrophobic [ˌhaidrəu'fəubik] adj. 疏水的，狂犬病的，恐水病的，患恐水病的
ionic [ai'ɔnik] adj. 离子的
irreversible [ˌiri'və:səbl，- sib -] adj. 不能撤回的，不能取消的
isomeric form 同分异构
keratin ['kerətin] n. （生化）角蛋白
levorotatory isomer 左旋异构体
macromolecule [ˌmækrəu'mɔlikju:l] n. 大分子，高分子
malaria [mə'lɛəriə] n. 疟疾，瘴气
matrix ['meitriks] n. 矩阵；基质，衬质，间质
mercaptoethanol [məˌkæptəu'eθənɔl] n. （化）巯基乙醇
metabolism [me'tæbəlizəm] n. 新陈代谢，代谢作用
opiate ['əupiit] n. 鸦片剂；adj. 安眠的；v. 缓和
optical isomer n. 旋光异构体，旋光异构物
organism ['ɔ:gənizəm] n. 生物体，有机体
peptide ['peptaid] n. 多肽
performic acid 过氧甲酸
pitch [pitʃ] n. 螺距
plane-polarized light 平面偏振光
polarimeter [ˌpəulə'rimitə] n. 偏光计
primary structure （免疫❷）一级结构
protein ['prəuti:n] n．（生化）蛋白质；adj. 蛋白质的
quaternary structure 四级结构
reversal [ri'və:səl] n. 颠倒，反转，反向，逆转
ribonuclease [ˌraibəu'nju:klieis] n. （生化）核糖核酸酶
secondary structure 二级结构
sickle cell anaemia n. 镰刀形红细胞贫血症
side chain 侧链，（聚合物中）支链
sigmoidal n. S形曲线；adj. S形的
silk fibroin 蚕丝蛋白
sodium ['səudjəm，-diəm] n. （化）钠
solubility [ˌsɔlju'biliti] n. 溶度，溶性，溶解性

---

❶“医”表示“医学”，后同。
❷“免疫”表示免疫学，后同。

stereoisomer [ˌstiəriəu'aisəmə] n.（化）立体异构体
steric ['stiərik,'sterik] n.（化）空间的，立体的，位的
substituent [sʌb'stitjuənt] n. 取代；adj. 取代的
substitution [ˌsʌbsti'tju:ʃən] n. 代替，取代作用，代入法，置换
sulfhydryl [sʌlf 'haidril] n. 巯基，硫氢基
tendon ['tendən] n.（解）腱
tensile ['tensail] adj. 可拉长的，可伸长的
tertiary structure 三级结构
tetrahedral ['tetrə'hedrəl] adj. 有四面的，四面体的
thalassaemias adj. 地中海贫血的
trivial ['triviəl] adj. 琐细的，价值不高的，微不足道的
tropomyosin [ˌtrɔpəu'maiəsin] n.（生化）原肌球蛋白
valence bond 价键
viscosity [vis'kɔsiti] n. 黏质，黏性
α helix α螺旋
β pleated sheet β折叠片
β turn β转角

难句分析

1. The key to the structure of the thousands of different proteins is the group of relatively simple building-block molecules from which proteins are built.
   译文：构成成千上万种不同蛋白结构的关键是一组组成蛋白相对简单的单位分子。
2. All of the 20 amino acids found in proteins have as common denominators a carboxyl group and an amino group bonded to the same carbon atom.
   译文：在蛋白质中所发现的 20 种氨基酸存在的共性，是一个羧基和一个氨基连接在同一个碳原子上。
3. With the single exception of glycine，which has no asymmetric carbon atom，all of the 20 amino acids obtained from the hydrolysis of proteins under sufficiently mild conditions are optically active.
   译文：除去甘氨酸没有不对称碳原子外，温和条件下水解蛋白所发现的 20 种氨基酸都具有旋光性。
4. To answer this question，Pauling and Corey evaluated a variety of potential polypeptide conformations by building precise molecular models of them. They adhered closely to the experimentally observed bond angles and distances for amino acids and small peptides.
   译文：为了解答这个问题，Pauling 和 Corey 通过构建精确分子模型的方法对一系列可能的多肽构象进行了计算和分析。他们的结果非常接近于实验观察到的氨基酸和小肽的键角、键长的数据。
5. The elucidation of the structure of the α helix is a landmark in molecular biology because it demonstrated that the conformation of a polypeptide chain can be predicted if the properties of its components are rigorously and precisely known.
   译文：α螺旋结构的阐明是分子生物学上的一块里程碑，因为它证明了如果清楚地知道了一条多肽组成成分的性质，这条多肽的构象是可以被预测出来的。
6. Analyses of the three-dimensional structures of numerous proteins have revealed that many of these chain reversals are accomplished by a common structural element called the β turn.

译文：对大量蛋白的三维结构的分析揭示出这种肽链反向结构中有许多是由一种叫做β折叠片的共同结构元素所构成的。

7. Mutations that result in the amino acid residues lining the haem pocket being replaced by ones that are less hydrophobic or more bulky may result in a failure to bind haem or failure to bind oxygen properly (i. e. not at all or irreversibly).

   Mutations是主语，由that引导定语从句修饰主语，定语从句中lining引导现在分词短语作residues的定语。being replaced是现在分词作residues定语。that are less…bulky是定语从句修饰ones，may result in是谓语。全句可译为：组成血红素口袋的氨基酸残基被疏水性弱或体积大的残基所取代引起的变异，可能导致不能结合血红素或不能正常地与氧结合（即完全不结合或不可逆地结合）。

（选自：李青山，安玉华，刘永新，陶小娟编．生物化学和分子生物学英语．长春：吉林大学出版社，1994.）

The Structure and Function of Enzymes

Chemical reactions in biological systems rarely occur in the absence of a catalyst. These catalysts are specific proteins called enzymes. The striking characteristics of all enzymes are their catalytic power and specificity. Furthermore, the activity of many enzymes is regulated. In addition, some enzymes are intimately involved in the transformation of different forms of energy. Let us examine these highly distinctive and biologically crucial properties of enzymes.

Enzymes have enormous catalytic power

Enzymes accelerate reactions by factors of at least a million. Indeed, most reactions in biological systems do not occur at perceptible rates in the absence of enzymes. Even a reaction as simple as the hydration of carbon dioxide is catalyzed by an enzyme.

$$CO_2 + H_2O \rightleftharpoons H_2CO_3$$

Otherwise, the transfer of CO_2 from the tissues into the blood and then to the alveolar air would be incomplete. Carbonic anhydrase, the enzyme that catalyzes this reaction, is one of the fastest known. Each enzyme molecule can hydrate 10^5 molecules of CO_2 in one second. This catalyzed reaction is 10^7 times faster than the uncatalyzed reaction.

Enzymes are highly specific

Enzymes are highly specific both in the reaction catalyzed and in their choice of reactants, which are called substrates. An enzyme usually catalyzes a single chemical reaction or a set of closely related reactions. The degree of specificity for substrate is usually high and sometimes virtually absolute.

Let us consider proteolytic enzymes as an example. The reaction catalyzed by these enzymes is the hydrolysis of a peptide bond.

$$\sim N(H)-C(H)(R^1)-C(=O)-N(H)-C(H)(R^2)-C(=O)\sim + H_2O \rightleftharpoons$$

peptide

$$\sim N(H)-C(H)(R^1)-C(=O)O^- \quad + \quad {}^-H_3N-C(H)(R^2)-C(=O)\sim$$

carboxyl component   amino component

Most proteolytic enzymes also catalyze a different but related reaction, namely the hydrolysis of an ester bond.

$$R^1-C(=O)-O-R^2 + H_2O \rightleftharpoons R^1-C(=O)O^- \quad + \quad HO-R^2 + H^+$$

ester   acid   alcohol

Proteolytic enzymes vary markedly in their degree of substrate specificity. Subtilisin, which comes from certain bacteria, is quite undiscriminating about the nature of the side chains adjacent to the peptide bond to be cleaved. Trypsin is quite specific in which it splits peptide bonds on the carboxyl side of lysine and argintine residues only. Thrombin, an enzyme participating in blood clotting, is even more specific than trypsin. The side chain on the carboxyl side of the susceptible peptide bond must be arginine, whereas the one on the amino side must be glycine.

Another example of the high degree of specificity of enzymes is provided by DNA polymerase Ⅰ. This enzyme synthesizes DNA by linking together four kinds of nucleotide building blocks. The sequence of nucleotides in the DNA strand that is being synthesized is determined by the sequence of nucleotides in another DNA strand that serves as a template. DNA polymerase Ⅰ is remarkably precise in carrying out the instructions given by the template. The wrong nucleotide is inserted into a new DNA strand less than once in a million times.

The activities of some enzymes are regulated

Some enzymes are synthesized in an inactive precursor form and are activated at a physiologically appropriate time and place. The digestive enzymes exemplify this kind of control. For example, trypsinogen is synthesized in the pancreas and is activated by peptide-bond cleavage in the small intestine to form the active enzyme trypsin. This type of control is also repeatedly used in the sequence of enzymatic reactions leading to the clotting of blood. The enzymatically inactive precursors of proteolytic enzymes are called zymogens.

Another mechanism that controls activity is the covalent insertion of a small group on an enzyme. This control mechanism is called covalent modification. For example, the activities of the enzymes that synthesize and degrade glycogen are regulated by the attachment of a phosphoryl group to a specific serine residue on these enzymes. This modification can be reversed by hydrolysis. Specific enzymes catalyze the insertion and removal of phosphoryl and other modifying groups.

A different kind of regulatory mechanism affects many reaction sequences resulting in the synthesis of small molecules such as amino acids. The enzyme that catalyzes the first step in such a biosynthetic pathway is inhibited by the ultimate product. The biosynthesis of isoleucine in bacteria illustrates this type of control, which is called feedback inhibition. Threonine is converted into isoleucine in five steps, the first of which is catalyzed by threonine deaminase. This enzyme is inhibited when the concentration of isoleucine reaches a sufficiently high level. Isoleucine binds to a regulatory site on the enzyme, which is distinct from its catalytic site. The inhibition of threonine deaminase is mediated by an allosteric interaction, which is reversible. When the level of isoleucine drops sufficiently, threonine deaminase becomes active again, and consequently isoleucine is again synthesized.

The specificity of some enzymes is under physiological control. The synthesis of lactose by the mammary gland is a particularly striking example. Lactose synthetase, the enzyme that catalyzes the synthesis of lactose, consists of a catalytic subunit and a modifier subunit. The catalytic subunit by itself cannot synthesize lactose. It has a different role, which is to catalyze the attachment of galactose to a protein that contains a covalently linked carbohydrate chain. The modifier subunit alters the specificity of the catalytic subunit so that it links galactose to glucose to form lactose. The level of the modifier subunit is under hormonal control. During pregnancy, the catalytic subunit is formed in the mammary gland, but little modifier subunit is formed. At the time of birth, hormonal levels change drastically, and the modifier subunit is synthesized in large amounts. The modifier subunit then binds to the catalytic subunit to form an active lactose synthetase complex that produces large amounts of lactose. This system clearly shows that hormones can exert their physiological effects by altering the specificity of enzymes.

Enzymes transform different kinds of energy

In many biochemical reactions, the energy of the reactants is converted into a different form with high efficiency. For example, in photosynthesis, light energy is converted into chemical-bond energy. In mitochondria, the free energy contained in small molecules derived from food is converted into a different currency, that of adenosine triphosphate (ATP). The chemical-bond energy of ATP is utilized in many different ways. In muscular contraction, the energy of ATP is converted into mechanical energy. Cells and organelles have pumps that utilize ATP to transport molecules and ions against chemical and electrical gradients. These transformations of energy are carried out by enzyme molecules that are integral parts of highly organized assemblies.

Enzymes do not alter reaction equilibria

An enzyme is a catalyst and consequently it cannot alter the equilibrium of a chemical reaction. This means that an enzyme accelerates the forward and reverse reaction by precisely the same factor. Consider the interconversion of A and B. Suppose that in the absence of enzyme the forward rate (K_F) is 10^{-4}/s and the reverse rate (K_R) is 10^{-6}/s. The equilibrium constant K is given by the ratio of these rates:

$$A \underset{10^{-6}/s}{\overset{10^{-4}/s}{\rightleftharpoons}} B$$

$$K=\frac{[B]}{[A]}=\frac{K_F}{K_R}=\frac{10^{-4}}{10^{-6}}=100$$

The equilibrium concentration of B is 100 times that of A, whether or not enzyme is present. However, it would take several hours to approach this equilibrium without enzyme, whereas equilibrium would be attained within a second when enzyme is present. Thus, enzymes accelerate the attainment of equilibria but do not shift their positions.

Enzymes decrease the activation energies of reactions catalyzed by them

A chemical reaction, A→B, goes through a transition state that has a higher energy than either A or B. The rate of the forward reaction depends on the temperature and on the difference in free energy between that of A and the transition state, which is called the Gibbs free energy of an activation and symbolized by ΔG.

$$\Delta G^{\neq}=G_{\text{transition state}}-G_{\text{substrate}}$$

The reaction rate is proportional to the fraction of molecules that have a free energy equal to or greater than $\Delta G^{\neq}$. The proportion of molecules that have an energy equal to or greater than $\Delta G^{\neq}$ increases with temperature.

Enzymes accelerate reactions by decreasing $\Delta G^{\neq}$, the activation barrier. The combination of substrate and enzyme creates a new reaction pathway whose transition-state energy is lower than it would be if the reaction were taking place in the absence of enzyme.

Formation of an enzyme-substrate complex is the first step in enzymatic catalysis

The making and breaking of chemical bonds by an enzyme are preceded by the formation of an enzyme-substrate (ES) complex (Fig. 1. 2). The substrate is bound to a specific region of the enzyme called the active site. Most enzymes are highly selective in their binding of substrates. Indeed, the catalytic specificity of enzymes depends in large part on the specificity of the binding process. Furthermore, the control of enzymatic activity may also take place at this stage.

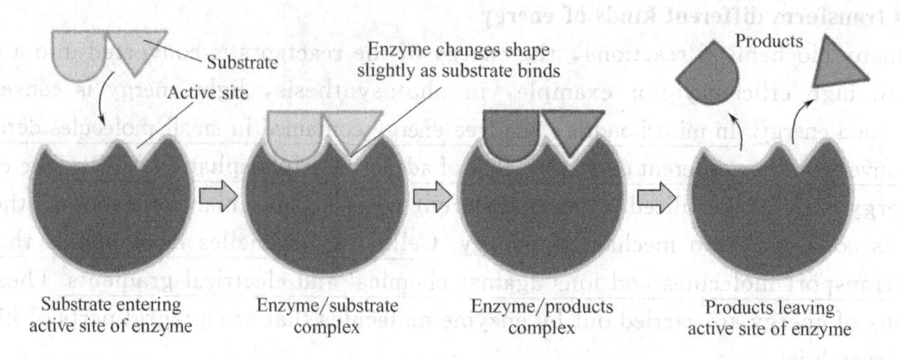

Fig. 1. 2 Induced-fit model of enzyme-substrate binding
In this model, the enzyme changes shape on substrate binding. The active site forms a shape complementary to the substrate only after the substrate has been bound

The existence of ES complexes has been shown in a variety of ways.

ES complexes have been directly visualized by electron microscopy and X-ray crystallography. Complexes of nucleic acids and their polymerase are evident in electron micrographs. Detailed information concerning the location and interactions of glycyl-L-tyrosine, a substrate of carboxypeptidase A, has been obtained from X-ray studies of that ES complex.

The physical properties of an enzyme, such as its solubility or heat stability, frequently change upon formation of an ES complex.

The spectroscopic characteristics of many enzymes and substrates change upon formation of an ES complex just as the absorption spectrum of deoxyhemoglobin changes markedly when it binds oxygen or when it is oxidized to the ferric state, as described previously. These changes are particularly striking if the enzyme contains a colored prosthetic group. Tryptophan synthetase, a bacterial enzyme that contains a pyridoxal phosphate prosthetic group, affords a nice illustration. This enzyme catalyzes the synthesis of L-tryptophan from L-serine and indole. The addition of L-serine to the enzyme produces a marked increase in the fluorescence of the pyridoxal phosphate group. The subsequent addition of indole, the second substrate, quenches this fluorescence to a level lower than that of the enzyme alone. Thus, fluorescence spectroscopy reveals the existence of an enzyme-serine complex and of an enzyme-serine-indole complex. Other spectroscopic techniques, such as nuclear and electron magnetic resonance, also are highly informative about ES interactions.

A high degree of stereospecificity is displayed in the formation of ES complexes. For example, D-serine is not a substrate of tryptophan synthetase. Indeed, the D-isomer does not even bind to the enzyme. This implies that the substrate-binding site has a very well defined shape.

ES complexes can sometimes be isolated in pure form. For an enzyme that catalyzes the reaction A+B=C, it is sometimes possible to isolate an ES complex. This can be done if the enzyme has a sufficiently high affinity for A and if B is absent from the mixture.

At a constant concentration of enzyme, the reaction rate increases with increasing substrate concentration until a maximal velocity is reached. In contrast, uncatalyzed reactions do not show this saturation effect. In 1913, Leonor Michaelis interpreted the maximal velocity of an enzyme-catalyzed reaction in terms of the formation of a discrete ES complex. At a sufficiently high substrate concentration, the catalytic sites are filled and so the reaction rate reaches a maximum. This is the oldest and most general evidence for the existence of ES complexes.

Some features of active sites

The active site of an enzyme is the region that binds the substrates (and the prosthetic group, if any) and contributes the residues that directly participate in the making and breaking of bonds. These residues are called the catalytic groups. Although enzymes differ widely in structure, specificity, and mode of catalysis, a number of generalizations concerning their active sites can be stated.

The active site takes up a relatively small part of the total volume of an enzyme. Most of the amino acid residues in an enzyme are not in contact with the substrate. This raises the intriguing question of why enzymes are so big. Nearly all enzymes are made up of more than 100 amino acid residues, which give them a mass greater than 10 kD and a diameter of more than 25Å.

The active site is a three-dimensional entity. The active site of an enzyme is not a point, a line, or even a plane. It is an intricate three-dimensional form made up of groups that come from different parts of the linear amino acid sequence-indeed, residues far apart in the linear sequence may interact more strongly than adjacent residues in the amino acid sequence, as has already been seen for myoglobin and hemoglobin. In lysozyme, and enzyme that will be discussed in more detail, the important groups in the active site are contributed by residues numbered 35, 52, 62, 63, and 101 in the linear sequence of 129 amino acids.

Substrates are bound to enzymes by relatively weak forces. ES complexes usually have equilibrium constants that range from 10^{-2} to 10^{-8} mol/L, corresponding to free energies of interaction ranging from -3 to -12kcal/mol. These values should be compared with the strengths of covalent bonds, which are between -50 and -110 kcal/mol.

Active sites are clefts or crevices. In all enzymes of known structure, substrate molecules are bound to a cleft or crevice from which water is usually excluded unless it is a reactant. The cleft also contains several polar residues that are essential for binding and catalysis. The nonpolar character of the cleft enhances the binding of substrate. In addition, the cleft creates a microenvironment in which certain polar residues acquire special properties essential for their catalytic role.

The specificity of binding depends on the precisely defined arrangement of atoms in an active site. A substrate must have a matching shape to fit into the site. Emil Fischer's metaphor of the lock and key, stated in 1894, has proved to be an essentially correct and highly fruitful way of looking at the stereospecificity of catalysis. However, recent work suggests that the active sites of some enzymes are not rigid. In such an enzyme, the shape of the active site is modified by the binding of substrate. The active site has a shape complementary to that of the substrate only after the substrate is bound. This process of dynamic recognition is called induced fit. Furthermore, some enzymes preferentially bind a strained from the substrate corresponding to the transition state.

Glossary

active site 活性位点，活性部位，活化部位
adenosine triphosphate (ATP) n. 腺苷三磷酸
allosteric [ˌæləˈsterik] adj. 变构效应的，异构的，别位的，活性中心外的
alveolar [ælˈviələ, ælviˈəulə] adj. 肺泡的，牙槽的，小泡的
atom [ˈætəm] n. 原子
carbohydrate chain 碳水化合物，糖类
carbon dioxide 碳（酸）酐，二氧化碳
carbonic anhydrase 碳酸酐酶
carboxyl [kɑːˈbɔksil] n. 羧基
carboxypeptidase [kɑːˌbɔksiˈpeptideis] n. 羧肽酶
catalyst [ˈkætəlist] n. 触媒剂，催化剂
catalytic group 催化基团
cleft [kleft] n. 裂口；adj. 裂，半裂的
crevice [ˈkrevis] n. 缝，缝隙
currency [ˈkʌrənsi] n. 流通
deaminase [diːˈæmineis] n. 脱氨基酶
deoxyhemoglobin [diˌɔksiˌhiːməˈgləubin] n. 去氧血红蛋白
diameter [daiˈæmitə] n. 直径，径
dynamic [daiˈnæmik] adj. 动态的，动力的
enzyme [ˈenzaim] n. 酶
equilibria n. 平衡（equilibrium 的复数）
equilibrium [ˌiːkwiˈlibriəm] n. 平衡，饱和，长期平衡
ester bond 酯键
feedback inhibition 反馈抑制
ferric state 高铁状态
forward reaction 正向反应
galactose [gəˈlæktəus] n. 半乳糖
Gibbs free energy 吉布斯自由能
glycine [ˈglaisiːn] n. 甘氨酸
glycogen [ˈglaikəudʒen] n. 糖原，肝淀粉，牲粉，动物淀粉
glycyl-L-tyrosine n. 甘氨酰（基）-L-酪氨酸
gradient [ˈgreidiənt] n. 梯度，陡度，斜度；adj. 倾斜的
hydration [haiˈdreiʃən] n. 水合（作用）
hydrolysis [haiˈdrɔlisis] n. 水解作用
indole [ˈindəul] n. 吲哚，苯并吡咯
induced fit 诱导契合
interconversion [ˌintə(ː)kənˈvəːʃən] n. 互变
intestine [inˈtestin] n. 肠
isoleucine [ˌaisəuˈluːsiːn, -sin] n. 异亮氨酸，异白氨酸（氨基酸类药）
lactose [ˈlæktəus] n. 乳糖
lysine [ˈlaisiːn] n. 赖氨酸，氨基已氨酸
lysozyme [ˈlaisəzaim] n. 溶菌酶，细胞壁溶解酶
mammary [ˈmæməri] adj. 乳房的，乳腺的
metaphor [ˈmetəfə] n. 比喻

mitochondria [ˌmaitəˈkɔndriə] n.（单 mitochondrion）线粒体
myoglobin [ˌmaiəˈgləubin] n. 肌红蛋白
nonpolar [ˈnɔnˈpəulə] n. 非极性的
nuclear and electron magnetic resonance 核磁共振
organelle [ˌɔːgəˈnel] n. 细胞器，类器官
pancreas [ˈpænkriəs, ˈpæŋ-] n. 胰腺
peptide bond n. 肽键
phosphoryl [ˌfɔsˈfɔːril] n. 磷酰基
photosynthesis [ˌfəutəuˈsinθəsis] n. 光合作用
prosthetic [prɔsˈθetik] adj. 假体的，修复的
pyridoxal [ˌpiriˈdɔksæl] n. 维生素 B_6，抗皮肤炎维生素，吡哆醛
quench [kwentʃ] n. 淬火，淬炼，熄灭，聚冷
reactant [riːˈæktənt] n. 试剂，反应物
spectroscopy [spekˈtrɔskəpi] n. 光谱学，波谱学，光谱法
spectrum [ˈspektrəm] n.（复 spectra）光谱，光系，波谱
stereospecificity [ˌstiəriəuˌspesiˈfisiti] n. 立体定向性，立体特异性，立体专一性
substrate [ˈsʌbstreit] n. 酶作用物，底物，基质，基层
subtilisin [ˌsʌbˈtilisin] n. 枯草杆菌溶素，枯草杆菌蛋白酶
subunit [sʌbˈjuːnit] n. 亚单位，亚单元，子群，亚基
threonine [ˈθriːəniːn] n. 苏氨酸，2-氨基-3-羟基丁酸
thrombin [ˈθrɔmbin] n. 凝血酶
transition state 过渡态，中间状态，迁移状态
trypsinogen [tripˈsinədʒən] n. 胰蛋白酶原
trypsin [ˈtripsin] n. 胰蛋白酶，结晶胰蛋白酶（蛋白水解酶）
tryptophan [ˈtriptəfæn] n. 色氨酸
velocity [viˈlɔsiti] n. 速度，速率
X-ray crystallography X射线衍射晶体分析法
zymogen [ˈzaimədʒen] n. 酶原

难句分析

1. Lactose synthetase，the enzyme that catalyzes the synthesis of lactose，consists of a catalytic subunit and a modifier subunit. The catalytic subunit by itself cannot synthesize lactose. It has a different role，which is to catalyze the attachment of galactose to a protein that contains a covalently linked carbohydrate chain.
   It 代表上句的 the catalytic subunit。which 引导的是非限定性定语从句。that 引导的定语从句修饰 a protein。全句可译为：乳糖合成酶催化乳糖的合成，它由一个催化亚单位以及一个调节亚基组成。催化亚单位不能独立合成乳糖。它具有一种不同的功能，可以催化半乳糖附着于一种蛋白，该蛋白包含一个共价结合的糖链。
2. In mitochondria，the free energy contained in small molecules derived from food is converted into a different currency，that of adenosine triphosphate（ATP）.
   that of adenosine triphosphate（ATP）是同位语，that 是代词，代替 currency。全句可译为：在线粒体中来源于食物的小分子中的自由能被转化为一种不同的流通物——腺苷三磷酸（ATP）。
3. Cells and organelles have pumps that utilize ATP to transport molecules and ions against chemical and electrical gradients.
   against（介词）在此作“逆着”“顶着”“对着”解释。全句可译为：细胞以及细胞器具

有能够利用腺苷三磷酸逆化学梯度以及电荷梯度运输分子和离子的泵的性质。

4. The spectroscopic characteristics of many enzymes and substrates change upon formation of an ES complex just as the absorption spectrum of deoxyhemoglobin changes markedly when it binds oxygen or when it is oxidized to the ferric state, as described previously.

   upon formation of an ES complex 中的介词 on 或 upon 后面跟随具有动作意义的名词时，可译作“在……以后，立即……”，“一……就……”。例如，On being heated, these two gases from a new compound.（这两种气体一经加热就形成了一种新的化合物。）just as 引导方式状语从句，译为“就像……那样……”，“恰似……那样……”。全句可译为：如前所述，许多酶和底物的分光镜检表征在形成一个 ES 复合物时会发生变化，就像还原血红蛋白结合氧或被氧化为三价状态时吸收光谱的变化一样。

5. It is an intricate three-dimensional form made up of groups that come from different parts of the linear amino acid sequence-indeed, residues far apart in the linear sequence may interact more strongly than adjacent residues in the amino acid sequence, as has already been seen for myoglobin and hemoglobin.

   as has already been seen for myoglobin and hemoglobin 是非限定性定语从句，修饰前面的整个句子，关系代词 as 在从句中作主语。全句可译为：它是一个由来自于线性氨基酸序列的不同部分的基团组成的复杂的三维立体结构，实际上线性序列中远隔着的残基之间的相互作用可以比氨基酸序列中邻近的残基的相互作用更强，正如肌球蛋白和血红蛋白中已经展示的那样。

（选自：李青山，安玉华，刘永新，陶小娟编．生物化学和分子生物学英语．长春：吉林大学出版社，1994.）

Polysaccharides and Glycoconjugates

Distinguishing features of polysaccharides

Do polysaccharides possess properties necessary for cell viability that are absent in other molecules? This is not an easy question to answer. Cells invest a lot of energy and metabolites in producing macromolecules. As much as 80%-90% of the dry weight of a cell is made up of proteins, nucleic acids, and polysaccharides. Each addition of an amino acid, nucleotide or sugar requires the hydrolysis of at least one, and usually more, phosphoanhydride bonds during the synthesis of the relevant polymer. Most cells (microbe, plant or animal) include polysaccharides as part of their composition. This text mainly concerns the structures and functions of these polysaccharides.

It might be asked why cells should produce polysaccharide even to the extent that some cells perform gluconeogenesis.

If polysaccharides fulfill unique functions then they would be expected to possess structures that are different from other cellular polymers such as the proteins. An obvious difference is that proteins contain nitrogen whereas polysaccharides frequently lack it. That is probably why plants, which often grow in nitrogen-limited environments, produce polysaccharides to support and protect their cells rather than a nitrogen-rich fibrous protein. Apart from chemical composition (Table 1. 1), there are other distinguishing features, all to do with shape and size. Firstly, a particular protein always has a specific number and sequence of amino acid residues. This number might alter for a variety of reasons, such as the modification of preproteins, but changes are always precise. Individual polysaccharides, in contrast, rarely have a specific M_r, but rather show a range or distribution of M_r. Nevertheless, they can be thought of as large or small, depending on the average M_r.

Table 1. 1 Periodic polysaccharides

name	repeating sequence	conformation
amylose	—Glc-α-1,4-Glc-α-1,4-Glc-α-1,4-Glc-	helix
cellulose	—Glc-β-1,4-Glc-β-1,4-Glc-β-1,4-Glc-	ribbon
chitin	—GlNAc-β-1,4-GlNAc-β-1,4-GlNAc-β-1,4-Glc-NAc	ribbon
xylan	—Xyl-β-1,4-Xyl-β-1,4-Xyl-β-1,4-Xyl-	extended helix
pectate	—GalUA-α-1,4-GaluA-α-1,4-GalUA-α-1,4-GalUA-	puckered ribbon
hyaluronate	—GlcNA-β-1,4-GlcUA-β-1,3-GlNAc-β-1,4-GlcUA-	helix
lipopolysaccharides of *Salmonella* sp.	Abc-α-1,3	Abc-α-1,3 | —Msan-α-1,4-Rha-α-1,3-Gal-α-1,3-Man-	helix

Secondly, the difference is in the variety of components. Each type of polysaccharide comprises very few types of sugars (Table 1. 1). Frequently there is but one. Thus, a polysaccharide may consist of only glucose residues (a glucan) or mannose residues (a mannan), etc. Many polysaccharides have repeating disaccharide units. Occasionally there are larger blocks, e. g. the tetra-, penta and hexasaccharides of lipopolysaccharides found on the surface of Gram-negative bacteria. Examples of these periodic or repeating sequences are shown in Table1. 1. The small number of sugar types in any one polysaccharide means that their structures are more uniform in shape than those of proteins. A sequence of repeating blocks of mono-, di-, trisaccharides, etc. produces polysaccharides of simple structures, such as ribbons or helices. This means that there is a limited array of groups on the polysaccharide surface capable of interaction with other molecules. Contrast this with the much greater range of chemical or physical interactions displayed by amino acid side-chains on the surface of a protein. This is probable why polysaccharides cannot act as catalysts. They simply do not have the subtle array of chemical groups oriented in particular, specific ways, as found at the active site of an enzyme. However, their very simplicity, especially in extended linear conformations, allows them to interact with other polymers to form large molecular networks with important and desirable properties.

Thirdly, although there is a limited range of sugar types, the glycosidic bond between adjacent residues is not confined to just one group in the sugar ring. In contrast to proteins, where identical peptide bonds join amino acid residues, a number of glycosidic bond arrangements are found between sugar residues. A carbon atom in the sugar ring, at position 1, often may be linked through the oxygen of the glycosidic bond to any of the other carbon atoms in the adjoining sugar residue.

The availability of different linkages means that the preferred conformation (or shape) of the polysaccharide is dependent on the glycosidic bond as well as the nature of the α-1, 4-produce a polymer with a shape and solubility quite different from one with identical residues, linked 1, 4, but with a β-configuration.

There is also the possibility of producing branched structures. Small but extensively branched chains, say 10-20 sugar residues, create unique molecular shapes that can be recognized specifically by proteins. In this manner, attached to proteins and lipids, carbohydrates aid in cellular recognition processes. In other examples, short branches or side chains are added to regularly shaped polysaccharides, thus producing local modifications to the contour. Such alterations in the shape of polysaccharides change the way in which they interact with other molecules. For example, some polysaccharides, because of their complementary surfaces, align with each other strengthen the association through hydrogen, hydrophobic and ionic bonds, and form polysaccharide aggregates. Local changes to the shape of the polysaccharide hinder the association. These changes produce polysaccharide hinder the association. These changes produce polysaccharide networks that trap clusters of water molecules to form gels, and bridging polysaccharides that act as cross-links among polysaccharide microfibrils.

The rest of this text will describe examples of polysaccharides and concentrate on their structures and functions. Like all macromolecules, polysaccharides have been tailored by evolutionary pressures to interact with other molecules, large and small. The questions are what do they do, how do they do it?

Functions

Sugar-bases structures have a variety of biological functions. These functions can usually be described as structural, energy storage and recognition.

Structural polysaccharides are molecules that protect and support biological structures. An important function of polysaccharides, especially in cells of microbes and plants, is to provide support. To accomplish this, a molecular net is thrown around the fragile plasma membrane containing the cytoplasm. A consequence of this is that the extracellular skeleton then largely dictates the shape and size of the cell. Typical constituents of this array are insoluble fibrous polysaccharides cross-linked to one another to form a net. Polysaccharides that participate cross-linking in this way are the other major class of skeletal cell-wall polymer.

Plant cell walls contain a number of these polysaccharides, which are known as hemicellulose and which contain a variety of sugars. Cross-linking of the fibrous polysaccharides β-1, 3-glucan, chitin and cellulose that support yeast and fungal cell walls is less well understood but is probably through hydrogen and hydrophobic bonds with branched β-1, 3/1, 6-D-glucans or formation of covalent bonds with proteins. The structure of the molecular net surrounding the cell is not permanent but is continually undergoing change to accommodate growth or morphological change.

The formation of extracellular support around both Gram-positive and Gram-negative bacterial cytoplasmic membranes is achieved in a different fashion to that in plant and yeast cell walls but with the same result. Fibrous polysaccharides are structurally modified to accommodate peptides that covalently cross-link them. The bacterial cell is thus surrounded by an enormous glycopeptide bag-shaped molecule that contains unique amino acid and sugar residues linked in a unique fashion. This giant biopolymer is called a peptidoglycan.

Polysaccharide or peptidoglycan nets, however, are not the universal solution to extracellular support. A number of mammalian cell types are surrounded, albeit with a much looser structure, by cross-linked fibrous proteins such as collagen and elastin. The requirements for support of these cells are the same, the response is similar, but the materials used are different.

Another important function of polysaccharides found in the cell walls of yeast, fungal, plant and some mammalian cells is to act as in-fill between the supportive skeletal fibers. This is matrix or amorphous material. Unlike the insoluble skeletal polymers, these polysaccharides provide protection rather than support. They produce a compact molecular barrier protecting the fragile plasma membrane from a potentially harmful environment, yet are sufficiently porous to allow small nutrient molecules access for binding to transport proteins lodged in the membrane. The matrix is a layer of partially immobilized water, a gel, surrounding the cell. Mannoproteins and branched β-1, 3/1, 6-D-glucans in yeast and fungal cell walls and pectins in plant cell walls are good examples of these types of carbohydrates.

Surrounding some mammalian cells, particularly those comprising particular cartilage tissue between bone surfaces, is a gel layer which has the specialized and curious property of altering viscosity in response to forces applied to it. Cartilage itself has the property of absorbing shock. Energy is absorbed through deformation of the cartilage which then returns to its original shape. These properties of lubrication and resilience are provided by a combination of proteoglycans (specialized forms of glycoproteins), glycosaminoglycans (polysaccharides), and proteins (such as collagen). The ability of proteoglycan solutions to become less viscous when shaken, stirred or forced through a nozzle is found in other solutions of biopolymers, solutions of compounds possessing this pseudoplastic or thixotropic behavior have many industrial uses.

The cell walls of bacteria do not appear to contain gel-forming polymers, though there are other major components besides peptidoglycan, e. g. the teichoic acids of Gram-positive cells, which contribute to the extracellular architecture. However, some bacteria and yeasts are surrounded by highly hydrated capsules of polysaccharide gel that are bound, often loosely, to the outer surface of the wall.

Storage polysaccharides are molecules rich in easily released chemical potential energy. As well forming components of the extracellular framework, polysaccharide structures are found within the cell, in the cytoplasm, and organelles. With few exceptions most cells are capable of producing and storing large ($M_r>10^6$) branched α-D-glucans. Bacteria, fungi and mammalian cells produce glycogen. Plant cells contain starch, a mixture of amylopectin and amylase. These polymers are stores of potential metabolic energy and are therefore referred to as storage or reserve polysaccharides. In most case this means that the glucosyl residues comprising the polymer may be converted to glucose-6-phosphate, which can enter the glycolytic pathway. Some ATP is produced by substrate-level phosphorylation and considerably more is produced through oxidative phosphorylation under aerobic conditions. Conversely, in times of plenty, whether carbohydrate is taken up, or manufactured within the cell (by photosynthesis or by gluconeogenesis), glucose-6-phosphate can be converted to glycogen or starch and the carbohydrate stored.

Glycogen and starch (Fig. 1. 3), though by far the most common, are not the only storage polymers. Some plants have unusual reserve polysaccharides. Jerusalem artichokes and dahlia tubers contain inulin, a fructan, while some grasses contain levans. All of these polysaccharides are smaller than glycogen or starch, with M_r of 3500-8000 (about 20-50 sugar residues per molecule). The considerable advantages of storing sugar residues as polysaccharides are described later, but they are not the only stores of cellular energy. Fats, which are more efficient energy stores than polysaccharides, and some proteins perform the same function.

Complex oligosaccharides are carbohydrates, participating in recognition between molecules. Later in the text it will be seen how polysaccharides form extended fibres for cellular support, water-trapping gels, and highly branched polymers for storage of metabolic energy. A common feature of all of these is that, for the most part, structures are simple and repetitive. There are, however, smaller carbohydrate polymers termed oligosaccharides, Which are covalently attached to lipid and protein and where the structure is anything but simple. There may be many such carbohydrate groupings attached to a single protein molecule. The structure of these carbohydrate or oligosaccharide units is often branched and lacks repeating sequences. What is the function of these sidechains?

Earlier in the text it was seen that one of the structural features distinguishing polysaccharides from proteins is the great variation possible in the configuration of the glycosidic bond between adjacent sugar residues. The many possibilities for linking allow a variety of shapes and surfaces to be formed from just a few sugar residues. Using a limited range of monosaccharides, such as mannose, galactose, *N*-acetylglucosamine, fucose and sialic acid, which are some of the commonest, a range of unique molecular shapes may be generated. These complex oligosaccharides act as tailored keys that assist a glycoprotein circulating in some vascular system, or a glycoprotein or glycolipid located on a cell surface, specifically to adhere to another cell surface. The oligosaccharide side chain is the key, the cell-surface receptor protein is

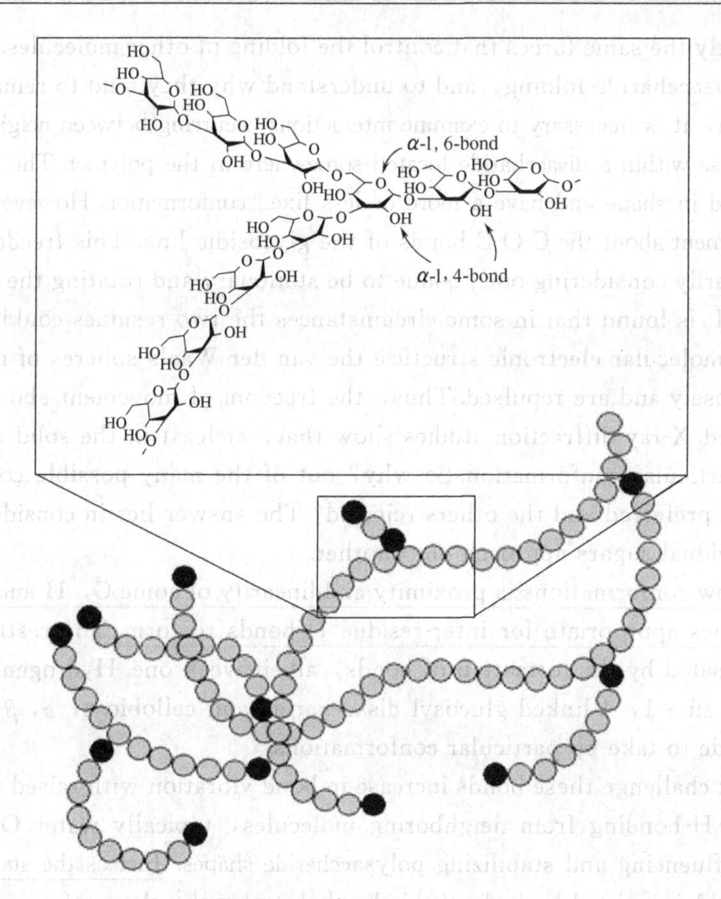

Fig. 1. 3 Glycogen structure

the lock. The combination of the two entities is very similar to the formation of an enzyme-substrate complex, but without the subsequent transformation of the substrate. In this way some circulating hormones are targeted to specific cells and cell-cell adhesion is made selective. Another role is in the control of selective cleavage of proteins. The same glycoprotein can be cut at different points by different proteases to produce a range of peptides with different activities and targets. Proteins that bind to specific sugar residues within specific glycosidic linkages are termed lectins.

Structures

The size of polysaccharides varies enormously. In contrasting the structure of proteins and polysaccharides, attention has already been drawn to the enigma of polysaccharide size: its variability compared with the preciseness of proteins. Apart from the seemingly carefully controlled structure of oligosaccharides of glycoproteins and glycoplipids, the sizes of polysaccharides show a distribution of M_r within the limits: 10^3-10^4 for inulin and 10^7-10^9 for glycogen, for example. With some exceptions, the functions of polysaccharide structure lie in the uniform and repeating shapes and surfaces rather than size itself. In contrast, the result of protein folding is to produce a unique but irregular surface, a feature that requires a precise sequence and fixed number of amino acids.

Knowing the shapes of polysaccharides is of great importance if the way they carry out their functions is to be understood. Whilst little is known about the termination of polysaccharide extension, more is known about the influences controlling their folding. These are, of course, exactly the same forces that control the folding of other molecules. In constructing a scheme of polysaccharide folding, and to understand why they tend to remain in a particular conformation, it is necessary to examine interactions occurring between neighboring residues, for instance, those within a disaccharide located somewhere in the polymer. The individual sugars are relatively rigid in shape and have a more or less fixed conformation. However, there is some freedom of movement about the C-O-C bonds of the glycosidic link. This freedom can be illustrated by arbitrarily considering one residue to be stationary and rotating the other about one of these bonds. It is found that in some circumstances the two residues could potentially collide. In term of molecular electronic structure the van der Waals spheres of individual atoms approach too closely and are repulsed. Thus, the freedom of movement about the glycosidic bond is restricted. X-ray diffraction studies show that, at least in the solid state, disaccharides adopt a particular conformation. So why, out of the many possible conformations remaining, is one preferred and the others rejected? The answer lies in considering what happens when individual sugars approach one another.

In only a few conformations is proximity and linearity of some C, H and O atoms in the individual residues appropriate for inter-residue H-bonds to form. The restriction in movement is compensated by the formation of bonds, albeit weak one. Hydrogen bonds can constrain maltose, an α-1, 4-linked glucosyl disaccharide and cellobiose, α, β-1,4-linked glucosyl dissacharide to take up particular conformations.

Forces that challenge these bonds increase in bond vibration with raised temperature and competition for H-bonding from neighboring molecules, typically water. Other bonds also contribute to influencing and stabilizing polysaccharide shapes. Just as the stacking of bases in double-helical DNA is assisted by hydrophobic bonds between the planes of purine and pyrimidine rings so hydrophobic bonds assist in the stacking of sugar chains on top of one another in, for example, cellulose and chitin microfibrils. The mutual repulsion of negative ion charges prevents anionic polysaccharides from coiling up in, for example, the glycosaminoglycans and pectins.

If maltose or cellobiose is now extended by glucose residues linked in the same fashion (α for maltose and β for cellobiose), then two different patterns begin to appear. The α-1,4-linked residues adopt a helical conformation and the β-1,4-linked take up an extended ribbon or linear conformation. Nevertheless, these conformations are maintained by rather weak bonds and they require stabilization for permanence. In the case of the β-1,4-linked ribbons, this occurs when they align to form bundles or microfibrils.

It is not unexpected to find that regular repeating sequences of residues and linkages form regular shapes. This presents a problem if polysaccharides are to produce net-like structures for the containment of cells. There must be molecular devices for cross-linking the fibrous molecules. For example, the bacterial cell wall peptidoglycan comprises a series of glycan ribbons that are covalently attached to one another through short peptides to form a bag-shaped molecule.

In the walls of plant cells the cross-linking is through hemicelluloses that bind via non-covalent bonds to the cellulose microfibrils. Polysaccharides must be adhesive in some parts but not in others. As might be expected, polymers with alternating properties have alternating sequences, producing alternating conformations. The regularity of periodic structure is occasionally altered by replacement of one sugar residue by another, covalent attachment of sugars producing side chains, or covalent modification that changes the conformation of a residue. Alterations tend to occur in groups and so produce domains. By these means, variations in folding patterns are introduced into the polymer at selected points. The consequence is a change in the ability of polymers to associate with one another. Helical sequences that were interweaving can no longer associate. Alternatively, lengths of helical polymers that become modified now extend into ribbon conformations and can associate with other linear polymers.

These structural changes, which need only be minor in occurrence, give rise to the term interrupted to describe this important group of polysaccharides.

Other, minor, interruptions in regular sequences can have a profound influence on structure. Thus, the occasional branching of polysaccharide chains allows linear molecules to become spherical. The cross-linking of polymers through mutual dissociation and reassociation induced by structural interruptions can trap and immobilize water molecules forming gels.

From this general introduction to polysaccharide function and structure it can be seen that, like proteins, they adopt helical, linear and extended coil conformations. Specific polysaccharides will now be described in more detail to illustrate how the structure of the molecule accounts for the role it plays in the protection and survival of the cell.

Glycolipids, glycoproteins and cell-cell recognition

The term glycoprotein describes a wide range of biopolymers. In general, glycoproteins are proteins to which sugar residues have been covalently attached after assembly of the peptide chain. The size, number of attachment points and variety of sugar residues varies enormously. However, only a limited number of types of amino acid residues are utilized: Ser, Thr and Asn are common sites of carbohydrate attachment. The structure of mucins consists of about 200 disaccharides attached to a single protein of some 1200 amino acid residues through serine residues. Viscous solutions of mucins lubricate and protect epithelial surfaces of the digestive and urinogenital tracts. Mucins from other sources carry additional larger oligosaccharide side-chains. The principal components of this complex are proteoglycans, the structure of which contains up to 150 polysaccharide chains attached to serine residues of a single core protein. Some biopolymers that were thought to be pure polysaccharides are now believed to have protein covalently attached to them. Glycogen is such an example.

Mucins, proteoglycans and glycogen are biopolymers, that contain carbohydrate as the principal structural component. The term glycoprotein, however, is more often used to describe proteins, such as glycophorin referred to in this section, which are mainly protein but contain between 5% and 10% by weight of carbohydrate. There is a range of structures. The saccharide side chains vary in number between 1 and about 20 and are linked to the protein at Asn, Ser and Thr residues. The saccharide moiety contains anything up to about 20 sugar residues and may be branched in structure. The range of oligosaccharyl moieties bound to protein can also be found bound to lipid, giving rise to glycolipids.

A more comprehensive term, describing proteins and lipids that contain covalently bound carbohydrate, is glycoconjugate.

There are over 100 different types of glycolipid structure, reflecting variations in the fatty acyl chains of the hydrophobic ceramide portion and the different combinations and types of sugar residue in the hydrophilic carbohydrate portion. The latter falls into five groups depending on the structure of the outer portion. They are ganglion, globo, lacto, type Ⅰ and type Ⅱ, and a remainder of disparate structure. A galactosyl-glucosyl disaccharide bridges these outer structures to the ceramide.

The term blood group antigen or blood group substance is misleading for it suggests that these molecules, glycoproteins or glycoplipids, are only associated with cells found in the bloodstream, such as the glycoprotein glycophorin which is on the surface of erythrocytes. This was, indeed, where they were first discovered but they are abundant elsewhere in organisms. They occur in relatively high concentrations on the surface of many cells. It is now clear that glycoplipids and glycoproteins play major roles in many cell-cell recognition processes.

Specific interactions can occur between the surface proteins of one cell and the carbohydrate moieties of glycoplipids or glycoproteins of the other. The cell density presumably influences the frequency of contact and leads to the promotion or inhibition of cell growth. In other words, glycoplipids and glycoproteins participate in transcytoplasmic membrane-signalling mechanisms.

Glycolipids occur in high concentrations on the surface of epithelial cells that constitute the mucous lining of many organs. Over 80% of human cancers are epithelial in origin, and it is perhaps the malfunction of the cell density sensing mediated by glycoplipids that is one of the causes contributing to the excessive cell proliferation that is characteristic of cancerous tissues.

During the early stages of embryo development changes occur in cell surface glycoplipids, This process enables the dividing and differentiating cells to be sorted into the correct three-dimensional cellular patterns, so the embryo grows and develops appropriately.

An example of binding between cells of different species is seen in N_2 fixation. Specific strains of the N_2-fixing bacteria *Rhizobium* adhere to the root-hair cells of selected plants because a lectin (that is, a glycoprotein that binds to specific sugars) on the plant cell surface binds to carbohydrate on the bacterial cell surface. The binding is specific, *R. trifolii* bind to clover and *R. phaseolus* to beans.

Glossary

N-acetylglucosamine *N*-乙酰氨基葡萄糖
aerobic [ˌeiə'rəubik] adj. 依靠氧气的，与需氧菌有关的
aggregate ['ægrigeit] n. 合计，集合体；adj. 集合的，聚合的；v. 聚集，集合
amorphous [ə'mɔːfəs] adj. 无定形的，无组织的
amylase ['æməˌleis] n. 淀粉酶
amylopectin [ˌæmiləu'pektin] n. 支链淀粉，胶淀粉
amylose ['æmələus] n. 直链淀粉，多糖
capsule ['kæpsjuːl] n. （植物）荚膜，胶囊，瓶帽，太空舱
cartilage ['kaːtilidʒ] n. （解）软骨
cellobiose [ˌselə'baiəus] n. 纤维二糖
cellulose ['seljuləus] n. 纤维素
ceramide ['serəmaid] n. 神经酰胺
chitin ['kaitin] n. 壳质，壳多糖，角素，几丁质
contour ['kɔntuə] n. 轮廓，周线，等高线
disaccharide [dai'sækəraid] n. 双糖，二糖
elastin [i'læstin] n. 弹性蛋白
erythrocyte [i'riθrəusait] n. 红细胞

fructan [ˈfræktən] n. 果聚糖
fucose [ˈfjuːkəus] n. 海藻糖
galactose [gəˈlæktəus] n. （生化）半乳糖
ganglion [ˈgæŋgliən] n.（复 ganglia）神经节，神经节，腱鞘囊肿
globo lacto [ˈlæktəu] 表示“乳酸，乳糖”之义
glucan [ˈgluːkən] n. 葡聚糖
gluconeogenesis [ˌgluːkəˌniːəuˈdʒenisis] n.（生化）糖原异生
glucosyl [ˈgluːkəsil] n. 葡萄糖基
glycoconjugate [ˌglaikəsaiˈæmidiːn] n. 糖结合物，配糖体
glycolipid [ˌglaikəˈlipid] n. 糖脂类
glycolytic pathway 糖分解途径
glycophorin [ˌglaikəˈfɔrin] n. 血型糖蛋白
glycosaminoglycan [ˈglaikəusəˈmiːnəuˈglaikæn] n. 黏多糖
glycosidic bond 糖苷键
Gram-negative bacteria 革兰阴性菌
hemicellulose [ˌhemiˈseljuləus] n. 半纤维素
hyaluronate [haiəˈlurəˌneit] n. 透明质酸盐（酯）
inulin [ˈinjulin] n. 菊粉
levan [ˈleˌvæn] n. 果聚糖
lipopolysaccharide [ˈlipəupɔliˈsækəraid] n. 脂多糖
lubrication [ˌluːbriˈkeiʃən] n. 润滑油
maltose [ˈmɔːltəus] n. 麦芽糖
mannan [ˈmænæn] n. 甘露聚糖
mannopyranose [ˌmænəpairəˌnəus] n. 吡喃甘露糖
mannose [ˈmænəus] n. 甘露糖
metabolite [miˈtæbəlait] n. 代谢物
microfibril [maikrəuˈfaibril] n. 微纤维
moiety [ˈmɔiəti] n. 二分之一，一部分，半族
morphological [ˌmɔːfəˈlɔdʒikəl] adj. 形态学（上）的，生态学的
mucin [ˈmjuːsin] n. 黏液素
nitrogen [ˈnaitrədʒən] n.（化）氮
oligosaccharide [ˌɔligəuˈsækəraid] n. 低聚糖，寡糖
pectate [ˈpekteit] n. 果胶酸盐（酯）
pectin [ˈpektin] n. 胶质
peptidoglycan [peptidəuˈglaikæn] n. 肽聚糖
polysaccharide [pɔliˈsækəraid] n.（生化）多糖，聚糖，多聚糖
porous [ˈpɔːrəs] adj. 多孔渗水的
preprotein [priːˈprəutiːn] n. 前蛋白
proteoglycan [prəutiəˈglaikæn] n. 蛋白聚糖，蛋白多糖
ribbon [ˈribən] n. 缎带，丝带，带状物
skeleton [ˈskelitən] n. 骨架，骨骼，基干
spherical [ˈsferikəl] adj. 球的，球形的
starch [stɑːtʃ] n. 淀粉
teichoic acid 磷酸质，磷壁酸
thixotropic [ˌθiksəˈtrɔpik] adj. 触变的，具有触变作用的；摇溶的
urinogenital [ˌjuərinəuˈdʒenitl] adj. 泌尿生殖器的
viability [ˌvaiəˈbiliti] n. 生存能力，发育能力
xylan [ˈzailæn] n. 木聚糖

难句分析

1. It might be asked why cells should produce polysaccharide even to the extent that some cells perform gluconeogenesis.
句中why cells should produces polysaccharide…为主语从句，even to the extent在从句中作状语，意为“甚至在某种程度上”，其后that引导的从句是extent的同位语。might是情态动词，含可能或不确定之意；should也是情态动词，此处表示惊奇，意为“竟然会”。全句可译为：有可能会有人问为什么细胞会产生多糖，甚至有的细胞会发生糖原异生作用。

2. That is probably why plants，which often grow in nitrogen-limited environments，produce polysaccharides to support and protect their cells rather than a nitrogen-rich fibrous protein.
句中why plants，…，produce polysaccharides to support and protect their cells为表语从句，从句中which引导非限定性定语从句，修饰plants。rather than为连词，此处连接名词，意为“而不是”。全句可译为：那就是经常在氮有限的环境下生长的植物产生多糖，以供养或保护其他细胞，而不是产生富含氮的纤维状蛋白的原因。

3. Apart from chemical composition，there are other distinguishing features，all to do with shape and size.
句中apart from为复合介词，意为“除……之外”。all为features的同位语，其后的to do with shape and size为动词不定式短语，作all的定语，其中短语to do with…意为“与……有关系”。全句可译为：除了化学组成，还有其他的显著特征，如形状和大小。

4. The availability of different linkages means that the preferred conformation（or shape）of the polysaccharide is dependent on the glycosidic bond as well as the nature of the α-1,4-produce a polymer with a shape and solubility quite different from one with identical residues，linked α-1,4，but with a β-configuration.
that引导的是主语从句，从句中quite different from one with…为形容词短语作a shape and solubility的定语，one为代词，代替a polymer。全句可译为：将会看到α-1,4连接的葡糖残基生成的高聚物其形状和溶解度与α-1,4连接、构型为β的相同葡糖碱基的高聚物完全不同。

5. A number of mammalian cell types are surrounded，albeit with a much looser structure，by cross-linked fibrous proteins such as collagen and elastin.
句中的albeit为连词，意为“即使”或“尽管”，此处引导一个省略形式的状语。全句可译为：许多类型的哺乳动物细胞，尽管它们的结构相当松散，但也受到交联的纤维状蛋白，如胶原蛋白和弹性蛋白的包围。

6. They produce a compact molecular barrier protecting the fragile plasma membrane from a potentially harmful environment，yet are sufficiently porous to allow small nutrient molecules access for binding to transport proteins lodged in the membrane.
句中protecting…from…为现在分词短语作barrier的定语，to allow…为不定式短语作结果状语。全句可译为：它们（这些多糖）形成一个紧密的分子屏障，可使易脆的质膜免遭潜在有害环境的影响，而且它们又是充分多孔的，有利于小的营养分子接近并与膜中的传递蛋白结合。

7. There are，however，smaller carbohydrate polymers termed oligosaccharides，which are covalently attached to lipid and protein and where the structure is anything but simple.

which 和 where 分别引导非限定性定语从句。“anything but…”作“根本不……”解。全句可译为：然而，有一些小的碳水化合物聚合物，与脂或蛋白共价相连，结构并不简单。

8. In only a few conformations is proximity and linearity of some C，H and O atoms in the individual residues appropriate for inter-residue H-bonds to form.

   该句为倒装句，其主语 proximity and linearity 可看作是一个词，意为“近程线性”，类似的例子如 iron and steel（钢铁），表语为 in only a few conformations。appropriate 为形容词作后置定语修饰 residues，for inter-residue H-bonds to form 为不定式复合结构作 appropriate 的逻辑主语。全句可译为：在适合形成残基间氢键的各个残基中某些 C、H 和 O 原子的近程线性排列仅存在于少数构象中。

9. Just as the stacking of bases in double-helical DNA is assisted by hydrophobic bonds between the planes of purine and pyrimidine rings so hydrophobic bonds assist in the stacking of sugar chains on top of one another in，for example，cellulose and chitin microfibrils.

   Just as 引出方式状语从句，so hydrophobic bonds assist in…为主句，so 为副词，在两个句子间起承上启下的作用。全句可译为：正像双螺旋 DNA 中碱基堆积受到嘌呤平面与嘧啶环之间的疏水键的帮助一样，纤维素和几丁质纤维糖中疏水键也有助于糖链的互相堆积。

10. In general，glycoproteins are proteins to which sugar residues have been covalently attached after assembly of the peptide chain.

    句中 to which 引导定语从句，其中 to 是介词，它引出的介词短语在从句中作状语。全句可译为：一般来说，糖蛋白是糖残基与肽链的共价装配以后形成的蛋白质。

11. The principal components of this complex are proteoglycans，the structure of which contains up to 150 polysaccharide chains attached to serine residues of a single core protein.

    the structure of which…引导非限定性定语从句，修饰 proteoglycans。从句中的attached to serine residues of a single core protein 为过去分词短语，修饰 polysaccharide chains。全句可译为：这种复合体主要成分是蛋白聚糖，其中包含由 10 个多聚糖链黏附于唯一核心蛋白质的丝氨酸残基。

12. The term blood group antigen or blood group substance is misleading for it suggests that these molecules，glycoproteins or glycoplipids，are only associated with cells found in the bloodstream，such as the glycoprotein glycophorin which is on the surface of erythrocytes.

    句中的 for 为并列连词，其词义可理解为“因为”；句中 glycoproteins and glycolipids 是 molecules 的同位语；found in the blood stream 为过去分词短语作 cells 的定语。全句可译为：血型抗原或血型物质这一术语会使人误解，因为它的意思是糖蛋白或糖脂这些分子只是与血液中细胞有关系，如位于血红细胞表面的糖蛋白——血糖蛋白。

13. Over 80% of human cancers are epithelial in origin，and it is perhaps the malfunction of the cell density sensing mediated by glycoplipids that is one of the causes contributing to the excessive cell proliferation that is characteristic of cancerous tissues.

    in origin 为介词短语作状语。句中两个 that 均为关系代词，引导定语从句，前者修饰 malfunction，后者修饰 proliferation。全句可译为：人的癌症有 80%以上都源于上皮，这可能由于糖脂调节所引起的细胞密度传感不正常所致，这种不正常是导致细胞过度增殖的原因之一，细胞过度增殖是癌组织的特征。

（选自：李青山，安玉华，刘永新，陶小娟编．生物化学和分子生物学英语．长春：吉林大学出版社，1994.）

The Structure of DNA

Structurally DNA is a polynucleotide. A formal analogy between polynucleotides and proteins may therefore be perceived. Polynucleotides are the products of nucleotide condensation, just as proteins are produced by the polymerization of amino acids. This similarity of structures is an important element which facilitates the transfer of genetic information between these two distinct classes of macromolecules. The structure of nucleotides and their constituent purine and pyrimidine bases are examined in the text.

The base composition of DNA varies considerably among species, particularly prokaryotes, which have a range of 25%-75% in adenine-thymine content. This range narrows with evolution, reaching limiting values of 45%-53% in mammals.

In addition to the four common bases, adenine, guanine, thymine, and cytosine, which occur in DNA from all sources, DNA isolated from many plant and animal tissues (e. g. , wheat germ, thymus gland) contains small amounts of the base 5-methylcytosine. Methylated derivatives of the bases are also present in all DNA molecules examined to date. In addition, the DNA of certain bacteriophages (the T-even coliphages) contains 5-hydroxymethyl-cytosine in place of cytosine, and this derivative occurs in a glucosylated form. Even uracil, a base constituent of RNA, has been found in certain *Bacillus subtilis* phages, instead of thymine.

Polynucleotides

Polynucleotides (Fig. 1. 4) are formed by the joining of nucleotides by phosphodiester bonds. The phosphodiester bond is the formal analog of the peptide bond in proteins. It serves to join, as a result of the esterification of two of the three hydroxyl groups of phosphoric acid, two adjoining nucleotide residues. Two free hydroxyl groups are present in deoxyribose on the C3′ and C5′ atoms. Therefore these are the only hydroxyl groups that can participate in the formation of a phosphodiester bond. Indeed, it turns out that the nucleotide residues in DNA polynucleotides are joined together by 3′, 5′-phosphodiester bonds.

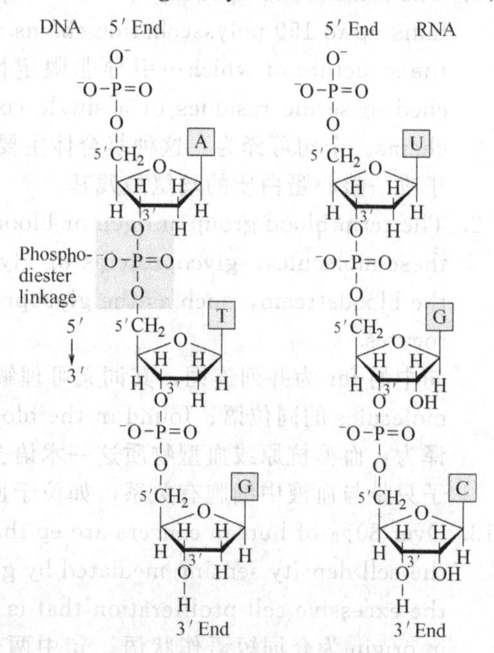

Fig. 1. 4 Phosphodiester linkages in the covalent backbone of DNA and RNA

The phosphodiester bonds (one of which is shaded in the DNA) link successive nucleotide units. The backbone of alternative pentose and phosphate groups in both types of nucleic acid is highly polar. The 5′end of the macromolecules lacks a nucleotide at the 5′position, and the 3′end lack a nucleotide at the 3′postion

In some instances, polynucleotides are linear polymers. The last nucleotide residue at each of the opposite ends of the polynucleotide chain serves as the two terminals of the chain. It is apparent that these terminals are not structurally equivalent, since one of the nucleotides must terminate at a 3′-hydroxyl group and the other at a 5′-hydroxyl group. These ends of the polynucleotides are referred to as the 3′ and the 5′ termini, and they may be viewed as corresponding to the amino and carboxyl termini in proteins. Polynucleotides also exist as cyclic structures, which contain no free terminals. Esterification between the 3′-OH terminus of a polynucleotide with its own 5′-phosphate terminus can produce a cyclic polynucleotide.

In this discussion long polymers of nucleotides joined by phosphodiester bonds are referred to as polynucleotides, in accordance with the prevailing nomenclature. A distinct name, oligonucleotide, is reserved for shorter nucleotide containing polymers. According to formal rules of nomenclature, however, polynucleotides must be named by using roots derived from the names of the corresponding nucleotides, and using the ending ylyl. For example, the polynucleotide segment, in which the 5′terminal is on the left of each nucleotide residue, should be named from left to right.

It is apparent, however, that the result of this approach is so cumbersome that abbreviations are generally preferred. For example, the oligonucleotides shown in the text is usually referred to as dAdCdGdT, and a polynucleotide containing only one kind of nucleotide, for example, dA, may be written as poly (dA) . Oligo-and polynucleotide structures are also written out in shorthand. In every instance the sequence is written starting on the left with the nucleotide of the 5′ terminus.

DNA is made of polynucleotides, and it is the specific sequence of bases along a polynucleotide chain that determines the biological proteins of the polymer. Although the structure of the nucleic acid building blocks, the bases, had been correctly known for many years, the polymeric structure initially proposed for DNA turned out to be one of the classical errors in the history of biochemistry. Experimental data obtained from what appears to have been partially degraded samples of DNA, and several other misconceptions, led to the erroneous conclusion that DNA consisted of repeating tetranucleotide units. Each tertranucleotide supposedly contained equimolar quantities of the four common bases. These impressions persisted to some degree until the late 1940s and early 1950s, when they were clearly shown to be in error. In the interim, however, these misconceptions were responsible for setting back the acceptance of the concept that the DNA of chromosomes carried genetic information. The monotonous structure of repeating tetranucleotides appeared incapable of having the versatility to encode for the enormous number of messages necessary to convey hereditary traits. Instead proteins, which can be ordered in an almost unlimited number of amino acid sequences, were favored as the most suitable candidates for a hereditary function. The transformation experiment carried out in the mid-1940s, and the subsequent finding the DNA consists of polynucleotide rather than tetranucleotide chains, were responsible for the general acceptance of the hereditary role of DNA that followed.

Hydrolysis of the phosphodiester bond: nucleases

The nature of the linkage between nucleotides to form polynucleotides was elucidated primarily by the use of exonucleases, which are enzymes that hydrolyze these polymers in a selective manner. Exonucleases cleave the last nucleotide residue in either of the two terminals of an oligonucleotide. Oligonucleotides can thus be degraded by the stepwise removal of individual nucleotides or small oligonucleotides from either the 5′ or the 3′ terminus. Nucleases sever the bonds in one of two nonequivalent positions as proximal (p) or distal (d) to the base which occupies the 3′ position of the bond. For example, the treatment of an oligodeoxyribonucleotide with venom diesterase, an enzyme obtained from snake venom, yields deoxyribonucleoside 5′-phosphates. In contrast, treatment with a diesterase isolated from animal spleen produces deoxyribonucleoside 3′-phosphates.

It should be noted that other nucleases, which cleave phosphodiester bonds located in the interior of polynucleotides and are designated as endonucleases, behave similarly in this respect. For instance, DNase Ⅰ cleaves only p linkages. While DNase Ⅱ cleaves d linkages. The points of cleavage along an oligonucleotides chain are indicated by arrows. Some endonucleases have been particularly useful in the development of early methodologies for sequencing of RNA polynucleotides. More recently other endonucleases, know as restriction endonucleases, have provided the basis for the development of recombinant DNA techniques.

Many nucleases do not exhibit any specificity with respect to the base adjacent to the linkage that is hydrolyzed. Certain nucleases, however, act more discriminately next to specific types of bases or even specific individual bases. Restriction nucleases act only on sequences of bases specifically recognized by each restriction enzyme. Nuclease also exhibit specificities, with respect to the overall structure of polynucleotides. For instance, some nucleases act on either single-stranded or double-stranded polynucleotides, whereas others discriminate between these two types of structures. In addition, some nucleases exclusively designated as phosphodiesterase will act on either DNA or RNA, whereas other nucleases will limit their activity to only one type of polynucleotide. The nuclease listed above illustrates some of the diverse properties of these enzymes.

Secondary structures of DNA

As has been emphasized previously, the polypeptide chains of protein are often arranged in space in a manner that leads to the formation of periodic structures (Fig. 1. 5). For instance, in the α helix each residue is related to the next by a translation of 1. 5 Å along the helix axis and a rotation of 100°. This arrangement places 3. 6 amino acid residues in each complete turn of the polypeptide helix. The property of periodicity is also encountered with polynucleotides, which usually occur in the form of helices.

Such preponderance of helical conformations among macromolecules is not surprising. The formation of helices tends to accommodate the effects of intermolecular forces, which in a helix can be distributed at regular intervals. The precise geometry of the polynucleotide helices varies, but the helical structure invariable results from the stacking of bases along the helix axis. In many instances stacking produces helices in which the bases are more or less perpendicularly oriented along the helix and

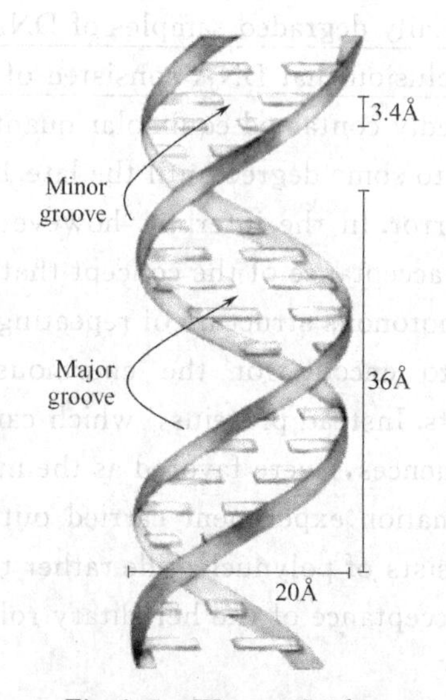

Fig. 1. 5 Waston-Crick model for the structure of DNA

The original model proposed by Waston and Crick had 10 bases pairs, or 34Å (3. 4nm), per turn of the helix; subsequent measurements revealed 10. 5 bases pairs, or 36Å (3. 6nm), per turn

touch one another. This arrangement, which obviously leaves no free space between two successive neighboring bases, is illustrated in the text. Such stacked helices, however, are not commonly encountered in nature. Rather, as it will become apparent from the subsequent discussion, polynucleotide helices tend to associate with one another to form double helices.

Forces that determine polynucleotide conformation

The hydrophobic properties of the bases are, to a large extent, responsible for forcing polynucleotides to adopt helical conformations. Examination of molecular models of the bases reveals that the edges of the rings contain polar groups (i. e. , amino and hydroxyl group residues) that are able to interact with other polar groups or surrounding water molecules. The faces of the rings, however, are unable to participate in such interactions and tend to avoid any contact with water. Instead they tend to interact with one another, producing the stacked conformation. The stability of this arrangement is further reinforced by an interchange between the electrons that circulate in the π orbitals located above and below the plane of each ring.

Clearly then, single-stranded polynucleotide helices are stabilized by both hydrophobic as well as stacking interactions involving the π orbitals of the bases. The stability of the helical structures is also influenced by the potential repulsion among the charged phosphate residues of the polynucleotide backbone. These repulsive forces introduce a certain degree of rigidity to the structure of the polynucleotide. Under physiological conditions, that is, at neutral pH and relatively high concentrations of salts, the charges on the phosphate residues are partially shielded by the cations present, and the structure can be viewed as a fairly flexible coil. Under more extreme conditions the stacking of the bases is disrupted and the helix collapses. A collapsed helix is commonly described as a random coil. A conversion, between a stacked helix and a random-coil conformation is depicted in the text.

DNA double helix

Although certain forms of cellular DNA exist as single-stranded structures, the most widespread DNA structure is the double helix. The double helix can be visualized as resulting from the interwinding around a common axis of two right-handed helical polynucleotide strands. The two strands achieve contact through hydrogen bonds, which are formed at the hydrophilic edges of their bases. These bonds extend between purine residues in one strand and pyrimidine residues in the other, so that the two types of resulting pairs are always adenine-thymine and guanine-cytosine. A direct consequence of these hydrogen-bonding specificities is that double stranded DNA contains equal amounts of purines and pyrimidines. Examination of space-filling models clearly indicates the structural compatibility of these bases in forming linear hydrogen bonds.

This relationship between bases in the double helix is described as complementarity. The bases are complementary because every base of one strand is matched by complementary hydrogen-bonding base on the other strand. For instance, for each adenine projecting toward the common axis of the double helix, a thymine must be projected from the opposite chain so as to fill exactly the space between the strands by hydrogen bonding with adenine. Neither cytosine nor guanine fits precisely in the available space in a manner that allows the formation of hydrogen bonds across strands. These hydrogen-bonding specificities ensure that the entire base sequence of one strand is complementary to that of the other strand.

The conventional double helix exists in various geometries designed as forms A, B, and C. These forms, however, share certain common characteristics. Specifically, the phosphate backbone is always located on the outside of the helix. Also, because the diesters of phosphoric acid are fully ionized at neutral pH, the exterior of the helix is negatively charged. The bases are well packed in the interior of the helix, where their faces are protected from contact with water. In this environment the strength of the hydrogen bonds that connect the bases can be maximized. The interwinding of the two polynucleotide strands produces a structure having two deep helical grooves that separate the winding phosphate backbone ridge.

However, the precise geometry of the double helix varies among the different forms. The original X-ray data obtained with highly oriented DNA fibers suggested the occurrence of a form, later designated as B, which appears to be the one commonly found in solution and in vivo. A characteristic of this form is that one of its grooves is wider than the other, and referred to as the major groove to distinguish it from the second, or minor groove. The nucleotide sequence of the polynucleotides can be discerned without dissociating the double helix by looking inside these grooves. As each of the four bases has its on orientation with respect to the rest of the helix, each base always shows the same atoms through the grooves. For instance, the C-6, N-7, and C-8 of the purine rings and the C-4, C-5, and C-6 of the pyrimidine rings line up in the major groove. The minor groove is paved with the C-2 and N-3 of the purine and the C-2 of the pyrimidine rings. Forms A and C differ from B in the pitch of the base pairs relative to the helix axis as well as in other geometric parameters of the double helix.

A new form of DNA was discovered recently, which has geometric characteristics radically different from those of the conventional forms (Fig. 1. 6). In this DNA, termed Z-DNA, the polynucleotide phosphodiester backbone assumes a "zig-zag" arrangement rather than the smooth conformation that characterizes other double-stranded forms. The Z-DNA structure forms a single groove as opposed to the two grooves that characterize B-DNA. Therefore, the conformation of Z-DNA structure forms a single groove as opposed to the two grooves that characterize B-DNA. Therefore, the conformation of Z-DNA may be viewed as the result of the major groove of B-DNA having "popped out" in order to form the outer convex surface of Z-DNA. This change places the stacked bases on the outer part of Z-DNA rather than in their conventional positions in the interior of the double helix. Another highly unusual property of

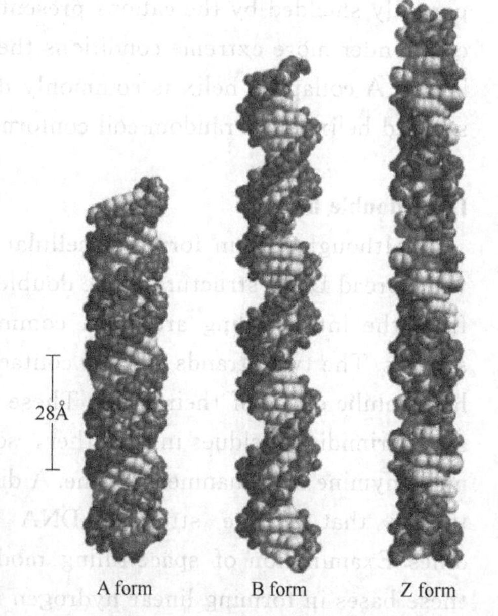

Fig. 1. 6 Comparison of A, B, and Z forms of DNA

Each structure shown here has 36 base pairs. The bases are shown in gray, the phosphate atoms in white, and the riboses and phosphate oxygens in dark

the Z structure is that it consists of left-handed rather than right-handed helices, which characterize the conventional forms. These major structural differences between the B-DNA and the Z-DNA, which are illustrated in the text, are partly the result of different conformations in the nucleotide residues between the two forms.

The biological function of Z-DNA is not known with certainty. Some evidence exists suggesting that Z-DNA influences gene expression and regulation. Apparently, Z-DNA is incorporated in small stretches, normally containing approximately one or two dozen nucleotide residues, in regions of the gene that regulate transcriptional activities. These stretches consist of alternating purines and pyrimidines in the sequence, which is a condition favoring the formation of the Z conformation. The Z form of DNA is stabilized by the presence of cations or polyamines and by methylation of either guanine residues in the C-8 and N-7 positions or cytosine residues in the C-5 position.

An important structural characteristic of all double-stranded DNA is that its strands are antiparallel. Polynucleotides are asymmetric structures with an intrinsic sense of polarity built into them. As it may be concluded from inspection, the two strands are aligned in opposite directions, that is, if two adjacent bases in the same strand, for example thymine and cytosine, are connected in the 5′-3′ direction, their complementary bases adenine and guanine will be linked in the 3′-5′direction (directions are defined by linking the 3′ and 5′ positions within the same nucleotide) . This antiparallel alignment produces a stable association between strands to the exclusion of the alternate parallel arrangement. Just as peptide preferred conformation of proteins, the formation of hydrogen bonds between complementary bases on antiparallel polynucleotide strands leads to the formation of the double helix.

The double-stranded structure for DNA was proposed in 1953. The proposal was partly based on the results of previously available X-ray diffraction studies, which suggested that the structures of DNA from various sources exhibited remarkable similarities. These studies also suggested that DNA had a helical structure containing two or more polynucleotides. An additional piece of evidence of central importance to the proposal was the clarification of the quantitative base composition of DNA, which was obtained independently in 1950. These results indicated the existence of molar equivalence between purines and pyrimidines, which turned out to be the essential observation suggesting the existence of complementarity between the two strands.

Stability of the DNA structure

The same factors that stabilize single-stranded polynucleotide helices, hydrophobic and stacking forces, are also instrumental in stabilizing the double helix. The separation between the hydrophobic core of the stacked bases and the hydrophilic exterior of the charged sugar-phosphate groups is even more striking in the double helix than with single-stranded helices. This arrangement, which produces substantial stabilization of the double stranded structures over single-stranded conformations, explains the preponderance of the former. The stacking tendency of single-stranded polynucleotides may be viewed as resulting from a tendency of the bases to avoid contact with water. The double-stranded helix is by far a more favorable arrangement, as it permits the phosphate backbone to be highly solvated by water while the bases are essentially removed from the aqueous environment.

Additional stabilization of the double helix results from its extensive network of cooperative hydrogen bonding. Although this bonding makes only a relatively minor contribution to the free energy of stabilization of the double helix, the physiological importance of hydrogen bonds should not be underestimated. By contrast to hydrophobic forces, hydrogen bonds are highly directional and for this reason are able to provide a discriminatory function for choosing between correct and incorrect base pairs. In addition, because of their directionality, hydrogen bonds tend to orient the bases in a way that favors stacking. Therefore although hydrogen bonds make a minor contribution to the total energy of stabilization, their contribution is essential for the stability of the double helix.

In the past, the relative importance of hydrogen bonding and hydrophobic forces in stabilizing the double helix as not always appreciated. However, studies on the effect of various regents on the stability of the double helix have suggested that the destabilizing effect of a reagent is not related to the ability of the reagent to break hydrogen bonds. Rather, the stability of the double helix is determined by the solubility of the free bases in the reagent, the stability decreasing as the solubility increases. Some of these findings emphasize the importance of hydrophobic forces in maintaining the structure of double-stranded DNA.

Ionic forces also have an effect on the stability and the conformation of the double helix. At physiological pH the electrostatic intrastrand repulsion between negatively charged phosphates forces the double helix into a relatively rigid rod-like conformation. In addition the repulsion between phosphate groups located on opposite strands tends to separate the complementary strands. In distilled water, DNA strands will separate at room temperature; near the physiological salt concentration, cations (in addition of other charged groups, for example, the basic side chains of proteins) shield the phosphate groups and decrease repulsive forces. Therefore, the flexibility of the double helix is partially restored and its stability is enhanced.

Denaturation

The double helix is stabilized by approximate 1 kcal per base pair. Therefore a relatively minor perturbation can produce disruption in double strandedness, provided that only a short section of the DNA is involved. As soon as the relatively few base pairs have separated, they close up again and release free energy, and then the adjacent base pairs unwind. In this manner, minor disruptions of double strandedness can be propagated along the length of the double helix. Therefore, at any particular moment the large majority of the bases of the double helix remain hydrogen bonded, but all bases can pass through the single-stranded state, a few at a time. This dynamic state of the double helix is characterized by the movement of an "open-stranded" portion up and down the length of the helix, as indicated in the text. The "dynamic" nature of this structure is an essential prerequisite for the biological function of DNA and especially the process of DNA synthesis.

Furthermore, the strands of DNA can be completely separated by increasing the temperature in solution. At relatively low temperature a few base pairs will be disrupted, creating one or more "open-stranded bubbles". These "bubbles" form initially in sections that contain relatively higher proportions of adenine and thymine pairs. Adenine-thymine pairs are bound by two hydrogen bonds and are therefore less stable than guanine-cytosine pairs, which contain three such bonds per air. As the temperature is raised, the size of the "bubbles" increases and eventually the thermal motion of the polynucleotides overcomes the forces that stabilize the double helix. This transformation is depicted in this text. At even higher temperature the strands can separate physically and acquire a random-coil conformation, referred to as denaturation. The process of denaturation is accompanied by a number of physical changes, including a buoyant density increase, a reduction in viscosity, change in the ability to rotate polarized light and changes in absorbency.

Changes in absorbency are frequently used for following experimentally the process of denaturation, DNA absorbs in the UV region due to the heterocyclic aromatic nature of its purine and pyrimidine constituents. Although each base has a unique absorption spectrum, all bases exhibit maximal at or near 260 nm. This property is responsible for the absorption of DNA at 260 nm. However, this absorbency is almost 40% lower than that expected from adding up the absorbency of each of the base components of DNA. This property of DNA, referred to as hypo chromic effect, results from the close stacking of the bases along the DNA helices. In this special arrangement interactions between the electrons of neighboring bases produce decrease in absorbency. However, as the ordered structure of the double helix is disrupted at increasing temperature, stacking interactions are gradually decreased. Therefore, a totally disordered polynucleotide, a random coil, eventually approaches an absorbance not very different from the sum of the absorbency of its purine and pyrimidine constituents.

Slow heating of double-stranded DNA in solution is accompanied by a gradual change in absorbency as the strands separate. However, since the interactions between the two strands are cooperative, the transition from double stranded to random coil conformation occurs over a narrow range of temperature. Before the rise of the melting curve, DNA is double-stranded. In the rising section of the curve an increasing number of base airs is interrupted as the temperature rises. Strand separation occurs at a critical temperature corresponding to the upper plateau of the curve. However, if the temperature is decreased before the complete separation of the strands, the native structure is completely restored.

The midpoint temperature, T_m, of this process, under standard conditions of concentration an ionic strength, is characteristic of the base content of each DNA. The higher the guanine-cytosine content, the higher the transition temperature between the double-stranded helix and the single strands. This difference in T_m is attributed to the increased stability of guanine-cytosine pairs, as a result of the three hydrogen bonds that connect them in DNA, in contrast to only two hydrogen bonds that connect adenine and thymine pairs.

Rapid cooling of a heated DNA solution normally produces denatured DNA, a structure that results from the reformation of some hydrogen bonds either between the separate strands or between different sections of the same strand. The latter must contain complementary base sequences. By and large denatured DNA is a disordered structure containing substantial amounts of random coil and single-stranded regions.

DNA can also be denatured at a pH above 11. 3 as the charge on several substituents on the rings of the bases is changed preventing these groups from participating in hydrogen bonding. Alkaline denaturation is often used as an experimental tool in preference to heat denaturation to prevent breakage of phosphodiester bonds that can occur to some degree at high temperature. Denaturation can also be induced at low ionic strength, because of enhanced intrastrand repulsion between negatively charged phosphates, as well as by various denaturing reagents, that is, compounds that weaken or break hydrogen bonds. A complete denaturation curve similar to that shown above can be obtained at a relatively low constant temperature, for instance room temperature, by variation of the concentration of an added denaturant.

Renaturation

Complementary DNA strands, separated by denaturation, can reform a double helix if appropriately treated by a process referred to as renaturation or reannealing. Renaturation depends upon the meeting of complementary DNA strands in an exact manner that can lead to the reformation of the original structure, and it is therefore a slow, concentration-dependent process. As a rule, maintaining DNA at temperature 10-15℃ below its T_m under conditions of moderate ionic strength (about 0. 15mol/L), provides the maximum opportunity for renaturation. At lower salt concentrations, the charged phosphate groups repel one another and prevent the strands from associating. As renaturation begins, some of the hydrogen bonds formed are extended between short tracts of polynucleotides that might have been distant in the original native structure. Short sequences, consisting for example of four to six base pairs, are reiterated many times within every DNA strand. Furthermore eukaryotic DNA contains a large number of much longer nucleotide sequences reiterated many times within each genome. Such sequences provide sites, for initial bases pairing, which produces a partially hydrogen-bonded double helix. These randomly base-paired structures are short-lived because the bases that surround the short complementary segments cannot pair and lead to the formation of a stable fully hydrogen-bonded structure. However, once the correct bases begin to pair by chance, the double helix over the entire DNA molecule is rapidly reformed (Fig. 1. 7).

Hybridization

The self-association of complementary polynucleotide strands has also provided the basis for the development of the technique of hybridization. The principle of this determination is simple. For a DNA of a given size the rate of annealing depends on the frequency with which two complementary segments can collide with each other. Therefore, the larger the number of reiterated sequences in a given DNA, the greater is the chance that a particular collision will result in the formation of annealed polynucleotides. On this basis the extent to which annealing takes place within a unit of time can be used to determine the number of reiterated sequences in the DNA.

Determinations of the maximum amount of DNA that can be hybridized have been used to establish homologies between the DNA of different species. This is possible because the base sequences of the DNA in each organism are unique for this organism. Therefore the annealed helices represent the same unique sequences of DNA even if the individual annealed strands originate from different cells. On this basis annealing can be used to compare the degree to which DNA isolated from different species are related to one another. Consequently, the observed homologies serve as indices of evolutionary relatedness and have been particularly useful for defining phylogenies in prokaryotes. "Hybridization" studies between DNA and RNA have, in addition, provided very useful information about the biological role of DNA, particularly the mechanism of transcription.

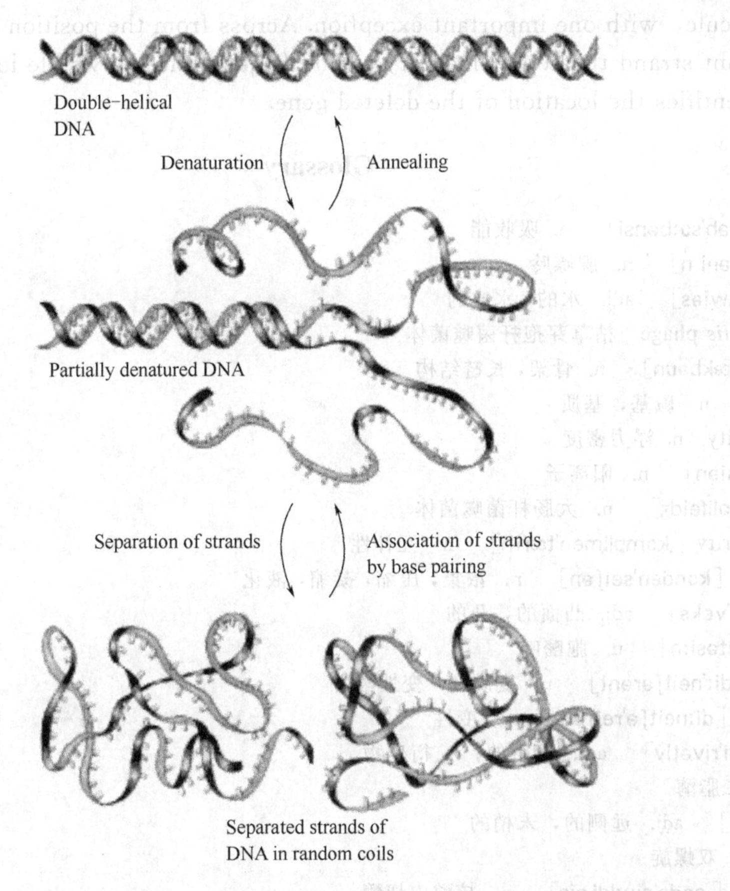

Fig. 1. 7 Reversible denaturation and annealing (renaturation) of DNA

In recent years hybridization techniques using membrane filters, usually made of nitrocellulose, have found increasing application. In general, hybridization can be quantitated by either measuring the amount of hybrid in equilibrium or the rate of hybrid formation under conditions in which one nucleic acid is present in large excess. The approach used for the latter determination is analogous to the Cot procedure and when it is used for DNA-RNA hybridization and RNA is present in excess it is referred to as the Rot method, or the Dot method when DNA is in excess.

A variant of filter hybridization, known as the Southern transfer, can be used for identifying the location of specific genes. Since a gene sequence represents a very small percent age of total DNA, the gene must be separated from the remaining DNA and amplified before hybridization.

Finally, the principle of hybridization has also served as the basis for the development of a technique that has permitted the construction of precise physical maps of DNA genes. This technique depends on the direct visualization under the electron microscope of single-stranded loops in the structures of artificially formed double-stranded DNA molecules known as heteroduplexes. The principle of the technique is simple. Heteroduplexes are constructed by hybridization of two complementary DNA strands. One of these strands, however, is selected on the basis that, as the result of a known mutation, it misses the gene being mapped. The complementary strands of the heteroduplex pair perfectly throughout the length of the molecule, with one important exception. Across from the position of the missing gene in the mutant strand the complementary strand forms a clearly visible loop. The position of the loop identifies the location of the deleted gene.

Glossary

absorbency [əb'sɔ:bənsi] n. 吸收能
adenine ['ædəni:n] n. 腺嘌呤
aqueous ['eikwiəs] adj. 水的，水成的
Bacillus subtilis phage 枯草芽孢杆菌噬菌体
backbone ['bækbəun] n. 骨架，长链结构
base [beis] n. 碱基，基质
buoyant density n. 浮力密度
cation ['kætaiən] n. 阳离子
coliphage ['kɔlifeidʒ] n. 大肠杆菌噬菌体
complementarity [ˌkɔmplimen'tæriti] n. 互补性
condensation [kɔnden'seiʃən] n. 浓集，压缩，凝缩，液化
convex ['kɔn'veks] adj. 凸面的，凸的
cytosine ['saitəsi:n] n. 胞嘧啶
denaturant [di:'neitʃərənt] n. 变质剂，变性药
denaturation [di:neitʃə'reiʃən] n. 变性
derivative [di'rivətiv] adj. 衍生的；n. 衍生物
diesterase 二脂酶
distal ['distəl] adj. 远侧的，末梢的
double helix 双螺旋
endonuclease [ˌendə'nju:kliˌeis] n. 核酸内切酶
esterification [esˌterifi'keiʃən] n. 酯化作用
exonuclease [ˌeksəu'nju:klieis] n. 核酸外切酶，外切核酸酶，水解核酸酶
flexibility [ˌfleksə'biliti] n. 灵活性，屈曲性，灵活度
free energy n. 自由能
genetic [dʒi'netik] adj. 遗传的，起源的
geometry [dʒi'ɔmitri] n. 几何形状，几何图
glucosylate n. 葡萄糖基化作用
guanine ['gwɑ:ni:n] n. 鸟嘌呤
heteroduplex n. 杂交双链，异型双链
hydrophobic force n. 疏水作用
hypochromic effect 减色效应，吸收减弱效应
interwind [ˌintə(:)'waind] v. 互相盘绕，互卷
ionic force n. 离子作用
methylation [ˌmeθi'leiʃən] n. 甲基化
methylcytosine [ˌmeθil'saitəsin] n. 甲基胞嘧啶
midpoint temperature, T_m 中点温度
oligodeoxyribonucleotide 寡脱氧核糖核苷酸
oligonucleotide [ˌɔligəu'nju:kliəutaid] n. 寡聚核苷酸
perturbation [ˌpə:tə:'beiʃən] n. 微扰，干扰，扰动
phosphodiester 磷酸二酯
phosphoric acid n. 磷酸
plateau ['plætəu, plæ'təu] n. 平高线，高地

polymerization [ˌpɔliməraiˈzeiʃən] n. 聚合作用
polynucleotide [ˌpɔliˈnjuːkljəˌtaid] n. 多聚核苷酸
prokaryote [prəuˈkæriəut] n. 原核生物
proximal [ˈprɔksiməl] adj. 接近的，临近的
purine [ˈpjuəriːn] n. 嘌呤
pyrimidine [ˌpaiəˈrimidiːn,ˈpiri-] n. 嘧啶
reannealing n. 重退火
renaturation [ˈriːˌneitʃəˈreiʃən] n. 退火，复性，复性作用
solvate [ˈsɔlveit] n. 溶血物，溶化物；vt. (使) 成溶剂化物
stacking [ˈstækiŋ] n. 堆积，堆叠
tetranucleotide [ˌtetrəˈnjuːkliətaid] n. 四核苷酸
thymine [ˈθaimiːn] n. 胸腺嘧啶
thymus gland 胸腺
uracil [ˈjuərəsil] n. 尿嘧啶
venom [ˈvenəm] n. 毒，毒物
viscosity [visˈkɔsiti] n. 黏（滞）性，黏（滞）度
wheat germ 麦芽胚
X-ray diffraction X射线衍射，X射线绕射

难句分析

1. Experimental data obtained from what appears to have been partially degraded samples of DNA, and several other misconceptions, led to the erroneous conclusion that DNA consisted of repeating tetranucleotide units.

   obtained from what appears to have been partially degraded samples of DNA 为过去分词短语，作主语 data 的定语，to have been 是 to be 的完成时态，degraded 为过去分词，是 sample 的定语。led to 为谓语。that 引导的从句作 conclusion 的同位语。全句可译为：从似乎已经降解的 DNA 样品得到的实验数据和其他一些错觉得出了 DNA 分子由重复的四种核苷酸单位组成的结论。

2. It should be noted that other nucleases, which cleave phosphodiester bonds located in the interior of polynucleotides and are designated as endonucleases, behave similarly in this respect.

   It 是形式主语，that 引导的从句作真正主语，其中 other nucleases 为主语，behave 为谓语，which 引导的定语从句修饰主语，定语从句中 located in the interior of polynucleotides 是过去分词短语，作 phosphodiester bonds 的定语。全句可译为：应该注意，另一些能够判断多核苷酸链内部磷酸二酯键的内切核酸酶类，在这方面具有相似性。

3. For instance, for each adenine projecting toward the common axis of the double helix, a thymine must be projected from the opposite chain so as to fill exactly the space between the strands by hydrogen bonding with adenine.

   for 引导介词短语作状语，so as to 引导不定式短语作目的状语。全句可译为：例如，由于每个腺嘌呤都指向双螺旋的轴，这样，另一条链的胸腺嘧啶必须通过与腺嘌呤形成氢键的方式指向双螺旋轴，以便正好充满两条链之间的空间。

4. The original X-ray data obtained with highly oriented DNA fibers suggested the occurrence of a form, later designated as B, which appears to be the one commonly found in solution and in vivo.

   obtained 引导的过去分词短语作 data 的定语，过去分词短语 later designated as B 和

which 引导的定语从句都为 a form 的定语。全句可译为：采用方向性强的 DNA 纤维得到的 X 射线衍射原始数据表明，有一种后来称为 B 型的 DNA 分子存在，这种形式的 DNA 在溶液状态和活体内都有发现。

5. DNA can also be denatured at a pH above 11.3 as the charge on several substituents on the rings of the bases is changed preventing these groups from participating in hydrogen bonding.

as 引导原因状语从句，原因状语从句中 preventing 引导分词短语作结果状语。全句可译为：当 pH 值高于 11.3 时，碱基环上的几个取代基电荷的改变使这些基因不能参与氢键的形成，因此 DNA 分子也可以变性。

（选自：李青山，安玉华，刘永新，陶小娟编．生物化学和分子生物学英语．长春：吉林大学出版社，1994.）

Cells—Discovery and Basic Structure

In 1655, the English scientist Robert Hooke made an observation that would change basic biological theory and research forever. While examining a dried section of cork tree with a crude light microscope, he observed small chambers and named them cells. Within a decade, researchers had determined that cells were not empty but instead were filled with a watery substance called cytoplasm.

Over the next 175 years, research led to the formation of the cell theory, first proposed by the German botanist Matthias Jacob Schleiden and the German physiologist Theodore Schwann in 1838 and formalized by the German researcher Rudolf Virchow in 1858. In its modern form, this theorem has four basic parts.

① The cell is the basic structural and functional unit of life; all organisms are composed of cells.

② All cells are produced by the division of preexisting cells (in other words, through reproduction). Each cell contains genetic material that is passed down during this process.

③ All basic chemical and physiological functions—for example, repair, growth, movement, immunity, communication, and digestion, are carried out inside of cells.

④ The activities of cells depends on the activities of subcellular structures within the cell (these subcellular structures include organelles, the plasma membrane, and, if present, the nucleus).

The cell theory leads to two very important generalities about cells and life in general.

a. Cells are alive. The individual cells of your organs are just as "alive" as you are, even though they cannot live independently. This means cells can take energy (which, depending on the cell type, can be in the form of light, sugar, or other compounds) and building materials (proteins, carbohydrates and fats), and use these to repair themselves and make new generations of cells (reproduction).

b. The characteristics and needs of an organism are in reality the characteristics and needs of the cells that make up the organism. For example, you need water because your cells need water.

Most of the activities of a cell (repair, reproduction, etc.) are carried out via the production of proteins. Proteins are large molecules that are made by specific organelles within the cell using the instructions contained within the genetic material of the cell.

Cytology is the study of cells, and cytologists are scientists that study cells. Cytologists have discovered that all cells are similar. They are all composed chiefly of molecules containing carbon, hydrogen, oxygen, nitrogen, phosphorus, and sulfur. Although many nonliving structures also contain these elements, cells are different in their organization and maintenance of a boundary, their ability to regulate their own activity, and their controlled metabolism.

All cells contain three basic features.

① A plasma membrane consisting of a phospholipid bilayer, which is a fatty membrane that houses the cell. This membrane contains several structures that allow the cell to perform necessary tasks—for example, channels that allow substances to move in and out of the cell, antigens that allow the cell to be recognized by other cells, and proteins that allow cells to attach to each other.

② A cytoplasm containing cytosol and organelles. Cytosol is a fluid consisting mostly of water and dissolved nutrients, wastes, ions, proteins, and other molecules. Organelles are small structures suspended in the cytosol. The organelles carry out the basic functions of the cell, including reproduction, metabolism, and protein synthesis.

③ Genetic material (DNA and RNA), which carries the instructions for the production of proteins.

Apart from these three similarities, cell structure and form are very diverse and are therefore difficult to generalize. Some cells are single, independent units and spend their entire existence as individual cells (these are the single-celled organisms such as amoebas and bacteria) . Other cells are part of multicellular organisms and cannot survive alone.

One major difference among cells is the presence or absence of a nucleus, which is a subcellular structure that contains the genetic material. Prokaryotic cells (which include bacteria) lack a nucleus, whereas eukaryotic cells (which include protozoans, animal and plant cells) contain a nucleus.

There are other major differences in cell structure and function between different types of organisms. For example, the cells of autotrophic organisms (most plants and some protozoans), which can produce their own food, contain an organelle called the chloroplast that contains chlorophyll and allows the cell to produce glucose using light energy in the process known as photosynthesis.

- The cells of plants, protists, and fungi are surrounded by a cell wall composed mostly of the carbohydrate cellulose; the cell wall helps these cells maintain their shape. Animal cells lack a cell wall but instead have a cytoskeleton, a network of long fibrous protein strands that attach to the inner surface of the plasma membrane and help them maintain shape.

There are even major differences in cells within the same organism, reflecting the different functions the cells serve within the organism. For example, the human body consists of trillions of cells, including some 200 different cell types that vary greatly in size, shape, and function. The smallest human cells, sperm cells, are a few micrometers wide (1/12000 of an inch) whereas the longest cells, the neurons that run from the tip of the big toe to the spinal cord, are over a meter long in an average adult! Human cells also vary significantly in structure and function. For example:

- Only muscle cells contain myofilaments, protein-containing structures that allow the cells to contract (shorten) and therefore cause movement.
- Specialized cells called photoreceptors within the eye have the ability to detect light. These cells contain special chemicals called pigments that can absorb light, and special organelles that can then turn the absorbed light into electrical current that is sent to the brain and is perceived as vision.

All plants and animals are composed of microscopic cells, which are the smallest basic organic units capable of carrying out the functions that we normally define as life. These essential cell functions include:

① taking in nutrients;

② combining the nutrients into substances for cell growth and repair;

③ reproducing themselves;

④ excreting waste matter.

Cells in people and other multicellular organisms reproduce themselves by dividing to form new cells. When the division process begins, long thread-like structures called chromosomes contract and become visible in the nucleus. At other times, chromosomes are, in effect, not visible because of their stretched out thin profile. Chromosomes (literally "colored bodies") were given this name, when they were discovered in the late 19th century, due to the fact that they become visible when stained dark purple with a synthetic dye used to enhance the observation of specific cell components.

Chromosomes contain almost all of the genetic information that determines inheritance. Different plant and animal species have different shapes and numbers of chromosomes. For instance, humans normally have 46 and our nearest living relatives, the chimpanzees and gorillas, have 48. Having more chromosomes does not necessarily mean that an organism is structurally more complex. For instance, chickens have 78, and there is a fern species that has 1260. More extraordinary still is a microscopic single-celled organism named *Oxytricha*. It has approximately 46 million chromosomes. Having the same number of chromosomes does not mean that two animals are the same species.

Within a species, different chromosomes are visually distinguishable on the basis of their size and the form of their components. The contracted strands form arms or chromatids. The point of attachment of two or more chromatids is called a centromere.

An individual's chromosomes can be photographed when they become visible during the cell division process. Then, all of the pictures of individual chromosomes in a cell may be cut out and organized into a karyotype. In a karyotype, the chromosomes are placed into homologous pairs. These are pairs that contain genes for the same traits at the same location (or locus) on the chromosomes. However, homologous chromosomes may have different alleles, or alternate forms of the same gene. Homologous chromosomes are paired early in the cell reproduction process.

The 46 chromosomes in normal humans consist of 22 pairs of autosomes and one pair of sex chromosomes. The autosomes carry the genes that determine most body characteristics. The sex chromosomes primarily determine sexual traits. If they are X and Y, the individual is almost always a male. In contrast, if both sex chromosomes are X's, the gender is female.

In fact, maleness is largely determined by the presence or the absence of a specific gene on the Y chromosome known as the *SRY* (sex-determining region Y) gene. Beginning in the 5th to 7th week after conception in humans, it triggers other genes that cause the undifferentiated embryonic sex organs to become testes. These, in turn, produce the hormone testosterone which stimulates the development of other masculine physical traits. If the *SRY* gene is not functioning or absent, an XY individual will very likely develop a female appearing body with breasts and a vagina but will not have a uterus or ovaries and subsequently will be sterile. Instead of ovaries, there will be rudimentary internal testes. Late or partial functioning of the *SRY* gene can result in individuals being sexually intermediate between male and female. However, this is not the only process that can cause ambiguous sexual characteristics.

Sometimes, in the production of sperm cells, an error can occur that results in a portion of the Y chromosome detaching and becoming part of an X chromosome. If that portion carries the *SRY* gene and it is inherited, the result can be a chromosomal female (XX) who will develop as sterile phenotypic male. About 1 in 20000 normal appearing males are chromosomal females (XX). Roughly the same number of phenotypic females are chromosomal males (XY). Both are sterile.

All autosomes and female sex chromosomes exist in homologous pairs. However, the X and Y chromosomes of males are mostly not homologous—they only share a few genes. As a result, the X and Y chromosomes are said to be hemizygous, or only partially homologous. The X chromosome is much larger and has many more genes not related to sexual characteristics.

One of the X chromosomes in the cells of human females is partly inactivated early in embryonic life. This is a self-preservation action to prevent a potentially harmful double dose of genes. Recent research points to the "*Xist*" gene on the X chromosome as being responsible for a sequence of events that silences one of the X chromosomes in females. The inactivated X chromosomes become highly compacted structures known as Barr bodies. The presence of Barr bodies has been used at the Olympics and other international sports competitions as a test to determine whether an athlete is a man or a woman. The measurement of testosterone levels has also been used for this purpose. Every person has about 30000 genes in each cell that determine the specific details of what he or she is like biologically. The chromosomes are made up of genes. More accurately, the chromosomes are mainly DNA molecules and the genes are sections of them. The 23 pairs of human chromosomes do not each have the same number of genes. Chromosome 1 is the largest, containing about 8% of all of our genes. Chromosome 21 is the smallest and has only about 1/6 as many genes as chromosome 1.

Animal cell structure

Animal cells are typical of the eukaryotic cell (Fig. 1. 8), enclosed by a plasma membrane and containing a membrane-bound nucleus and organelles. Unlike the eukaryotic cells of plants and fungi, animal cells do not have cell walls. This feature was lost in the distant past by the single-celled organisms that gave rise to the kingdom animalia. Most cells, both animal and plant, range in size between 1 and 100 micrometers and are thus visible only with the aid of a microscope.

The lack of a rigid cell wall allowed animals to develop a greater diversity of cell types, tissues, and organs. Specialized cells that formed nerves and muscles tissues impossible for plants to evolve have these organisms mobility. The ability to move about by the use of specialized muscle tissues is a hallmark of the animal world, though a few animals, primarily sponges, do not possess differentiated tissues. Notably, protozoans locomote, but it is only via nonmuscular means, in effect, using cilia, flagella, and pseudopodia.

The animal kingdom is unique among eukaryotic organisms because most animal tissues are bound together in an extracellular matrix by a triple helix of protein known as colla-

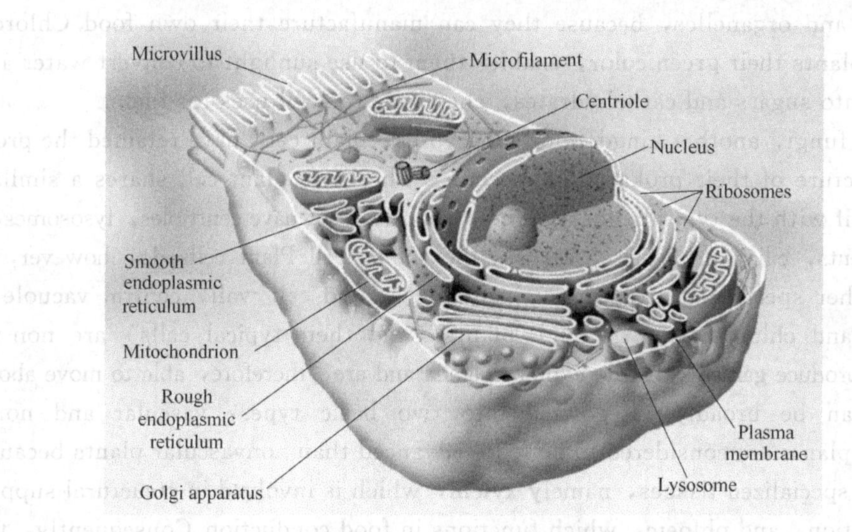

Fig. 1. 8 An animal cell

gen. Plant and fungal cells are bound together in tissues or aggregations by other molecules, such as pectin. The fact that no other organisms utilize collagen in this manner is one of the indications that all animals arose from a common unicellular ancestor. Bones, shells, spicules, and other hardened structures are formed when the collagen-containing extracellular matrix between animal cells becomes calcified.

Animals are a large and incredibly diverse group of organisms. Making up about three-quarters of the species on Earth, they run the gamut from corals and jellyfish to ants, whales, elephants, and, of course, humans. Being mobile has given animals, which are capable of sensing and responding to their environment, the flexibility to adopt many different modes of feeding, defense, and reproduction. Unlike plants, however, animals are unable to manufacture their own food, and therefore, are always directly or indirectly dependent on plant life.

Most animal cells are diploid, meaning that their chromosomes exist in homologous pairs. Different chromosomal ploidies are also, however, known to occasionally occur. The proliferation of animal cells occurs in a variety of ways. In instances of sexual reproduction, the cellular process of meiosis is first necessary so that haploid daughter cells, or gametes, can be produced. Two haploid cells then fuse to form a diploid zygote, which develops into a new organism as its cells divide and multiply.

The earliest fossil evidence of animals dates from the Vendian Period (650 million to 544 million years ago), with coelenterate-type creatures that left traces of their soft bodies in shallow-water sediments. The first mass extinction ended that period, but during the Cambrian Period which followed, an explosion of new forms began the evolutionary radiation that produced most of the major groups, or phyla, known today. Vertebrates (animals with backbones) are not known to have occurred until the early Ordovician Period (505 million to 438 million years ago).

Plant cell structure

Plants are unique among the eukaryotes, organisms whose cells have membrane-enclosed nuclei and organelles, because they can manufacture their own food. Chlorophyll, which gives plants their green color, enables them to use sunlight to convert water and carbon dioxide into sugars and carbohydrates, chemicals the cell uses for fuel.

Like the fungi, another kingdom of eukaryotes, plant cells have retained the protective cell wall structure of their prokaryotic ancestors. The basic plant cell shares a similar construction motif with the typical eukaryote cell, but does not have centrioles, lysosomes, intermediate filaments, cilia, or flagella, as does the animal cell. Plant cells do, however, have a number of other specialized structures, including a rigid cell wall, central vacuole, plasmodesmata, and chloroplasts. Although plants (and their typical cells) are non-motile, some species produce gametes that do exhibit flagella and are, therefore, able to move about.

Plants can be broadly categorized into two basic types: vascular and nonvascular. Vascular plants are considered to be more advanced than nonvascular plants because they have evolved specialized tissues, namely xylem, which is involved in structural support and water conduction, and phloem, which functions in food conduction. Consequently, they also possess roots, stems, and leaves, representing a higher form of organization that is characteristically absent in plants lacking vascular tissues. The nonvascular plants, members of the division Bryophyta, are usually no more than an inch or two in height because they do not have adequate support, which is provided by vascular tissues to other plants, to grow bigger. They also are more dependent on the environment that surrounds them to maintain appropriate amounts of moisture and, therefore, tend to inhabit damp, shady areas.

It is estimated that there are at least 260000 species of plants in the world today. They range in size and complexity from small, nonvascular mosses to giant sequoia trees, the largest living organisms, growing as tall as 330 feet (100 meters). Only a tiny percentage of those species are directly used by people for food, shelter, fiber, and medicine. Nevertheless, plants are the basis for the Earth's ecosystem and food web, and without them complex animal life forms (such as humans) could never have evolved. Indeed, all living organisms are dependent either directly or indirectly on the energy produced by photosynthesis, and the byproduct of this process, oxygen, is essential to animals. Plants also reduce the amount of carbon dioxide present in the atmosphere, hinder soil erosion, and influence water levels and quality.

Plants exhibit life cycles that involve alternating generations of diploid forms, which contain paired chromosome sets in their cell nuclei, and haploid forms, which only possess a single set. Generally these two forms of a plant are very dissimilar in appearance. In higher plants, the diploid generation, the members of which are known as sporophytes due to their ability to produce spores, is usually dominant and more recognizable than the haploid gametophyte generation. In Bryophytes, however, the gametophyte form is dominant and physiologically necessary to the sporophyte form.

Animals are required to consume protein in order to obtain nitrogen, but plants are able to utilize inorganic forms of the element and, therefore, do not need an outside source of protein. Plants do, however, usually require significant amounts of water, which is needed for the photosynthetic process, to maintain cell structure and facilitate growth, and as a means of bringing nutrients to plant cells. The amount of nutrients needed by plant species varies significantly, but nine elements are generally considered to be necessary in relatively large amounts. Termed macroelements, these nutrients include calcium, carbon, hydrogen, magnesium, nitrogen, oxygen, phosphorus, potassium, and sulfur. Seven microelements, which are required by plants in smaller quantities, have also been identified: boron, chlorine, copper, iron, manganese, molybdenum, and zinc.

Thought to have evolved from the green algae, plants have been around since the early Paleozoic era, more than 500 million years ago. The earliest fossil evidence of land plants dates to the Ordovician Period (505 million to 438 million years ago). By the Carboniferous Period, about 355 million years ago, most of the Earth was covered by forests of primitive vascular plants, such as lycopods (scale trees) and gymnosperms (pine trees, ginkgos). Angiosperms, the flowering plants, didn't develop until the end of the Cretaceous Period, about 65 million years ago just as the dinosaurs became extinct.

Bacteria cell structure

They are as unrelated to human beings as living things can be, but bacteria are essential to human life and life on planet Earth. Although they are notorious for their role in causing human diseases, from tooth decay to the Black Plague, there are beneficial species that are essential to good health. For example, one species that lives symbiotically in the large intestine manufactures vitamin K, an essential blood clotting factor. Other species are beneficial indirectly. Bacteria give yogurt its tangy flavor and sourdough bread its sour taste. They make it possible for ruminant animals (cows, sheep, goats) to digest plant cellulose and for some plants, (soybean, peas, alfalfa) to convert nitrogen to a more usable form.

Bacteria are prokaryotes, lacking well-defined nuclei and membrane-bound organelles, and with chromosomes composed of a single closed DNA circle. They come in many shapes and sizes, from minute spheres, cylinders and spiral threads, to flagellated rods, and filamentous chains. They are found practically everywhere on Earth and live in some of the most unusual and seemingly inhospitable places.

Evidence shows that bacteria were in existence as long as 3.5 billion years ago, making them one of the oldest living organisms on the Earth. Even older than the bacteria are the archeans (also called archaebacteria) tiny prokaryotic organisms that live only in extreme environments: boiling water, super-salty pools, sulfur-spewing volcanic vents, acidic water, and deep in the Antarctic ice. Many scientists now believe that the archaea and bacteria developed separately from a common ancestor nearly four billion years ago. Millions of years later, the ancestors of today's eukaryotes split off from the archaea. Despite the superficial resemblance to bacteria, biochemically and genetically, the archean are as different from bacteria as bacteria are from humans.

In the late 1600s, Antoni van Leeuwenhoek became the first to study bacteria under the microscope. During the nineteenth century, the French scientist Louis Pasteur and the German physician Robert Koch demonstrated the role of bacteria as pathogens (causing disease). The twentieth century saw numerous advances in bacteriology, indicating their diversity, ancient lineage, and general importance. Most notably, a number of scientists around the world made contributions to the field of microbial ecology, showing that bacteria were essential to food webs and for the overall health of the Earth's ecosystems. The discovery that some bacteria produced compounds lethal to other bacteria led to the development of antibiotics, which revolutionized the field of medicine.

There are two different ways of grouping bacteria. They can be divided into three types based on their response to gaseous oxygen. Aerobic bacteria require oxygen for their health and existence and will die without it. Anerobic bacteria can't tolerate gaseous oxygen at all and die when exposed to it. Facultative anaerobes prefer oxygen, but can live without it.

The second way of grouping them is by how they obtain their energy. Bacteria that have to consume and break down complex organic compounds are heterotrophs. This includes species that are found in decaying material as well as those that utilize fermentation or respiration. Bacteria that create their own energy, fueled by light or through chemical reactions, are autotrophs.

Cell digestion and the secretory pathway

The primary sites of intracellular digestion are organelles known as the lysosomes, which are membrane-bounded compartments containing a variety of hydrolytic enzymes. Lysosomes maintain an internal acidic environment through the use of a hydrogen ion pump in the lysosomal membrane that drives ions from the cytoplasm into the lumenal space of the organelles. The high internal acidity is necessary for the enzymes contained in lysosomes to exhibit their optimum activity. Hence, if the integrity of a lysosomal membrane is compromised and the enzymatic contents are leaked into the cell, little damage is done due to the neutral pH of the cytoplasm. If numerous lysosomes rupture simultaneously, however, the cumulative action of their enzymes can result in autodigestion and the death of the cell.

Lysosomal hydrolytic enzymes are manufactured in the rough endoplasmic reticulum (Rough ER), from whence they are transferred in a transport vesicle to the *cis* face of the Golgi apparatus or complex. Inside of the Golgi complex, the enzymes undergo additional processing and are transformed from an inactive to an active state. Lysosomes may then bud from the *trans* face of the Golgi apparatus, though they may also form via other mechanisms (such as the gradual transformation of endosomes).

The majority of the membrane and secretory proteins involved in cellular digestion are synthesized on the surface of the endoplasmic reticulum (rough ER) or are translocated to the organelle after being produced in the cytoplasm. In addition to serving as a center for protein processing and secretion, the endoplasmic reticulum forms a portion of the nuclear envelope. Newly synthesized proteins are sorted at ER exit sites and enter vesicles for trafficking to intermediate compartments (VTC), where proteins are sorted for transport to the Golgi complex or returned to the endoplasmic reticulum.

Once they arrive in the cisternal stacks of the Golgi apparatus, proteins are sorted and transported to the *trans*-Golgi network or sent back to earlier compartments for re-processing. Within the *trans* network, proteins are sorted for trafficking to either the plasma membrane, endosomes, or lysosomes (for degradation). Those proteins destined for the plasma membrane can also be recycled through the endosomes and sent back to the membrane or the Golgi, as well as sorted to late endosomes and potentially to lysosomes. Protein transfer to the mitochondria and peroxisomes in many cases probably doesn't involve processing through vesicles. The mitochondria are critical in the production of energy (via adenosine triphosphate; ATP) and lipid biosynthesis, whereas peroxisomes play a role in lipid metabolism and help reduce the intracellular concentration of free radicals.

Several different varieties of macromolecules may be digested by lysosomes and arrive at the organelles by disparate pathways. For example, many unicellular organisms and certain cells in multicellular organisms consume particles of food and other items via a process called phagocytosis. When the food is engulfed by the cell during this process, a vacuole forms around it from an invagination in the cell to plasma membrane that pinches off to completely encase the foreign material. Once released into the cell, the food vacuole is able to fuse with a lysosome. The action of the hydrolytic enzymes upon the contents of the vacuole digests them, breaking them down into monomers, such as simple sugars and amino acids, which are able to traverse the lysosomal membrane via transport proteins and enter the cytosol, where they serve as cell nutrients.

In addition to providing nutrients, the phagocytic process is important to some cell types as a mode of defense. Certain components of the vertebrate immune system, for instance, utilize phagocytosis to fight off infections. Specifically, macrophages and neutrophils, which are often referred to as scavenger cells, actively ingest bacteria and other microorganisms they encounter in the body. These intruders are then digested with the help of lysosomes.

An alternate mechanism that materials may be acquired by lysosomes for digestion is strictly intracellular. Termed autophagy, the process entails the removal of membrane-bounded organelles or other cytoplasmic components through the action of lysosomes. This digestive pathway is perhaps best illustrated in the situation where a lysosome fuses to a fractured mitochondrion. Similar to phagocytosis, the enzymes present in the lysosome disassemble the mitochondrial components into monomers and the breakdown products from autophagy are released into the cytosol.

The autophagic process is often a means of cell renewal, allowing damaged macromolecules to be recycled into products the cell can readily utilize. The liver cells, where mitochondria have an average lifetime of only about 10 days, are a good example of this function. Microscopic examination of liver cells reveals the constant presence of lysosomes containing mitochondria. Moreover, additional smooth endoplasmic reticulum produced in liver cells due to exposure to certain drugs is broken down via autophagy when the chemical substance is removed, emphasizing the fact that autophagy is a selective process.

Lysosomes are also important for their role in the programmed death of certain cells. As previously discussed, if the enzymes of a single lysosome are released into a cell, there is little change in the cytosol, but a massive enzymatic discharge by many lysosomes can be fatal to the cell. Through the coordinated release of lysosomal enzymes a number of important developmental changes occur in various multicellular organisms. As a tadpole matures into a frog, for instance, it loses its tail due to the lysosomal destruction of the cells contained in the appendage.

Cell reproduction

Most human cells are frequently reproduced and replaced during the life of an individual. However, the process varies with the kind of cell. Somatic, or body cells, such as those that make up skin, hair, and muscle, are duplicated by mitosis. The sex cells, sperm and ova, are produced by meiosis in special tissues of male testes and female ovaries. Since the vast majority of our cells are somatic, mitosis is the most common form of cell replication.

Mitosis

The cell division process that produces new cells for growth, repair, and the general replacement of older cells is called mitosis. In this process, a somatic cell divides into two completely new cells that are identical to the original one. Human somatic cells go through the 6 phases of mitosis in 0.5 to 1.5 hours, depending on the kind of tissue being duplicated.

Some human somatic cells are frequently replaced by new ones and other cells are rarely duplicated. Hair, skin, and fingernails are replaced constantly and at a rapid rate throughout our lives. In contrast, brain and nerve cells in the central nervous system are rarely produced after we are a few months old. Subsequently, if they are destroyed later, the loss is usually permanent, as in the case of paraplegics. Liver cells usually do not reproduce after an individual has finished growing and are not replaced except when there is an injury. Red blood cells are also somewhat of an exception. While they are being constantly produced in our bone marrow, the specialized cells from which they come do not have nuclei nor do the red blood cells themselves.

Meiosis

Meiosis is a somewhat similar but more complex process than mitosis. This is especially true in females. While mitosis produces 2 daughter cells from each parent cell, meiosis results in 4 sex cells, or gametes in males and 1 in females. Unlike the cells created by mitosis, gametes are not identical to the parent cells. In males, meiosis is referred to as spermatogenesis because sperm cells are produced. In females, it is called oogenesis because ova, or eggs, are the main ultimate product.

Glossary

acidity [ə'siditi] n. 酸度，酸性
adenosine triphosphase [ə'denəsi:n trai'fɔsfəteis] 腺苷三磷酸（酯）酶
alfalfa [æl'fælfə] n.（植[1]）紫花苜蓿
algae ['ældʒi:] n. 藻类，海藻
amoebas [ə'mi:bəz] n. 阿米巴，变形虫
anerobic ['ænərəubik] adj. 厌氧的
angiosperm ['ændʒiəu,spə:m] n. 被子植物
animalia [,æni'meiliə] n. 动物界，动物类
antibiotics ['æntibai'ɔtiks] n. 抗生素，抗生学
archaebacteria [,ɑ:kibæk'tiəriə] n.（复）（微）原始细菌（一种不同于细菌和动植物细胞且要求完全厌氧条件并能产生甲烷的微生物）
archean [ɑ:'kiən] adj.（地质[2]）太古代的
autophagy [ɔ:'tɔfədʒi] n.（生[3]）（细胞的）自我吞噬（作用）
autosome ['ɔ:təsəum] n. 正染色体，常染色体
autotrophic [,ɔ:təu'trɔfik] adj. 自造营养物质的，自给营养的
autotroph ['ɔ:təutrɔf] n.（生）自养生物，靠无机物质生存的生物
bacteriology [bæk,tiəri'ɔlədʒi] n. 细菌学
Barr bodies 巴尔体（存在于雌性体细胞核内的一种无活性、浓集的X染色体）
boron ['bɔ:rən] n.（化）硼
bryophyte ['braiə,fait] n. 苔藓类的植物
calcify ['kælsifai] vi. 变成石灰质，钙化；vt. 使成石灰，使钙化
calcium ['kælsiəm] n.（化）钙
cell [sel] n. 单元，细胞
centriole ['sentriəul] n.（生）细胞中心粒，中心体

---

[1]“植”表示“植物学”，后同。
[2]“地质”表示“地质学”，后同。
[3]“生”表示“生物学”，后同。

centromere [ˈsentrəˌmiə] n.（生）着丝点，着丝粒
chimpanzee [ˈtʃimpənˈziː] n.（动）黑猩猩
chlorine [ˈklɔːriːn] n.（化）氯
chlorophyll [ˈklɔːrəfil] n.（生化）叶绿素
chloroplast [ˈklɔ(ː)rəplɑːst] n.（植）叶绿体
chromatid [ˈkrəumətid] n.（生）染色单体
chromosome [ˈkrəuməsəum] n.（生）染色体
cilia [ˈsiliə] n. 纤毛，睫
cisternal [sisˈtəːnə] n.（解❶）池，内胞浆网槽
coelenterate [siˈlentəreit] n.（动❷）腔肠动物；adj. 腔肠动物的
collagen [ˈkɔləˌdʒən] n. 胶原质
conception [kənˈsepʃən] n. 受孕
copper [ˈkɔpə] n.（化）铜
coral [ˈkɔrəl] n. 珊瑚，珊瑚虫
cytologist [saiˈtɔlədʒist] n.（生）细胞学家
cytology [saiˈtɔlədʒi] n.（生）细胞学
cytoskeleton [ˌsaitəˈskelitən] n. 细胞骨架，细胞支架
cytosol [ˈsaitəusɔl] n.（生）细胞溶质，胞液
degradation [ˌdegrəˈdeiʃən] n. 降解，递降，分解作用
digestion [diˈdʒestʃən, daiˈdʒestʃən] n. 消化（作用）
dinosaur [ˈdainəsɔː] n. 恐龙
diploid [ˈdiplɔid] n. 倍数染色体；adj. 双重的，倍数的，双倍的
ecology [i(ː)ˈkɔlədʒi] n. 生态学
ecosystem [iːkəˈsistəm] n. 生态系统
embryonic [ˌembriˈɔnik] adj.（生）胚胎的，开始的
endoplasmic reticulum [ˌendəˈplæzmik riˈtikjuləm] 内质网
endosome [ˈendəˌsəum] n. 内含体
erosion [iˈrəuʒən] n. 腐蚀，侵蚀
excrete [eksˈkriːt] vt. 排泄，分泌
fern [fəːn] n.（植）蕨类植物
flagellated [ˈflædʒəleitid] adj.（生）有鞭毛的
flagella [fləˈdʒelə] n. 鞭节，鞭毛
free radical [ˈrædikəl] n.（化）自由基，游离基
gamete [ˈgæmiːt] n.（生）接合体，配偶子
gametophyte [gəˈmiːtəˌfait] n.（植）配偶体
gamut [ˈgæmət] n. 全音阶，长音阶，全音域
ginkgo [ˈgiŋkgəu] n. 银杏树
Golgi apparatus [ˈgɔːldʒi ˌæpəˈreitəs] n.（生）高尔基体
gorillas [gəˈrilə] n. 大猩猩
gymnosperm [ˈdʒimnəuˌspəːm] n. 裸子植物
hallmark [ˈhɔːlmɑːk] n. 标志，标点
haploid [ˈhæplɔid] n.（生）单倍体，仅有一组染色体的细胞；adj. 单一的
hemizygous [ˌhemiˈzaigəs] adj.（生）半合子的
heterotroph [ˈhetərəutrɔf] n.（生）异养生物
hydrolytic [ˌhaidrəˈlitik] adj.（化）水解的
inheritance [inˈheritəns] n.（生）遗传
inhospitable [inˈhɔspitəbl] adj. 冷淡的

❶“解”表示“解剖学”，后同。
❷“动”表示“动物学”，后同。

invagination [inˌvædʒiˈneiʃən] n. 内陷
jellyfish [ˈdʒelifiʃ] n. 水母
karyotype [ˈkæriətaip] n.（生物）染色体组型
lipid [ˈlipid,ˈlaipid] n. 脂质，脂类
locomote [ˌləukəˈməut] v. 移动，行动
lumenal [ˈljuːminl] adj.（解）内腔的
lycopod [ˈlaikəpɔd] n. 石松属的植物
lysosome [ˈlaisəsəum] n.（生）（细胞中的）溶酶体
macroelement [ˈmækrəˈelimənt] n. 常量元素
macromolecule [ˌmækrəuˈmɔlikjuːl] n. 大分子
macrophage [ˈmækrəfeidʒ] n.（生）巨噬细胞
magnesium [mægˈniːzjəm] n.（化）镁
manganese [ˌmæŋgəˈniːz,ˈmæŋgəniːz] n.（化）锰
marrow [ˈmærəu] n.（生）髓，骨髓
masculine [ˈmɑːskjulin] adj. 男性的，雄性的
matrix [ˈmeitriks] n. 基质，衬质，间质
meiosis [maiˈəusis] n. 减数分裂，成熟分裂
metabolism [meˈtæbəlizəm] n. 新陈代谢，变形
microbial [maiˈkrəubiəl] adj. 微生物的，由细菌引起的
microelement [ˈmaikrəuˌelimənt] n.（化）微量元素
microvillus [ˌmaikrəuˈviləs] n. 微绒毛；微小突起物；指状突起，一些上皮细胞表面上突起的微小发状结构的总称，尤指小肠上的
mitochondria [ˌmaitəˈkɔndriə] n.（生）线粒体
mitosis [miˈtəusis] n. 有丝分裂
molybdenum [məˈlibdinəm] n.（化）钼
monomer [ˈmɔnəmə] n. 单体
moss [mɔs] n. 苔藓
multicellular [ˌmʌltiˈseljulə] adj. 多细胞的
myofilament [ˌmaiəuˈfiləmənt] n.（解）肌丝（肌原纤维的组成部分）
neuron [ˈnjuərɔn] n.（解）神经细胞，神经元
neutrophil [ˈnjuːtrəfil] adj.（细胞等）嗜中性的；n.（解）嗜中性粒细胞
nucleus [ˈnjuːkliəs] n.（nuclear 的复数）核，细胞核
nutrient [ˈnjuːtriənt] n. 营养素，养分，养料
oogenesis [ˌəuəˈdʒenisis] n.（生）卵子发生，成卵法
optimum [ˈɔptiməm] n. 最适条件；adj. 最适宜的
organelle [ˌɔːgəˈnel] n. 细胞器官
organ [ˈɔːgən] n.（生）器官
ovary [ˈəuvəri] n.（生）卵巢
ova [ˈəuvə] n. ovum 的复数
ovum [ˈəuvəm] n.（生）卵，卵子
Paleozoic era [pæliəˈzəuik ˈiərə] n. 古生代
paraplegic [ˌpærəˈpliːdʒik] adj.（医）（患）截瘫的；n. 截瘫患者
pathogen [ˈpæθədʒ(ə)n] n.（微[1]）病菌，病原体
pectin [ˈpektin] n. 胶质，果胶
peroxisome [pəˈrɔksisəum] n.（生化）过氧物酶体
phagocytosis [fægəˌsaiˈtəusis] n.（生）噬菌作用
phloem [ˈfləuem] n.（植）韧皮部
phospholipid [ˌfɔsfəuˈlipid] n.（生化）磷脂

[1]"微"表示"微生物学"，后同。

phosphorus [ˈfɔsfərəs] n. 磷
photosynthesis [ˌfəutəuˈsinθəsis] n. 光合作用
phyla [ˈfailə] n.（生物分类学上的）门（phylum 的复数）
pigment [ˈpigmənt] n.（生）色素，颜料
plasma [ˈplæzmə] n. 血浆，原浆，原生质
plasmodesmata [ˌplæzməuˈdezmeitə] n.（复）胞间连丝
ploidy [ˈplɔidi] n.（生）倍数性，倍性
potassium [pəˈtæsjəm] n.（化）钾
protist [ˈprəutist] n.（生）原生生物
pseudopodia [ˌsjuːdəuˈpɔdiə] n.（植）假足，（动）伪足（pseudopodium 的复数）
rudimentary [ruːdiˈmentəri] adj. 不发育的，退化的
ruminant [ˈruːminənt] adj. 反刍动物的；n. 反刍动物
rupture [ˈrʌptʃə(r)] v. 破裂；n. 破裂
scavenger [ˈskævindʒə] n. 清除剂，消除剂
sequoia [siˈkwɔiə] n.（植）美洲杉
somatic [səuˈmætik] adj. 躯体的，体壁的，体细胞的，菌体的，躯体性的
sourdough [ˈsauədəu] n. 酵母
spermatogenesis [ˌspəːmətəuˈdʒenisis] n.（生）精子发生
sperm [spəːm] n. 精液，精子
spicules [ˈspikjuːl;ˈspai-] n. 针状体，小穗状花序，骨针
spinal cord [ˈspainl kɔːd] n. 脊髓
sponge [spʌndʒ] n.（解）海绵体
spore [spɔː, spɔə] n. 孢子；vi. 长孢子
sporophyte [ˈspɔːrəˌfait] n. 孢子体；vi. 孢子形成体
sulfur [ˈsʌlfə] n.（化）硫，硫黄
testes [ˈtestis] n. testis 的复数 [testis（解）睾丸]
testosterone [tesˈtɔstərəun] n.（生化）睾丸激素，睾酮
theorem [ˈθiərəm] n.（数❶）定理，原理，法则
unicellular [ˈjuːniˈseljulə] adj. 单细胞的
uterus [ˈjuːtərəs] n.（解）子宫
vacuole [ˈvækjuəul] n. 空泡，（生）液泡
vagina [vəˈdʒainə] n.（解）阴道
vascular [ˈvæskjulə] adj.（解）（动）脉管的，有脉管的，血管的
vesicle [ˈvesikl] n.（解）（动）囊，泡；（植）小泡
vitamin [ˈvaitəmin,ˈvi-] n. 维生素
xylem [ˈzailem] n.（植）木质部
yogurt [ˈjɔgət] n.（=yoghurt）酸奶酪，酵母乳
zinc [ziŋk] n.（化）锌

难句分析

Recent research points to the "*Xist*" gene on the X chromosome as being responsible for a sequence of events that silences one of the X chromosomes in females.
译文：最近的研究指出，位于X染色体上的被称作"*Xist*"的基因负责使一条雌性X染色体沉默。

[选自：Vision learning（www. visionlearning. com）and the Molecular Expressions website（http：//micro. magnet. fsu. edu/index. html）]

❶"数"表示"数学"，后同。

The Origins of Genetics and Molecular Biology

Genetics is the name given to the study of heredity, the process by which characteristics are passed from parents to offspring so that all organisms, human beings included, resemble their ancestors. The central concept of genetics is that heredity is controlled by a vast number of factors, called genes, which are discrete physical particles present in all living organisms.

The first geneticists were mainly interested in how genes are transmitted from parents to their offspring during reproduction and how different genes act together to control variable traits such as height and eye colour. A change in emphasis occurred during the 1930s when it was recognized that if genes are physical entities then, like other cell components, they must be made of molecules and it should therefore be possible to study them directly by biophysical and biochemical methods. This led to a new branch of genetics, called molecular biology, which had as one of its initial aims the identification of the chemical nature of the gene. This new approach led to new concepts, and soon biologists ceased to regard individual genes simply as units of inheritance and instead began to look on them as unit of biological information, with the entire complement of genes in an organism containing the total amount of information needed to construct a living, functioning example of the organism. The aim of geneticists and molecular biologists over the past 50 years has been to understand the way in which biological information is stored in genes and how that information is made available to the living cell.

Genetics and molecular biology are very closely related subjects, and although there are distinctions between them it is more constructive to treat them as one. For this reason the term molecular genetics is now often used to describe that branch of biology concerned with the study of all aspects of the gene. This book is about molecular genetics and as such covers some of the most fascinating discoveries and intriguing problems of modern science. However, before describing the discoveries and examining the problems we must first turn our attention to the way in which genetics and molecular biology developed during the late nineteenth and early twentieth centuries. This will set the scene for the subsequent chapters by providing the historical and intellectual backdrop against which our more recent endeavor to understand the gene have been set.

Mendel and the experimental approach to genetics

Genetics as we know it originated with Gregor Mendel, in particular with a paper he published in 1866 in the "Verhandlungen des naturforschenden Vereines in Brunn"—the Proceedings of the Society of Natural Sciences in Brno. Mendel was typical of many nineteenth century scientists in that he had an insatiable curiosity about the natural and physical world and was keenly interested in the diversity of living things. If Mendel's family had been able to afford for their son an education fitting his intellectual talents then Gregor could very well have risen to become a noted Professor in one of the great European universities. Unfortunately his parents were poor and although Mendel spent four terms at the University of Vienna his education was piecemeal and what he obtained was achieved by perseverance rather than privilege.

It is well known that Mendel joined a monastery near Brno, which was at the time in Austria but is now in the Czech Republic, and eventually in 1868 became abbot. This choice of career was a fortunate one, because it seems that his duties in the years 1856-1864 were not so onerous as to prevent him carrying out a lengthy series of experiments concerning the inheritance in garden peas of traits such as height, flower colour and seed shape (Fig. 1. 9). These experiments were presented to the Brno Society in papers read by Mendel on 8 February and 8 March 1865 and published the following year in the Society Proceedings.

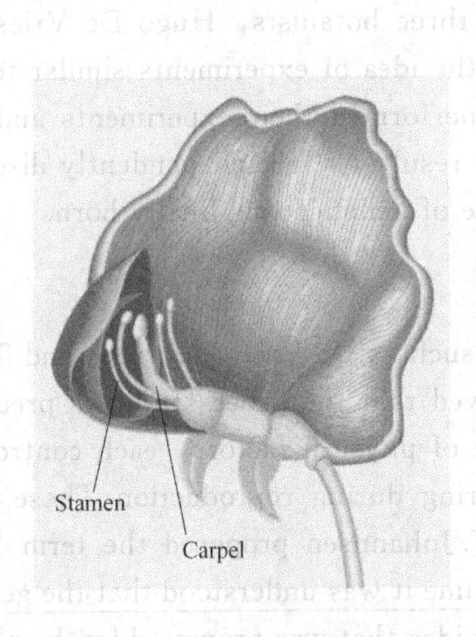

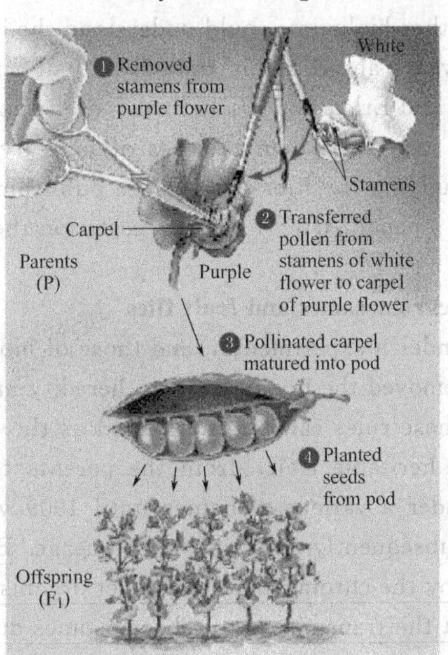

Fig. 1. 9 Mendel's experiments

What should be appreciated at the outset is that Mendel's contribution was not only the basic Laws of Heredity but also the demonstrations that heredity could be studied experimentally. Before Mendel, and indeed for the remaining 35 years of the nineteenth century, concepts about heredity were based almost entirely on simple observations of the living world. The idea that heredity could be studied systematically by straightforward experiments was totally alien to the mainstream of biological thought at that time. Mendel's experiments, which were carefully planned and meticulously executed, demonstrated that heredity works not by some mysterious process but in a predictable and logically consistent manner. His techniques and results provided a clear indication of how the problem of heredity could be unraveled by experimental means. Unfortunately Mendel remained until his death the only person who understood and appreciated the true significance of his work.

The birth of genetics-Mendel rediscovered

The suggestion that Mendel's paper was "lost" because it was published in an obscure journal by an unknown monk in a remote monastery is not really consistent with the facts. The Proceedings of the Brno Society were fairly well circulated and copies of the 1866 issue containing Mendel's paper were sent to at least 55 European libraries and learned societies, including the Royal Society and Linnean Society in London. Mendel sent reprints of his paper to many of the leading botanists of the day, including Professor Kerner of the University of Innsbruck (who evidently did not read it as the reprint was found after his death with the pages still uncut) and Professor Nageli of Munich, with whom Mendel corresponded regularly between 1866 and 1873. Mendel's work was noticed to the extent that he or his paper was mentioned in 16 publications in the late nineteenth century, including the ninth edition of the "Encyclopaedia Britannica".

The fact is that it was not until 1900 that biological thought had progressed to the point where other biologists could understand the importance of Mendel's work. Mendel was literally 35 years ahead of his time. Eventually in 1900 three botanists, Hugo De Vries, Carl Correns and Erich von Tschermak, each conceived the idea of experiments similar to Mendel's to test their own theories of heredity. Each performed their experiments and then, when studying the literature before publishing their results, each independently discovered Mendel's paper. After a lengthy gestation the science of genetics was finally born.

Genes, chromosomes and fruit flies

Mendel's experiments, and those of biologists such as De Vries, Correns and Tschermak, removed the mystique from heredity and showed that the process follows predictable rules. These rules can be rationalized as the passage of physical factors, each controlling a separate heritable trait, from the parents to offspring during reproduction. These factors went under a variety of names until 1909 when W. Johannsen proposed the term "gene" which subsequently entered common usage. By this time it was understood that the genes are carried by the chromosomes of higher organisms, an idea that was prompted by the observation that the transmission of chromosomes during cell division and reproduction exactly parallels the behavior of genes during these events. The chromosome theory was stated in its most convincing form in 1903 by W. S. Sutton, who at that time was a graduate student at Columbia University in New York and who subsequently became a surgeon before his early death in 1916.

Once the experimental basis of heredity had been established and the chromosome theory accepted, the way was open for a rapid advance in the understanding of genetics. That this advance occurred as rapidly as it did was mainly the result of the intuition and imagination of Thomas Hunt Morgan and the members of his research group, notably Calvin Bridges, Arthur Sturtevant and Hermann Muller. Morgan and his colleagues achieved something that many biologists dream about: they discovered an organism that was ideally suited for the particular research programme that they wished to carry out. This organism was *Drosophila melanogaster*, the fruit fly. *Drosophila* possesses several features that make it very suitable for genetic analysis but most important from Morgan's point of view was the fact that a large number of stable variant forms of the fly could be obtained. The differences between these variants involve features such as wing shape and body colour, some traits having several different varieties. These varieties enabled Morgan to investigate the way in which different combinations of genes work together in controlling the inheritance of an individual characteristic. Morgan's group developed many of the techniques that have now become standard methods in genetic analysis, including those that map the relative positions of different genes on a chromosome. Between 1911 and 1929 the "fly-room" at Columbia University provided the data that remain the foundation to our knowledge of the gene as a unit of inheritance.

What is life? —the advent of molecular biology

The rediscovery of Mendel's work and the remarkable advances made by Morgan and his colleagues attracted the attention of many biologists not actively engaged in genetic research, and during the first few decades of the twentieth century our understanding of the gene was advanced not only by geneticists but also by cytologists, physiologists, biochemists and biophysicists. But the contribution that in the long run has proved most influential was made not by biologists at all but by a group of physicists whose previous work had been in the area of quantum mechanics, far removed from Mendel and fruit flies. The involvement of physicists in biology was heralded by a lecture called "Light and Life" presented by Niels Bohr to an international congress in Copenhagen in August 1932, and subsequently reached its culmination with the publication of the book *What is Life?* by Erwin SchrÖdinger in 1944. Bohr and SchrÖdinger both attempted to interpret the central question in biology—the nature of life itself—in physical terms, but both were frustrated by the apparent limitations of trying to explain life by orthodox physical principles.

Paradoxically, neither Bohr's lecture nor SchrÖdinger's book stands up to present-day scrutiny: both contain errors and inconsistencies that today seem almost naive. Their contributions are important not so much for the ideas put forward but more for the influence that they had on other people. Several young physicists listened to Bohr's lecture, or read the text published the following year in the scientific journal *Nature*, and were fascinated by the challenge provided by biology. Among these was Max Dellbruck, a German who subsequently was forced by the political climate in Europe to emigrate to the USA. Delbruck learnt from Bohr about bacteriophages (viruses that attack bacteria) and decided that their infection cycle could be used as an experimental system with which to tackle the question of what genes are and how they work. He brought into existence in 1940 the "Phage Group", an informal association of physicists, biologists and chemists, all working in different laboratories but all with a common interest in the gene. The formation of this group stimulated the development of molecular biology as a discipline in its own right and led within 20 years to an understanding of the chemical nature of the gene. But we must now curtail our historical narrative as we have reached the point at which a detailed description of the gene must begin.

Glossary

bacteria [bæk'tiəriə] n. 细菌
bacteriophage [bæk'tiəriəfeidʒ] n. 噬菌体
biochemist ['baiəu'kemist] n. 生物化学家
biologist [bai'ɔlədʒist] n. 生物学家
biophysical ['baiəu'fizikəl] adj. 生物物理的
biophysicist ['baiəu'fizisist] n. 生物物理学家
botanist ['bɔtənist] n. 植物学家
chromosome ['krəuməsəum] n. (生) 生物染色体

culmination [kʌlmi'neiʃ(ə)n] n. 极度，极点
curtail [kəː'teil] vt. 缩减，减少
cytologist [sai'tɔlədʒist] n. (生) 细胞学家
geneticist [dʒi'netisist] n. 遗传学者
genetics [dʒi'netiks] n. 遗传学
gestation [dʒes'teiʃən] n. 怀孕，酝酿，妊娠
heredity [hi'rediti] n. 遗传，形质遗传
insatiable [in'seiʃəbl] adj. 不知足的，贪求无厌的
meticulously [mi'tikjuləsli] adv. 过细地，谨慎地
mystique [mis'tiːk] n. 神秘性，奥秘，秘法
onerous ['ɔnərəs] adj. 繁重的，费力的
orthodox ['ɔːθədɔks] adj. 正统的，传统的
paradoxical [ˌpærə'dɔksikəl] adj. 荒谬的
physicist ['fizisist] n. 物理学家，唯物论者
physiologist [ˌfizi'ɔlədʒist] n. 生理学家
piecemeal ['piːsmiːl] adj. 一点一点的，逐渐的
proceedings [prə'siːdiŋs] n. 会议录，活动，学报，议程，记录，项目，年 (学) 报
quantum ['kwɔntəm] n. 量，额，(物) 量子，量子论
scrutiny ['skruːtini] n. 细察
unravel [ʌn'rævəl] v. 阐明，解释，解开

难句分析

1. This choice of career was a fortunate one, because it seems that his duties in the years 1856-1864 were not so onerous as to prevent him carrying out a lengthy series of experiments concerning the inheritance in garden peas of traits such as height, flower colour and seed shape.
   译文：这个（修道士的）职业的选择是值得庆幸的，因为（在那里）从1856年至1864年他并不很忙，所以没有影响到他从事一系列的长时间的遗传学实验，这些实验是通过观察豌豆的高度、花的颜色和种子的形状来完成的。
2. By this time it was understood that the genes are carried by the chromosomes of higher organisms, an idea that was prompted by the observation that the transmission of chromosomes during cell division and reproduction exactly parallels the behavior of genes during these events.
   译文：到这个时期，人们理解了基因是由高等生物的染色体运载的，这种观点是通过观察染色体在细胞分裂和增殖过程中与基因的行为总是一致的这一现象来确立的。
3. Their contributions are important not so much for the ideas put forward but more for the influence that they had on other people.
   译文：他们的贡献之所以重要并不是因为所提出的思想，而是他们对其他人的影响。

（选自：Brown TA. Genetics a molecular approach. 3rd ed. London：chapman and hall，1998.）

The Genome Jigsaw

To understand why high-throughput gene-sequencing technology often produces frustrating results, says Titus Brown, imagine that 1000 copies of Charles Dickens' novel A Tale of Two Cities have been shredded in a wood chipper. "Your job is to put them back together into a single book," he says.

That task is relatively easy if the volumes are identical and the shreds are large, says Brown, a microbiologist and bio-informatician at Michigan State University in East Lansing. It is harder with smaller shreds, he says, "because if the sentence fragments are too small, then you can't uniquely place them in the book". There are too many ways they might fit together. "And it's harder still if the original pile of books includes multiple editions," he says.

Researchers in genetic sequencing today face a similar task. An organism's DNA—made up of four basic building blocks, or bases, denoted by the letters A, T, C and G—is chopped into short snippets, sequenced to determine the order of its bases and reassembled into what researchers hope is a good approximation of the organism's actual genome.

Today's high-throughput sequencing technology is remarkably powerful and has led to an explosion of sequencing projects in laboratories around the world, says Jay Shendure, a molecular biologist who develops sequencing methods at the University of Washington School of Medicine in Seattle. Thousands of patient tumours and more than 10000 vertebrate species have been or are being sequenced. High-throughput sequencing is now an essential tool for basic and clinical research, with applications ranging from detection of microbial "bio-threats" to finding better biofuels.

But some types of genomic DNA cannot be sequenced by high-throughput methods, leaving many frustrating gaps in data (see "What makes a tough genome?"). For example, a genome might contain long stretches in which the sequence simply repeats—as if Dickens had filled whole pages with a word or sentence written over and over—making that passage hard, if not impossible, to reconstruct by the usual technologies. And the widespread adoption of next-generation sequencing has meant that the quality of genome assemblies has declined significantly over the past six years, says Evan Eichler, a molecular biologist also at the University of Washington. Although "we can generate much, much more sequence, the short sequence-read data translate into more gaps, missing data and more incomplete references," he says.

Incomplete genomes make it harder for researchers to identify and interpret sequence variations. "Instead," Eichler says, "we focus only on the accessible portions, creating a biased view," which in turn hinders efforts to study the genetic basis of disease or how species have evolved. For example, the human-genome sequence, used as a reference by scientists around the world, has more than 350 gaps, says Deanna Church, a genomicist at the US National Center for Biotechnology Information. An updated reference genome is filling in much of the missing data, but "even with the release of the new assembly, there will still be gaps and regions that aren't well represented", she says. "It is definitely a work in progress."

More than 900 human genes are in regions where there is much repetition. About half of these genes are in areas so poorly understood that they are often excluded from biomedical study, says Eichler. Certain regions of chromosomes, notably those near centromeres (where the two halves of a chromosome connect) and telomeres (the ends of chromosomes) are especially incomplete in the reference genome.

This lack of information can have medical consequences. For example, researchers have known for more than a decade that medullary cystic kidney disease—a rare disorder that occurs in mid-life—can be caused by mutations in a gene hidden somewhere along a 2-million-base-pair stretch of chromosome 1. Early detection of the mutation is the first step towards preventative therapies, but would require a DNA test. The gene, however, lies within a region rich in sequence repeats as well as in the bases guanine (G) and cytosine (C). Such "GC-rich" regions, like repetitions, are difficult to sequence.

Only by reverting to Sanger sequencing—a classic but more laborious approach—and combining it with special assembly methods were researchers able to decipher the DNA region involved in the disease. The results, which were published in February, mean that a test to screen younger members of families affected by the disorder is now a possibility.

Sanger sequencing is a painstaking process in which each type of DNA base is labelled with a different compound. The labelled DNA is then separated and the sequence is read. For the Human Genome Project, researchers combined Sanger sequencing with techniques to establish markers that locate where the sequences fit. The approach, which has been in use for decades, delivers a read accuracy and contiguity of sequence that are unmatched by current technology, Shendure says. "I couldn't do anything remotely approaching the quality of what resulted from the project." But the art of Sanger sequencing and its associated methods cannot be scaled up for the high- throughput sequencing projects done today. "We need to think about how to 'next-generation-ify' all of this," he says.

Research to do just that is well under way, with a variety of methodologies that address problems such as repetitive sequences and GC-rich regions, as well as the knotty task of assembling complete genomes for organisms that have four or even eight copies of each chromosome, for example, as opposed to humans' two.

Some of the technologies on the horizon promise to deliver longer reads and, possibly, fewer headaches for researchers trying to assemble them. But until those instruments are on bench tops, scientists are combining new and old approaches to refine sequencing.

Rich is poor

Some of the newer approaches aim to tackle GC-rich regions. For high-throughput sequencing, DNA is often first chopped into short fragments, which are then amplified by polymerase chain reaction (PCR). But the enzyme used in PCR "has trouble getting through" GC-rich regions, says Shendure. As a result, GC-rich stretches can end up poorly represented in the DNA sample delivered to the sequencer, thus skewing the data. Some sequencing technologies, such as those made by Illumina, based in San Diego, California, use amplification before and during the sequencing process, causing further bias against GC regions.

A number of sample-preparation approaches reduce this GC bias. The amplification step is cut out completely in platforms made by Pacific Biosciences, based in Menlo Park, California, and in a method being developed at Oxford Nanopore Technologies in Oxford, UK. And although DNA read lengths differ among platforms, the most widely used bench top sequencers—which are made by Illumina—generate short reads, of around 150 base pairs.

"The killer with short reads is that they're very sensitive to repeated content," says Brown.

If the read length is shorter than a repeat—or, to draw on the book analogy, if the shreds of the novel are only a fraction as long as a repeated paragraph—it is hard or even impossible to uniquely place. "That's where things like long reads or other technologies can be helpful," says Shendure. Long DNA fragments can bridge repetitive regions and thus help to map them. As another way to ease assembly, researchers in Shendure's group and elsewhere are exploring different methods to tag and group DNA fragments before sequencing. "There are more on the horizon," says Shendure, but he prefers to divulge the details in research publications.

The terms "short" and "long" are in a state of flux in this fast-moving industry. The first generation of Illumina machines generated reads of around 25 base pairs in length; the latest ones have upped that to around 150 base pairs (see "Extended sequence"). But it is still hard to assemble a complete genome from reads of this length.

Geoff Smith, who directs technology development at Illumina in Cambridge, UK, acknowledges the drawbacks of short-read technology for sequencing repetitive regions and various types of genomic rearrangements. He says that the company aims to address issues that crop up as researchers compare genomes they sequence to reference genomes, or sequence organisms from scratch without references.

Illumina has launched a service to allow longer reads with its current short-read technology. Last year the firm bought Moleculo, a company based in San Francisco, California, which has developed a process to create long reads by stitching together short ones through a proprietary sample-preparation and computational process. In July Illumina began offering Moleculo's process as a service for customers.

The Moleculo process first creates DNA fragments about 10000 bases (10 kilobases) in length. The fragments are sheared and amplified, then grouped and tagged with a unique barcode that helps to identify which larger fragment they originated from and aids in assembly.

At present, sample preparation for the Moleculo process takes around two days. Smith says that he and his team are refining the process and that by the end of the year Illumina will launch Moleculo as a stand-alone sample-preparation kit. He says that company scientists are now evaluating the kit's performance by sequencing a well-known genome, but he prefers not to say which one.

"We suspect you will be able to uncover a lot more of the genome with 10-kilobase reads versus the [150-base-pair] read length that we currently have," says Smith. He adds that the company plans to increase the fragment length to 20 kilobases. He and his team hope to "develop better molecular-biology tools to allow us to reach into these difficult- to-sequence parts of the genome but also use those tools on well-characterized genomes," he says. The team is also tuning the Illumina software to better distinguish between false and correct reads.

The company's initiative comes at a time of intense commercial and academic activity around long-read sequencing technology and new assembly methods. Finished genomes have taken a back seat, leaving many highly fragmented assemblies that need completing, says Jonas Korlach, chief scientific officer of sequencing manufacturer Pacific Biosciences in Menlo Park, California, whose sequencer generates read lengths of around 5 kilobases.

Korlach agrees that long reads will help to sequence repetitive regions, such as those that characterize many plant genomes, for example. They will also help with the challenge of distinguishing between copies of chromosomes, important in identifying the tiny variants that can affect biological function. Humans are diploid, meaning they have two copies of each chromosome, but "many organisms, especially plants, have even more copies, which makes resolving all the different chromosomes so much harder", Korlach says.

Tough nuts

Plant sequencing, in particular, will benefit from improvements. The spruce genome is a "real nightmare", says Stefan Jansson, a plant biologist at the Umeå Plant Science Centre in Sweden. Jansson led a study that generated a draft assembly of the Norway spruce genome (*Picea abies*). In addition to being the largest genome yet sequenced, it also contains many repeats, and the differences between its chromosomes are larger than in the human genome. "Sequencing diploid spruce is like mixing human and chimpanzee DNA and then trying to assemble them simultaneously," Jansson says.

Many plants have more than two copies of chromosomes. Bread wheat (*Triticum aestivum*), for example, is hexaploid, and sequencing and assembling the six sets of chromosomes to completion has proven extremely difficult. And although some strawberry species are diploid, the commercial strawberry (*Fragaria×ananassa*) is octoploid: it has eight sets of seven chromosomes, four sets from each parent, says Thomas Davis, a plant biologist at the University of New Hampshire in Durham. "Good thing Mendel didn't use octoploid strawberries to try to understand heredity," he says.

Davis and his colleagues have published a draft genome of the diploid woodland strawberry (*Fragaria vesca*), and now want to apply their experience to the octoploid strawberry. Assembling this tough-nut genome will require high-quality reads longer than 500 base pairs, Davis says. He believes he can succeed, although he does not want to share his methodology just yet. "If anyone cracks that nut, he'll do it," says Kevin Folta, a molecular biologist at the University of Florida in Gainesville, who led the woodland-strawberry project.

The plant world has many other challenging genomes to offer. The onion genome is massive, Folta says, and sugarcane has 12 copies of each chromosome. "Those will take special techniques," he says.

Every platform has benefits and drawbacks, and scientists must weigh the costs, sample-preparation time and sequencing-error rates for each. To sequence the woodland strawberry, for example, the scientists used a combination of three platforms.

But for polyploid genomes, short-read sequencing is almost a waste of time, says Clive Brown, chief technology officer at Oxford Nanopore. "You don't know where your short read comes from, which chromosome it is from," he says. "It's very hard to piece that together." He believes that the problem will be helped by instruments, including those in development in his company, that can generate long reads without the need for special sample preparation or complex assembly. The longer the reads, the easier the assembly, because the overlapping sequences will help researchers to determine which sequence belongs to which chromosome.

Fresh approaches were needed to crack the genome of the oil palm (*Elaeis guineensis*), reported last month in Nature. The effort was more than a decade in the making. Oil palm is an important source of food, fuel and jobs in southeast Asia, and the industry is under pressure to produce it sustainably and avoid increased rainforest logging, says study co-leader Ravigadevi Sambanthamurthi, director of the advanced biotechnology and breeding centre at the Malaysian Palm Oil Board in Kajang, which works with the country's oil-palm industry.

With millions of repeats distributed throughout the plant's genome, short reads could fit in many possible spots in the assembled DNA sequence. "It is as if you were assembling a jigsaw puzzle in which most of the pieces are identical," says Robert Martienssen, a geneticist at Cold Spring Harbor Laboratory in New York, who co-led the project with Sambanthamurthi.

Classic sequencing methods were too laborious and expensive for the oil-palm project. So Martienssen suggested applying a technique based on a finding he had made in 1998—that repeats in plant genomes can be distinguished from genes because the cytosine bases in the repeats usually carry methyl groups. Before fragments are sent to the sequencer, they are treated with enzymes that digest methylated DNA and thereby remove the repeats from the samples.

To complete the oil-palm project, the scientists applied this methylation-filtration technique and then sequenced the DNA regions housing genes. The technique has now been commercialized through Orion Genomics, a company based in St Louis, Missouri, which Martienssen co-founded.

The researchers used a high-throughput sequencer made by 454 Life Sciences, a company owned by Roche and based in Branford, Connecticut, that generates short reads from longer, filtered fragments. In preparing the samples, the researchers used bacteria to amplify DNA in large chunks on bacterial artificial chromosomes—an approach also used in the Human Genome Project—to pin down hard-to-map regions by retaining them next to genes with known positions to act as signposts.

Assembly of the oil-palm genome called for extensive computational resources, which crashed multiple times, the researchers say. But now, with the genome in hand, they have located a gene that encodes the shell of the palm fruit, knowledge they hope to harness to increase the plant's yield.

Sambanthamurthi says that when the researchers finally pinned down the shell gene, they popped a bottle of champagne, then celebrated with a traditional Malaysian meal served on a banana leaf.

The long and the short

Bacterial genomes are smaller and less complex than those of plants and other multicellular organisms, but they, too, have regions that are tough to sequence. For example, Bordetella pertussis, which causes whooping cough, has hundreds of insertion sequence elements—stretches of mobile DNA inserted into various locations in the genome—each more than 1 kilobase long. Proponents of long-read technology say that spanning these regions with long reads will deliver sequencing efficiency gains.

Korlach points out that it took a team of more than 50 scientists to solve the bacterium's complete genome 7. But long-read technology can make assembly of highly repetitive genomes faster and easier, he says. He says that he and scientists in the Netherlands were able to assemble nine whooping-cough bacterial strains in one month.

Whether a read is classified as "long" or "short" is in great flux. Two years ago, scientists might have said that a long read was 1 kilobase, Korlach says. "Now [Pacific Bio- sciences] customers are generating an average of 5000 bases, with some reads longer than 20000 bases—and we are working to deliver even more than that." Ultimately, a "long read" will be as long as is needed to sequence a given genome, he says.

Korlach knows that some scientists say his company's sequencers are pricey, but he says that the newer versions have seen a significant drop in price and an increase in throughput. He says that the question of price is often raised "in the context of pure cost per sequenced base". And, he adds, if a certain sequencing technology is the only one that will work to solve a medically important question, "then there is no price tag that can be put on this medically relevant information".

Last year, researchers collaborating with Pacific Biosciences used the company's sequencer to distinguish the repetitive genomic region involved in fragile X syndrome, a developmental disorder that is caused by repeats in a particular region on the X chromosome, and that worsens in severity with higher numbers of repeats.

As technology developers get closer to instruments that produce longer reads, scientists will need longer DNA fragments at the beginning of their sequencing experiments. Several companies focus on helping researchers to prepare DNA fragments for sequencing. For example, Sage Science, based in Beverly, Massachusetts, has a platform that uses pulsed-field electrophoresis to select and sort DNA fragments of sizes ranging from 50 base pairs to 50000 base pairs. In May, the company began marketing its instrument to accompany the Pacific Biosciences sequencing platform.

Steve Siembieda, who is responsible for business development at Advanced Analytical Technologies in Ames, Iowa, says that his company sees the trend towards longer reads as writing on the wall. The company has licensed patents from Iowa State University, also in Ames, to develop an instrument to assess the integrity, fragment length and concentration of DNA samples.

With this instrument, an electric field is applied to a tiny amount of DNA so that it is pulled into a long, hair-thin capillary tube containing a gel with a fluorescent dye that binds to DNA molecules. As the DNA fragments move through the gel, they separate according to size. "Small molecules move fast, big molecules move slowly," Siembieda says. As the molecules pass by a window in the capillary, a flash of light excites the dye and a camera records the DNA fragment length.

The instrument's readout tells scientists whether the size distribution of the DNA fragments is in the range needed for a given sequencing platform and whether the DNA has the right concentration. Siembieda says that skipping these measurements can be the wrong experimental shortcut—if the concentration or fragment size is off, "a sequencer may run for nine days, it will cost them thousands of dollars, plus all the time wasted to not make sure they have the appropriate material". The instrument will possibly be used in developing the Moleculo process, but negotiations between the two companies are still under way.

Technology development at Advanced Analytical is focusing increasingly on long DNA fragments, which are challenging to resolve, Siembieda says. One solution is to customize gels for different applications. At present, the company's instrument can resolve lengths of up to 20 kilobases and the company is working on resolving longer fragments, he says.

Assembly required

Scientists are applying many methods and tricks to create longer fragments. "Unfortunately, these technology tricks create erroneous data at points, so now you're stuck with some data that may be wrong," says Michigan State's Titus Brown. He was part of an effort, published in April 9, to sequence the lamprey (Petromyzon marinus) genome, one-third of which is covered by long repeats. Obtaining an assembly even with Sanger sequencing, which generates 1-kilobase reads, was difficult, he says. In addition, the lamprey genome has many GC-rich regions. The team used several types of software to assemble the complete DNA sequence.

In July, scientists published a comparison of software programs used to assemble sequence reads. The researchers found that different assemblers give different results—even when fed the same sequence reads. Brown says that biologists should never forget that assemblies are not certainties. Every new sequencing technology—from how the DNA sample is prepared to the sequencing chemistry—has the potential for error and bias. "If you have short reads, or bad biology, you're going to have a very hard time getting a good assembly, even in theory," he says.

Ideally, a genome assembly should deliver end-to-end chromosomal sequences, says Shendure. What worries him more than the discordance among assemblers in the comparison study is that all of the assemblies were very fragmented. "That's not a fault of the assem blers, that's a fault of the data that we're putting into the assemblers and the fact that we're not capturing contiguity at these longer scales," he says. "The algorithms can only make do with the ingredients that they are provided by the technologies."

Brown is hopeful about the potential impact of longer-read technology. If Pacific Biosciences or Oxford Nanopore "deliver on many inexpensive long reads—more than 10 kilobases, I'd say—regardless of accuracy, you would end up revolutionizing the genome- assembly field, because it would give you so much more information to work with", he says. However, he adds, assembly software has to be compatible with each sequencing method. "So we're continually playing catch up, where new sequencing technologies lead to new sequence- analysis approaches a year or three later."

Eichler agrees that sequencing and assembly must continue to improve. Read lengths longer than 200 kilobases and with 99.9% accuracy rates will be needed to unpick repeats and other complications, he says. He says that the Pacific Biosciences instrument and what he knows of Moleculo "fall short of this, but are on the right track". All read-length requirements depend on the genome and complexity, he adds. For many bacterial genomes, current read lengths and accuracy are already sufficient, he says.

The next telescope

Oxford Nanopore plans to launch its new sequencing technology in the near future, but no date has been given. The technology expands on findings by researchers at Harvard University in Cambridge, Massachusetts, the University of Oxford, UK, and the University of California in Santa Cruz to harness the abilities of pore-forming proteins for DNA- sequencing devices.

One of the weaknesses of current high- throughput sequencing technology is amplification chemistry, says Oxford Nanopore's Clive Brown. Although DNA is made up of four bases, it is possible that more than those canonical four—such as bases that are methylated—should be detected, he says.

And in some sections of genomes, bases are naturally missing. But current sequencers do not capture such variations—instead, says Brown, they produce the equivalent of a four-colour photocopy of a picture with many more colours. "A lot of the detail is lost immediately, as soon as you make a four-colour copy," he says. Ideally, "you take a chromosome and run it through the sequencer. You can't quite do that yet." He, too, says that the next crucial phase of sequencing technology will be about long reads.

Brown says that to his mind, sequencers are just opening the door to characterizing the genome. People can get "very cosy about what they can see", with scientific instruments, he says. He likens today's sequencers to the first telescopes, which offered a view of the Moon's features and exploration of the visible spectrum. "It gets you a long way, you can count the stars, see the planets," he says. But the telescope does not show other celestial phenomena—such as dark matter or galactic movement.

Like astronomers with their telescopes, genome researchers will get a clearer picture of the genome as the sequencing technologies improve, he says. And, inspired by that picture, they will strive to see even more.

Glossary

algorithms [ˈælgərɪðəm] n. 算法
amplification [ˌæmplifiˈkeiʃn] n. 扩大，扩充，详述，引申，推广，增幅
base [beɪs] n. 碱基
chimpanzee [ˌtʃimpænˈziː] n. 黑猩猩
chromosome [ˈkrəuməsəum] n. 染色体
computational [ˌkəmpjuˈteiʃənl] adj. 计算的
cystic [ˈsistik] adj. 囊的，胆囊的，膀胱的
cytosine [ˈsaitəusiːn] n. 胞嘧啶，氧氨嘧啶
diploid [ˈdiplɔid] adj. 双重的；双倍性的，具有两套染色体的
electrophoresis [iˌlektrəufəˈriːsis] n. 电泳分离法
Elaeis n. 油棕属

encode [in'kəʊd] vt. 编码
fragmented ['frægmənt] adj. 成碎片的，片段的
Fragaria n. 草莓属
genome ['dʒiːnəʊm] n. 基因组，染色体组
guanine ['gwɑːniːn] n. 鸟嘌呤
hexaploid ['heksəplɔid] adj. 有六倍体的
high-throughput sequencing 高通量测序
kilobase ['kiləubeis] n. 千碱基
medullary ['medələri] adj. 骨髓的，髓质的
methylation [meθi'leiʃn] n. 甲基化作用
microbial [mai'krəubiəl] adj. 微生物的，由细菌引起的
molecular biologist 分子生物学
multicellular [mʌlti'seljʊlə] adj. 多细胞的
mutation [mjuː'teiʃn] n. 变化，突变，变异
octoploid ['ɔktəplɔid] n. 八倍体
organism ['ɔːgənizəm] n. 有机体，生物体
overlapping [əʊvə'læpiŋ] adj. 相互重叠的
polymerase chain reaction 聚合酶链反应
syndrome ['sindrəʊm] n. 症候群，综合征
Triticum aestivum n. 小麦
vertebrate adj. 脊椎的，脊椎动物的；n. 脊椎动物

难句分析

1. Researchers in genetic sequencing today face a similar task. An organism's DNA—made up of four basic building blocks, or bases, denoted by the letters A, T, C and G—is chopped into short snippets, sequenced to determine the order of its bases and reassembled into what researchers hope is a good approximation of the organism's actual genome.
   译文：现如今，基因测序研究人员都面临着相似的挑战。一个生物体的DNA，由四个基本模块或碱基（用字母A、T、C和G表示）组成，为了测定碱基的顺序，将DNA切成一些短片段并进行测序，随后重新组装成为研究者们希望的更接近生物体真实基因组的样子。
2. For example, a genome might contain long stretches in which the sequence simply repeats—as if Dickens had filled whole pages with a word or sentence written over and over—making that passage hard, if not impossible, to reconstruct by the usual technologies.
   译文：例如，一个基因组可能包含很长的简单重复序列，这类似于狄更斯曾用一个字或一句话不断地重复来填充整个文章，这使得段落变得很难，但用普通的技术来重建也并非不可能。
3. Research to do just that is well under way, with a variety of methodologies that address problems such as repetitive sequences and GC-rich regions, as well as the knotty task of assembling complete genomes for organisms that have four or even eight copies of each chromosome, for example, as opposed to humans' two.
   译文：目前正在进行的研究工作是应用不同的方式来解决出现的问题，例如重复序列、高GC含量区域，还有一个棘手问题是为生物体完整构建一个每个染色体有四至八个拷贝的基因组，而不是像与人类一样的两组拷贝的基因组。

4. The amplification step is cut out completely in platforms made by Pacific Biosciences, based in Menlo Park, California, and in a method being developed at Oxford Nanopore Technologies in Oxford, UK. And although DNA read lengths differ among platforms, the most widely used bench top sequencers—which are made by Illumina—generate short reads, of around 150 base pairs.

   译文：通过加利福尼亚州门洛帕克市的太平洋生物科学公司（Pacific Biosciences）制造的平台，同时采用在英国牛津的牛津纳米技术公司（Oxford Nanopore Technologies）开发的方法，扩增步骤被完全切除。虽然DNA阅读长度在不同的平台上有所不同，但使用由Illumina生产的台式测序仪可以产生约150个碱基对的短读数。

5. Humans are diploid, meaning they have two copies of each chromosome, but "many organisms, especially plants, have even more copies, which makes resolving all the different chromosomes so much harder", Korlach says.

   译文：人类是二倍体，这意味着他们每个染色体有两个拷贝，但是Korlach说"许多生物体，特别是植物，有更多的拷贝，这使得解决所有不同的染色体变得更加困难"。

6. And although some strawberry species are diploid, the commercial strawberry (*Fragaria* × *ananassa*) is octoploid: it has eight sets of seven chromosomes, four sets from each parent, says Thomas Davis, a plant biologist at the University of New Hampshire in Durham.

   译文：达勒姆新罕布什尔大学的植物生物学家托马斯·戴维斯说：虽然有些草莓品种是二倍体，但是商业草莓（*Fragaria*×*ananassa*）是八倍体，它有八套染色体组，每套各七条染色体，每个亲本提供四套染色体组。

7. Classic sequencing methods were too laborious and expensive for the oil-palm project. So Martienssen suggested applying a technique based on a finding he had made in 1998—that repeats in plant genomes can be distinguished from genes because the cytosine bases in the repeats usually carry methyl groups.

   译文：经典的测序方法对于油棕项目而言过于费力和昂贵。所以Martienssen建议应用一项于1998年发现的技术——在植物基因组中的重复基因可以被区分出来，因为重复基因中的胞嘧啶碱基通常带有甲基。

8. In preparing the samples, the researchers used bacteria to amplify DNA in large chunks on bacterial artificial chromosomes—an approach also used in the Human Genome Project—to pin down hard-to-map regions by retaining them next to genes with known positions to act as signposts.

   译文：在准备样品的过程中，工作人员利用细菌在人造染色体上大量扩增DNA（在人类基因组计划中也使用了这一种方法），通过将相邻的已知基因作为标记点来确定难以定位的区域。

9. For example, Sage Science, based in Beverly, Massachusetts, has a platform that uses pulsed- field electrophoresis to select and sort DNA fragments of sizes ranging from 50 base pairs to 50000 base pairs.

   译文：例如，Sage Science，位于马萨诸塞州贝弗利，有一个使用脉冲场电泳的平台，可以选择和分选大小范围从50个到50000个碱基对的DNA片段。

（选自：Marx V. Next-generation sequencing: The genome jigsaw. Nature. 2013; 501 (7466): 263-8.）

Functions of Cell Membranes

Plasma membranes

The plasma membrane of the eukaryotic cell defines the cell boundary. This membrane by weight is about half lipid and half protein. The plasma membrane of the cell exemplifies the basic compartment function of membranes; it separates and delineates intracellular from extracellular domains.

It is the role of the plasma membrane (Fig. 1. 10) to maintain the difference between the inside and the outside of the cell by controlling the entrance and exit of materials. In order to carry out this compartmental function, the plasma membrane is designed to be impermeable to substances other than water. All materials going in and out of the cell must encounter the plasma membrane. Water goes through the plasma membrane rapidly, but nothing else manages to cross the plasma membrane as readily. Therefore, to get substances in and out of the cell, the cell must have transport systems in the plasma membrane.

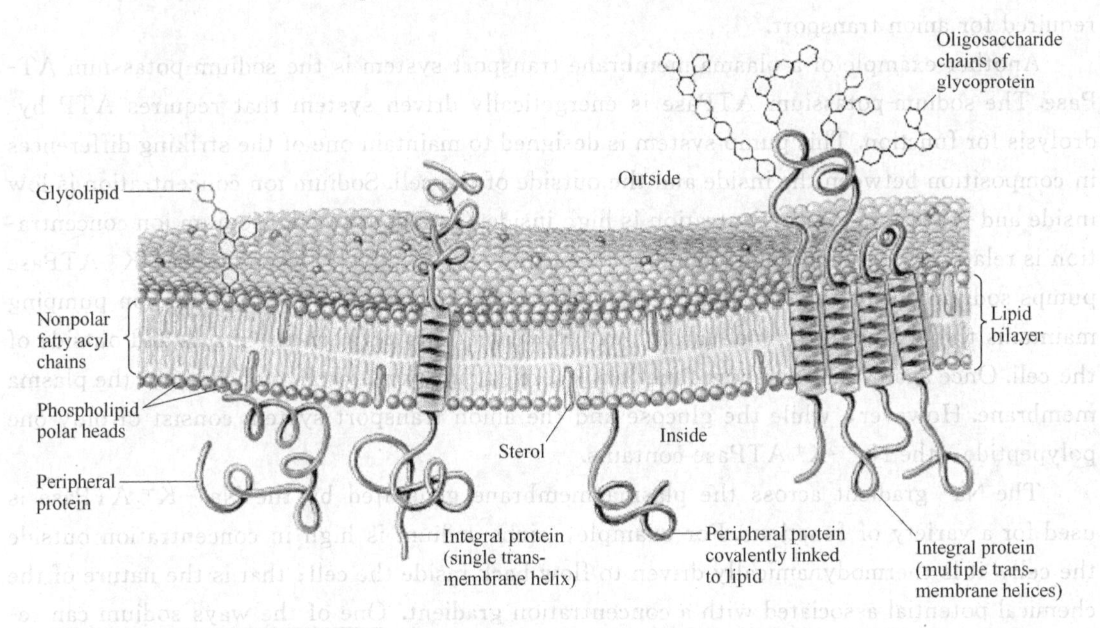

Fig. 1. 10 Fluid mosaic model for membrane structure

The fatty acyl chains in the interior of the membrane form a fluid hydrophobic region. Internal proteins float in this sea of lipid, held by hydrophobic interactions with their nonpolar amino acid side chains. Both proteins and lipids are free to move laterally in the plane of the bilayer, but movement of either from one face of the bilayers to the other is restricted. The carbohydrate moieties attached to some proteins and lipids of the plasma membrane are exposed on the extracellular surface of the membrane

Consider the plasma membrane of the human erythrocyte as an example. The simple structure of the human erythrocyte has made this cell a useful model for other cell membranes. Since no other cell membranes complicate the studies, a membrane study on the human erythrocyte is a study of the erythrocyte plasma membrane. So far, such studies have not led membrane biochemists seriously astray in attempts to understand other cell plasma membranes. The plasma membrane of the human red blood cell contains a protein specific for the transport of glucose across that membrane. Glucose is otherwise much less able to penetrate the intact erythrocyte. Because the human erythrocyte has only a plasma membrane, penetrating the cell involves transport across only the plasma membrane. The glucose transporter consists of a protein embedded in the plasma membrane. Membrane proteins responsible for transport usually are embedded in the membrane.

It is interesting to note that in contrast to the human erythrocytes, avian erythrocytes, such as chicken or turkey, are nucleated. Therefore, they have intracellular membranes, and the avian cell is, consequently, more complex in structure than the human cell.

Another one of the major membrane proteins contained in the human erythrocyte plasma membrane is the protein responsible for anion transport across that plasma membrane. This protein functions as a small hole in the membrane, which specifically allows anions to go back and forth rapidly across the membrane. In erythrocytes, rapid chloride-bicarbonate exchange across the plasma membrane is functionally integrated with the binding and absorption of oxygen by hemoglobin and with CO_2 transport. That exchange is made possible by the anion transport protein, a protein embedded in the plasma membrane. ATP hydrolysis is not required for anion transport.

Another example of a plasma membrane transport system is the sodium-potassium ATPase. The sodium-potassium ATPase is energetically driven system that requires ATP hydrolysis for function. This pump system is designed to maintain one of the striking differences in composition between the inside and the outside of the cell. Sodium ion concentration is low inside and potassium ion concentration is high inside. Outside the cell, sodium ion concentration is relatively high and potassium ion concentration is relatively low. The Na^+-K^+ ATPase pumps sodium out of the cell and potassium into the cell simultaneously. This ion pumping maintains the distinct difference in Na^+-K^+ concentrations between the inside and outside of the cell. Once again this transport function is carried out by a protein embedded in the plasma membrane. However, while the glucose and the anion transport system consist of only one polypeptide, the Na^+-K^+ ATPase contains.

The Na^+ gradient across the plasma membrane generated by the Na^+-K^+ ATPase is used for a variety of functions. For example, since sodium is high in concentration outside the cell, it is thermodynamically driven to flow back inside the cell; that is the nature of the chemical potential associated with a concentration gradient. One of the ways sodium can respond to the difference in chemical potential is to reenter the cell through a cotransport system involving a membrane protein distinct from the sodium pump. In cotransport, a nutrient the cell requires is transported across the plasma membrane along with a sodium ion. Sodium is moving with its concentration gradient in this system. Therefore, the energy stored in the sodium gradient can be used to drive a nutrient against its concentration gradient and into the cell. The sodium gradient that is set up by the sodium-potassium ATPase is essential for allowing the inward transport of some nutrients for the cell.

In addition to transport proteins, there are a number of receptor protein signalling systems in the plasma membrane. One of the enzymes influenced by plasma membrane receptors is the adenylate cyclase, which is located on the inside of the plasma membrane of the cell. This enzyme can be activated by hormones as they bind to cell surface receptors. Receptor-stimulated adenylate cyclase generates a signal or a second messenger, increasing cyclic AMP levels in the cell. Such an increase can cause a number of changes in the metabolic activity of the cell.

Another function of the plasma membrane of the cell is to control cell-cell interactions. During development cells differentiate and proceed to organize themselves into organs or organisms. This process depends on cells recognizing the right cell matrix and building the structure of the organism or organ according to the pattern based on their genetic material. This cell-cell recognition is mediated by the plasma membrane.

The opposite extreme, when the cells do not recognize that they are encountering other cells, and consequently, multiply without control, is characteristic of tumor cells.

The plasma membrane of the cell is involved in the conduction event of the nerve impulse. The plasma membrane of the axon carries information as a transient transmembrane electrical potential in the form of a potential spike that moves along the plasma membrane of the axon.

A related example of plasma membrane activity is synaptic function. The synapse describes a specialized region of the cell-cell contact, where the plasma membranes of two cells directly oppose each other. In the membrane of the cell that is initiating the piece of information to be conveyed, there are closed vesicles near the surface that contain, for example, acetylcholine as a neurotransmitter. Upon the change in electrical potential in the plasma membrane of the axon, these acetylcholine-containing vesicles will move to fuse with the plasma membrane and release the acetylcholine into the synaptic space. The acetylcholine diffuses across the synaptic space, where a receptor protein is located in the opposing cell plasma membrane. Upon binding the acetylcholine, this receptor initiates a new nerve impulse by modifying sodium permeability in the postsynaptic membrane.

The plasma membrane is also involved in other fusion events. The lumen of the ER, for example, is equivalent to the outside of the cell because of a fusion mediated transport process that carries membrane material, and material to be secreted, through the cell. The secretion of proteins to the outside of the cell begins with the synthesis and insertion of the protein to be secreted into the lumen of the ER. Vesicles pinch off from the ER and move to fuse with the Golgi, carrying with them the newly synthesized proteins to be secreted. Another vesicle-mediated process moves this material from the Golgi to the plasma membrane. Then these vesicles fuse with the plasma membrane. The membranes of the vesicles become part of the plasma membrane. The membranes of the vesicles become part of the plasma membrane, which results in the release of the vesicular contents to the outside the cell.

Another receptor function in the plasma membrane involves the recognition and binding of the serum low-density lipoprotein particle, LDL. Through a process called receptor-mediated endocytosis, the receptor and its bound LDL particle are incorporated into a vesicle, which pinches off from the plasma membrane and moves into the interior of the cell for further processing.

An area of communication between cells is the gap junction. This gap junction is particularly important in cell layers that function in a concerted fashion, such as an involuntary muscle tissue. In the gap junction, the plasma membranes of two cells are relatively close to each other but not touching. A narrow space remains between the cells, which is about 20-40Å across. Across this gap are connections between the two cells. These connections are apparently hollow protein tubes. Ions can pass through these tubes from one cell to another. This ion movement allows for cell communication on a primitive level, though at a rapid rate.

There are other examples of junctions between cells that do not necessarily fall under the category of communication. One example is the tight junction. This is a region in which the electron microscope does not reveal any significant intercellular space between the two cell plasma membranes. The tight junction allows epithelial tissue, such as in the lining of the intestine and the kidney, to form sealed cell layers. As a consequence, the two sides of the cell layer are not in direct communication with each other. The tight junctions force the plasma membranes of these cells to function as the primary pathway for transport through the cell layer. This pathway exploits the exquisite control of transport offered by the plasma membrane.

Another example of specialization of the plasma membrnne is found in the disk membrane of retinal cells. The rod cells are responsible for black and white vision. The rod outer segment is the part of the rod cell that contains stacked disks. The disk-shaped membranes hold the photopigment, rhodopsin. Rhodopsin is a membrane protein embedded in the membrane. The disks are specialized regions derived from plasma membrane. The plasma membrane evaginates and pinches off a disk-like structure that contains the photopigment. This disk then becomes part of the stack in the rod outer segment.

Yet another example of specialization can be found in epithelial cell layers. The plasma membrane of such an epithelial cell is roughly divided into two regions. One is the basal lateral region where transported molecules are, in many cases, secreted. The other is the brush border membrane. The name, brush border, derives from the microvilli that decorate that surface of the cell. These numerous projections of the plasma membrane greatly increase the surface area, which facilitates the transport carried out by the brush border membrane.

One other specialized membrane is provided by the Schwann cell. During development, the Schwann cell wraps itself around a nerve axon in concentric and multitudinous fashion. It creates a rather simple membrane containing mostly lipid and a couple of proteins. The many layers of the Schwann cell plasma membrane insulate the nerve electrically. When the myelin sheath breaks down, the nerve axon is no longer properly insulated. Nerve conduction thus becomes faulty in demyelinating diseases.

On the surface of the plasma membrane, there are large projections of material that are predominately carbohydrate. The carbohydrate structure attached to glycoproteins and glycoplipids in the cell makes a coat around the surface of the cell. This coat offers protection to the cell and a means by which one cell can adhere to another. This adherence phenomena is actually a recognition function whereby cells adhere to each other based on their surface carbohydrate structure. The carbohydrate also functions in some cells as a means of retaining water near the surface of the cell in dehydrating circumstances.

Extending out of the plasma membranes of some cells, particularly unicellular organisms, are flagella pellicles. Flagella are long projections of the plasma membrane with a number of cytoskeletal components inside. These flagella are involved in locomotion of the microorganism. These projections constitute another specialized piece of plasma membrane.

Basement membrane is not a membrane at all. It is a support structure largely consisting of fibrous proteins, like collagen, which give structural support on which cells can organize themselves into a tissue.

This completes an overview of the functions of the plasma membrane of eukaryotic cells. The plasma membrane creates a domain. By separating the extracellular medium from the inside of the cell, the plasma membrane controls the passage of materials into and out of the cell through its transport functions. The plasma membrane is a focal point for control of cellular activity. It regulates cell function through its receptors and through control of membrane transport. The plasma membrane provides the means for cell-cell communication, and for cellular organization during development (and dedifferentiation in tumor growth).

Intracellular membranes

Organelles within the eukaryotes are bound by either single or double membranes. These internal membranes not only enclose specialized regions within the cell, but also are involved in cellular processes, such as biosynthesis, transport, recycling, energy metabolism, and degradation.

Nuclear membrane

The nuclear membrane is actually a double membrane. Thus, one should properly refer to the inner and outer nuclear membranes. It would be anticipated that they would be functionally distinct and thus be different in composition. Unfortunately, almost no compositional information is available. Thin section electron micrographs suggest that the nuclear membrane has patches or pores. However, these are not likely to be holes in the nuclear membrane, since the nucleus is not freely permeable. An interesting speculation is that the "pores" may be processing centers that stain differently and thus are visualized under an electron microscope. There have to be places where nuclear information can be passed to the outside, in the form of mRNA, for example, and regulatory information can be passed to the inside of the nucleus, such as with steroid hormones. Some investigators suggest that the pores, which are rather large (10nm in diameter), may be areas where messenger RNA processing takes place prior to export to the cytoplasm. Chromosomal material may have some attachment to the nuclear membrane. Thus, one may speculate on a direct role of the inner nuclear membrane in gene regulation. More likely is an indirect role, in which the nuclear membranes function somewhat analogously to the plasma membrane. The mechanism of such a role might involve receptors for hormones on the nuclear membrane, which when occupied, stimulate an intranuclear "second messenger" process, similar to some plasma membrane receptors.

Endoplasmic reticulum

The endoplasmic reticulum is another internal membrane system of the eukaryotic cell. The ER membrane system encloses a specialized region referred to as the lumen. Because the lumen of the ER is separated from the cytosol of the cell, the ER allows for separation of function and separation of materials. This membrane is the site of biosynthesis of many components; in particular, many membrane components are synthesized on the ER, such as membrane proteins and lipids. The ER is subdivided into rough and smooth ER. The designation arises from the morphology observed in the electron microscope. Some parts of the ER have bumps, which give a rough appearance. The bumps are ribosomes, which are protein manufacturing complexes bound to the ER.

The smooth ER is simply where the ribosomes are not attached. These two regions of the ER apparently come in patches, thus, enabling their distinction in the electron microscope.

Ribosomes are involved in the synthesis of membrane proteins and proteins to be secreted by the cell into the extracellular medium. The membrane proteins are inserted into the membrane during or shortly after the time of their synthesis.

Concurrent with their synthesis, the secreted proteins are injected into the lumen of the ER. At this point, the protein is effectively secreted. The lumen of the ER is related morphologically to the exterior of the cell. Upon moving into the lumen of the ER, a protein to be secreted can progress through this membrane system and be transported to the Golgi. After the Golgi, the secreted proteins advance to the plasma membrane via vesicular transport. This transport occurs in such a fashion that the original cytoplasmic surface of the ER membrane becomes coincident with the cytoplasmic surface of the plasma membrane. By so doing, these membrane vesicles fuse with the plasma membrane, and the inside of the vesicle becomes continuous with the outside of the cell. The proteins to be secreted are thus transferred to the exterior of the cell.

Golgi

Golgi is a closed membrane system as is the ER. Communication between the ER and the Golgi membranes appears to be via a vesicular pathway. Therefore, the lumen of the Golgi is functionally connected with the lumen of the ER. The Golgi is also in communication with the plasma membrane of the cell, again by vesicular transport. There is considerable movement of material all around the cell, with material moving from the ER to the Golgi to the plasma membrane. One also observes movement of membrane and trapped material from the plasma membrane to the Golgi and to the ER. Furthermore, there is movement of material from the plasma membrane to the lysosome elsewhere. There also has to be movement from the ER membrane to the mitochondria. The Golgi is intimately involved in these complex pathways for movement of cellular material. The Golgi is also involved in synthesis, because it glycosylates lycoproteins, including secreted proteins.

Movement of material is a directional process in which the Golgi itself can be functionally subdivided. For example, there appears to be a progressive glycosylation as proteins move through the Golgi. The Golgi has been divided into *cis* and *trans* aspects to emphasize the fact that there is a progression of metabolic behavior. Primitive glycosylation takes place on one side (closest to the ER or even in the ER), and a more extensive glycosylation takes place on the side closest to the plasma membrane.

Vesicles of communication

One of the challenges for cell biologists is to understand the movement of material throughout the cell, considering the fact that the target sites for this movement are all morphologically and functionally different. What governs this movement of material? If one considers the protein composition of the various intracellular membranes, one finds, for example, that there is little in common between the protein composition of the plasma membrane and the protein composition of the mitochondria. Furthermore, there is little in common between the protein composition of the plasma membrane and that of the ER. How these differences in membrane composition are maintained in the fact of interorganelle communication is a fascinating, unsolved mystery.

It is worthwhile here to introduce one other example of membrane vesicular movement, receptor-mediated endocytosis. For example, the LDL receptor is one that operates using receptor-mediated endocytosis. In this process, the receptor, which is a protein embedded in the cell's plasma membrane, first binds the LDL particle. The LDL particle then moves to a specialized patch of plasma membrane, which makes a coated pit. The name, coated pit, is given to this patch of membrane because the plasma membrane that invaginates to form a vesicle for intracellular movement is coated by electron dense material, as seen with an electron microscope. The electron dense material turns out to be a protein that fabricates a matrix surrounding this invaginated region of the plasma membrane (This protein, which is called clathrin, is in fact capable of making hollow baskets by itself without the membrane. These baskets resemble the coat of the coated vesicle) . The clathrin-coated vesicle eventually pinches off so that a coated vesicle forms. This coated vesicle then can move to the Golgi, or depending on what is happening, can move to the lysosome, which apparently is what happens in case of LDL.

An interesting related phenomenon is that receptors can frequently recycle. Therefore, the system must work in such a way that it can sort itself, cycling those things that are appropriate to the plasma membrane back to the plasma membrane.

The transferring receptor is another system that works via a recycling mechanism. This receptor, however, according to current models, apparently employs an additional interesting twist. The transferring-containing endocytotic vesicle fuses with an endosome and the pH is lowed, likely by an ATP-driven proton pump. Because of the low pH, the iron of the transferring is releases (transferring affinity for the iron is diminished at low pH) . The apo-transferrin is then hypothesized to be recycled to the surface with its receptor. At the higher extracellular pH, apo-transferrin affinity for the receptor is diminished. This provides a likely mechanism for release of the apo-transferrin and binding of the transferring containing iron for another cycle of iron uptake.

Recycling is also observed for the receptor that is involved in the degradation of the Ashwell protein. This is a glycoprotein that is cycled into the cell via receptor-mediated endocytosis for degradation. Release of the glycoprotein by the receptor allows the receptor to be recycled to the cell surface for another uptake cycle.

Mitochondrial membranes

Another organelle in the cell is the mitochondrion, which has two membranes, the outer and the inner mitochondrial membranes. The inner mitochondrial membrane significantly differs from other membranes because of its high protein content. For example, the inner mitochondrial membrane has much less lipid than the plasma membrane and the ER membrane. ATP synthesis is perhaps the most important function of the mitochondria. The pathway of ATP synthesis involves an intricate set of enzymes located in the inner mitochondrial membrane. This pathway is good example of the earlier observation that some cellular functions are uniquely dependent on membranes. The inner membrane is a highly convoluted membrane with lots of surface area and has many enzymes of the oxidative phosphorylation pathway incorporated in it. Oxidative phosphorylation is a series of biochemical steps in which electrons are transported and ATP is synthesized, while oxygen is reduced to water.

Not only do all of these steps take place in and around the inner mitochondrial membrane, but oxidative phosphorylation would not take place without a membrane and its structural features. The process leading to ATP synthesis creates, through a mechanism involving the movement of electrons, a proton gradient across the inner mitochondrial membrane. Embedded in that membrane is an enzyme this is capable of ATP synthesis from ADP and inorganic phosphate. The enzyme obtains the energy for ATP synthesis from the proton gradient. Thus, protons get pumped across the inner mitochondrial membrane and then flow back through this membrane via the protein responsible for ATP synthesis. It is the flow of protons through this protein that produces the ATP. Such a pH gradient cannot exist without a membrane that is otherwise impermeable to protons. Therefore, the ATP synthesis system functions only because it is a membrane system.

Lysosome

Another important organelle in the eukaryotic cell is the lysosome. This organelle is bounded by a single membrane. The lysosome is a degradative organelle. In fact in the lysosome, where the pH is lower than in the rest of the cell, there are enzymes responsible for degradation; these include enzymes for cleaving proteins, phospholipids, cholesterol esters, and carbohydrates into their respective subunits. Through intracellular vesicles, the lysosome is in communication with other organelles and the plasma membrane.

Viruses

Enveloped viruses constitute an important class of membranes outside the area of eukaryotic and prokaryotic cells. Viruses contain genetic material. Some contain RNA and some contain DNA. Viruses differ in how that genetic material is packaged. It is usually packaged in connection with protein-nucleic acid interactions, which create a complex of protein and RNA or protein and DNA. This complex is coiled, and a tight structure is created, which disassembles inside the cell on infection and takes over the metabolic machinery of the cell on infection and takes over the metabolic machinery of the cell. Some viruses consist of only protein and nucleic acid. Other viruses have an envelope or a membrane surrounding the protein-nucleic acid complex (nucleocapsid).

Two examples of enveloped viruses will be considered. The first example is vesicular stomatitis virus, VS virus. This is a rhabdovirus. It has a bullet shape covered with spikes, as visualized in the electron microscope. On the inside is found a nucleocapsid, which contains a single strand of RNA that is tightly wrapped with protein. The viral proteins play both structural and functional roles. Between the nucleocapsid and the viral membrane there is a protein underlying the plasma membrane. This protein is called the M protein, which is the matrix or structural protein of the viral envelope membrane. M changes the properties of the membrane when it interacts with the viral envelope.

The envelope surrounding the nucleocapsid contains a phospholipids bilayer. Inserted in that phospholipids bilayer are glycoproteins of a single type, referred to as G protein. The properties of the envelope membrane are distinctly different from normal cell plasma membranes. This difference is of interest to the biologist and to the membrane biochemist because it is from the plasma membrane of the cell the virus buds when it replicates.

In the process of infection, the virus initially fuses its viral membrane with the plasma membrane of the target cell with the aid of the G protein. The viral envelope becomes one with the membrane of the cell, injecting the nucleocapsid into the cell. This is followed by the process of viral replication. The cell, under the influence of the viral RNA, starts making viral proteins and the associated RNA for the virus. Newly synthesized G protein is inserted into the plasma membrane. The nucleocapsids then migrate to the plasma membrane and, with the aid of the M protein, associate with a patch of the plasma membrane enriched in the G protein. This process eventually culminates in the production of new viral particles by budding from the plasma membrane.

Thus, the virus selects host cell membrane components to make its envelope. However, the viral membrane exhibits different properties than the membrane from which it buds. The cause of this difference is the protein in the membrane.

Another example is the paramyxovirus Sendai virus, which is also an enveloped virus. It is a spherical virus, exhibiting spikes protruding from its envelope. It contains RNA in its nucleocapsid. It also has an M protein. The spikes are different from those found in the vesicular stomatitis virion because there are two kinds of spikes in Sendai virus. There is a spike referred to as F and a spike referred to as HN, each kind of spike consisting of different proteins. This viral particle has to fuse with the plasma membrane of the cell for successful infection. Sendai virus more specialized than the VS virus in that it has the F protein, which appears to be designed to enhance the fusion event, Therefore, F stands for fusion protein. The stimulation of membrane fusion by F is exploited to promote artificially induced cell fusion.

The other surface glycoprotein, HN, has both hemaglutinin and neurominidase activity. The hemaglutinating activity is what causes Sendai virus to stick to cell surfaces. The ability of Sendai to agglutinate red cells provides an assay for the activity of the HN protein.

The envelope membrane of Sendai virus is even more different from cell plasma membranes than is the envelope of vesicular stomatitis virus. The surface of Sendai virus membranes appears to be at least partially dehydrated. Dehydration may be a part of some fusion pathways. Thus, the surface of this virus may be specially constructed to enhance the fusion activity of Sendai.

Glossary

acetylcholine [ə͵si:təlˈkəulin] n. （化）乙酰胆碱
agglutinate [əˈglu:tineit] adj. 黏着的，胶合的；vt. 使……黏着；vi. 成胶状，黏合
anion [ˈænaiən] n. 阴离子
avian [ˈeiviən] adj. 鸟类的；n. 鸟
axon [ˈæksɔn] n. （神经的）轴突
bicarbonate [baiˈkɑ:bənit] n. 碳酸氢盐
chloride [ˈklɔ:raid] n. 氯化物
chromosomal [͵krəuməˈsəuməl] adj. 染色体的
cyclase [ˈsikleis] n. （生化）环化酶
dedifferentiation n. 去分化
dehydrate [di:ˈhaidreit] vt. （使）脱水
electron micrograph 电子显微镜照片

endosome [ˈendəˌsəum] n. 内含体
erythrocyte [iˈriθrəusait] n. 红细胞
glycosylate [ˈglaikəsileit] vt. (生化)使(蛋白质)糖基化
Golgi [ˈgɔːldʒi] n. 高尔基体
gradient [ˈgreidiənt] n. 梯度，倾斜度，坡度；adj. 倾斜的
impermeable [imˈpəːmjəbl] adj. 不能渗透的，不渗透性的
impulse [ˈimpʌls] n. 推动，刺激，冲动，推动力
inorganic [ˌinɔːˈgænik] adj. 无生物的，无机的
intestine [inˈtestin] n. 肠
kidney [ˈkidni] n. 肾
low-density lipoprotein(LDL) 低密度脂蛋白
lumen [ˈljuːmin] n. (解)内腔
lysosome [ˈlaisəsəum] n. (生)(细胞中的)溶酶体
membrane [ˈmembrein] n. 膜
microvilli [ˌmaikrəuˈvilai] n. 微绒毛
mitochondrion [maitəuˈkɔndriən] n. 线粒体
myelin [ˈmaiəli(ː)n] n. 髓鞘
nerve [nəːv] n. 神经，胆量，勇气，叶脉
neurominidase [ˌnjuəˈminədeis] n. (生化)神经氨(糖)酸苷酶，唾液酸苷酶
nucleocapsid [njuːkliəˈkæpsid] n. 核壳体，核蛋白壳，核包核酸，病毒粒子
organelle [ˌɔːgəˈnel] n. 细胞器
paramyxovirus [ˌpærəˌmiksəˈvaiərəs] n. (微)副黏病毒
pellicle [ˈpelikl] n. 薄皮，薄膜；(动)表皮
photopigment [ˈfəutəuˌpigmənt] n. 感光色素
pit [pit] vt. 去……之核
potassium [pəˈtæsiəm] n. 钾
proton [ˈprəutɔn] n. 质子
retinal cell 视黄醛细胞，视网膜细胞
rhabdovirus [ˈræbdəuˌvaiərəs] n. (医)棒状病毒
rhodopsin [rəuˈdɔpsin] n. 视紫红质
rod cell 杆状核细胞
Schwann cell 神经鞘细胞，施万细胞
Sendai virus n. 仙台病毒，副流感病毒 1 型
sheath [ʃiːθ] n. 鞘，护套，外壳
sodium [ˈsəudiəm] n. 钠
steroid [ˈstiərɔid] n. 类固醇，甾体化合物
synapse [siˈnæps] n. (解)神经元的神经线连接，神经腱；vi. 形成突触
vesicle [ˈvesikl] n. 小囊泡，顶囊

难句分析

1. Water goes through the plasma membrane rapidly, but nothing else manages to cross the plasma membrane as readily.

   译文：水可以快速地通过质膜，但其他物质都不能像水那样容易地透过质膜。

2. One of the ways sodium can respond to the difference in chemical potential is to reenter the cell through a cotransport system involving a membrane protein distinct from the sodium pump.

译文：钠对化学位差响应的方式之一是通过共传递系统重新进入细胞，共传递系统中涉及的蛋白质和钠泵是不同的。

3. The opposite extreme，when the cells do not recognize that they are encountering other cells，and consequently，multiply without control，is characteristic of tumor cells.

   when 引导定语从句修饰 The opposite extreme，其中 that they are encountering other cells 是 recognize 的宾语从句。全句可译为：能识别其他细胞并且无控制的增殖是另一种极端，这种状态是瘤细胞的特征。

4. The mechanism of such a role might involve receptors for hormones on the nuclear membrane，which when occupied，stimulate an intranuclear "second messenger" process，similar to some plasma membrane receptors.

   which 引导的定语从句和形容词短语 similar to 都修饰 receptors for hormones。全句可译为：这一作用的机制可能包括核膜上的激素受体，当这一受体被占据时，能够启动核内"第二信使"过程，激素的受体与某些质膜受体相似。

5. The designation arises from the morphology observed in the electron microscope. Some parts of the ER have bumps，which give a rough appearance. The bumps are ribosomes，which are protein manufacturing complexes bound to the ER.

   译文：电子显微镜观察发现，内质网的某些部分有突起使得其表面变得粗糙。这些突起是核糖体，是它将包装好的蛋白复合体运送到内质网的。

6. What governs this movement of material? If one considers the protein composition of the various intracellular membranes，one finds，for example，that there is little in common between the protein composition of the plasma membrane and the protein composition of the mitochondria.

   译文：是什么控制这种物质的运动？如果认为是各种胞内膜的蛋白组分，将会发现例如胞浆膜和线粒体膜的蛋白组成有很少是相似的。

（选自：李青山，安玉华，刘永新，陶小娟编．生物化学和分子生物学英语．长春：吉林大学出版社，1994.）

General Properties of Immune Responses

The term immunity is derived from the Latin word immunitas，which referred to the protection from legal prosecution offered to Roman senators during their tenures in office. Historically，immunity meant protection from disease and，more specifically，infectious disease. The cells and molecules responsible for immunity constitute the immune system，and their collective and coordinated response to the introduction of foreign substances is called the immune response.

The physiologic function of the immune system is defense against infectious microbes. However，even noninfectious foreign substances can elicit immune responses. Furthermore，mechanisms that normally protect individuals from infection and eliminate foreign substances are themselves capable of causing tissue injury and disease in some situations. Therefore，a more inclusive definition of immunity is a reaction to foreign substances，including microbes，as well as to macromolecules such as proteins and polysaccharides，regardless of the physiologic or pathologic consequence of such a reaction. Immunology is the study of immunity in this broader sense and of the cellular and molecular events that occur after an organism encounters microbes and other foreign macromolecules.

Historians often credit Thucydides，in Athens during the fifth century BC，as having first mentioned immunity to an infection that he called "plague" (but that was probably not the bubonic plague we recognize today). The concept of immunity may have existed long before，as suggested by the ancient Chinese custom of making children resistant to smallpox by having them inhale powders made from the skin lesions of patients recovering from the disease. Immunology，in its modern form，is an experimental science，in which explanations of immunologic phenomena are based on experimental observations and the conclusions drawn from them. The evolution of immunology as an experimental discipline has depended on our ability to manipulate the function of the immune system under controlled conditions. Historically，the first clear example of this manipulation，and one that remains among the most dramatic ever recorded，was Edward Jenner's successful vaccination against smallpox. Jenner，an English physician，noticed that milkmaids who had recovered from cowpox never contracted the more serious smallpox. On the basis of the observation，he injected the material from a cowpox pustule into the arm of an 8-year-old boy. When this boy was later intentionally inoculated with smallpox，the disease did not develop. Jenner's landmark treatise on vaccination (Latin vaccines，of or from cows) was published in 1798. It led to the widespread acceptance of this method for inducing immunity to infectious diseases，and vaccination remains the most effective method for preventing infections. An eloquent testament to the importance of immunology was the announcement by the World Health Organization in 1980 that smallpox was the first disease that had been eradicated worldwide by a program of vaccination.

Since the 1960s，there has been a remarkable transformation in our understanding of the immune system and its functions. Advances in cell culture techniques (including monoclonal antibody production)，immunochemistry，recombinant DNA methodology，X-ray crystallography, and creation of genetically altered animals (especially transgenic and knockout mice) have changed immunology from a largely descriptive science into one in which diverse immune phenomena can be explained in structural and biochemical terms.

Innate and adaptive immunity

Defense against microbes is mediated by the early reactions of innate immunity and the later responses of adaptive immunity. Innate immunity (also called natural or native immunity) consists of cellular and biochemical defense mechanisms that are in place even before infection and poised to respond rapidly to infections. These mechanisms react only to microbes and not to noninfectious substances, and they respond in essentially the same way to repeated infections. The principal components of innate immunity are①physical and chemical barriers, such as epithelia and antimicrobial substances produced at epithelial surfaces; ②phagocytic cells (neutrophils, macrophages) and NK (natural killer) cells; ③blood proteins, including members of the complement system and other mediators of inflammation; and④proteins called cytokines that regulate and coordinate many of the activities of the cells of innate immunity. The mechanisms of innate immunity are specific for structures that are common to groups of related microbes and may not distinguish fine differences between foreign substances. Innate immunity provides the early lines of defense against microbes.

In contrast to innate immunity, there are other immune responses that are stimulated by exposure to infectious agents and increase in magnitude and defensive capabilities with each successive exposure to a particular microbe. Because this form of immunity develops as a response to infection and adapts to the infection, it is called adaptive immunity. The defining characteristics of adaptive immunity are exquisite specificity for distinct molecules and an ability to "remember" and respond more vigorously to repeated exposures to the same microbe. The adaptive immune system is able to recognize and react to a large number of microbial and nonmicrobial substances. In addition, it has an extraordinary capacity to distinguish among different, even closely related, microbes and molecules, and for this reason it is also called specific immunity, to emphasize which potent protective responses are "acquired" by experience. The components of adaptive immunity are lymphocytes and their products (Fig. 1. 11). Foreign substances that induce specific immune responses or are the targets of such responses are called antigens. By convention, the terms immune responses and immune system refer to adaptive immunity, unless stated otherwise.

Innate and adaptive immune responses are components of an integrated system of host defense in which numerous cells and molecules function cooperatively. The mechanisms of innate immunity provide effective defense against infections. However, many pathogenic microbes have evolved to resist innate immunity, and their elimination requires the powerful mechanisms of adaptive immunity. There are two important links between innate immunity and adaptive immunity. Firstly, the innate immune response to microbes stimulates adaptive immune responses and influences the nature of the adaptive responses. Secondly, adaptive immune responses use many of the effector mechanisms of innate immunity to eliminate microbes, and they often function by enhancing the antimicrobial activities of the defense mechanisms of innate immunity.

Innate immunity is phylogenetically the oldest system of host defense, and the adaptive immune system evolved later. In invertebrates, host defense against foreign invaders is medi-

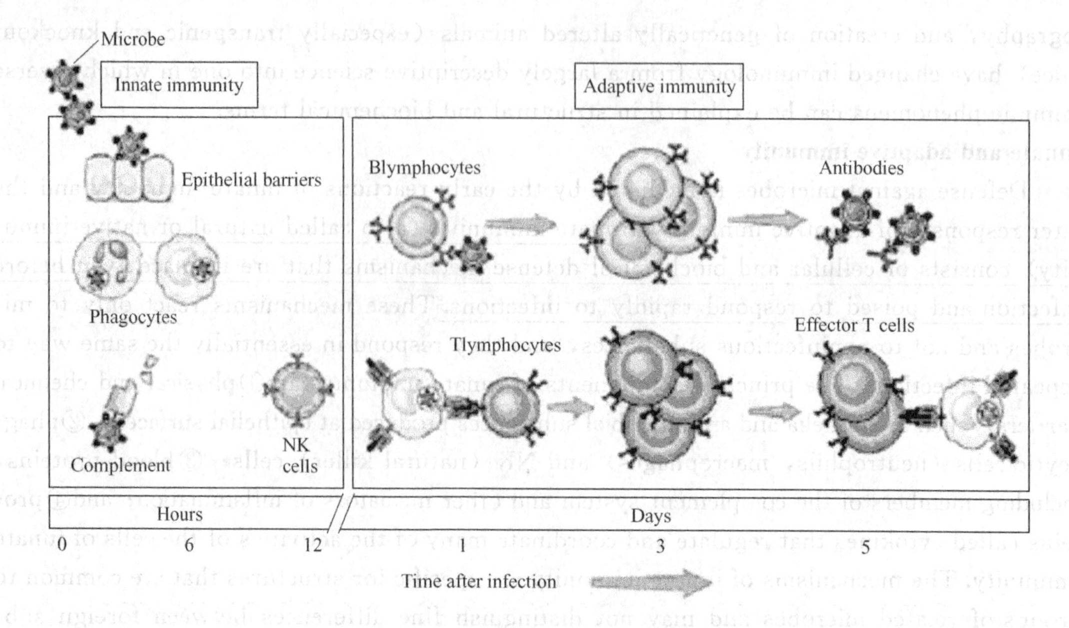

Fig. 1. 11 Innate and adaptive immunity

The mechanisms of innate immunity provide the initial defense against infections. Adaptive immune responses develop later and consist of activation of lymphocytes. The kinetics of the innate and adaptive immune responses are approximatons and may vary in different infections

ated largely by the mechanisms of innate immunity, including phagocytes and circulating molecules that resemble the plasma proteins of innate immunity in vertebrates. Adaptive immunity, consisting of lymphocytes and antibodies, first appeared in jawed vertebrates and became increasingly specialized with further evolution.

Types of adaptive immune responses

There are two types of adaptive immune responses, called humoral immunity and cell-mediated immunity, which are mediated by different components of the immune system and function to eliminate different types of microbes. Humoral immunity is mediated by molecules in the blood and mucosal secretions, called antibodies, which are produced by cells called B lymphocytes (also called B cells). Antibodies recognize microbial antigens, neutralize the infectivity of the microbes, and target microbes for elimination by various effector mechanisms. Humoral immunity is the principal defense mechanism against extracellular microbes and their toxins because secreted antibodies can bind to these microbes and toxins and assist in their elimination. Antibodies themselves are specialized, and different types of antibodies may activate different effector mechanisms. For example, some types of antibodies promote phagocytosis, and others trigger the release of inflammatory mediators from leukocytes such as mast cells. Cell-mediated immunity, also called cellular immunity, is mediated by T lymphocytes (also called T cells). Intracellular microbes, such as viruses and some bacteria, survive and proliferate inside phagocytes and other host cells, where they are inaccessible to circulating antibodies. Defense against such infections is a function of cell-mediated immunity, which promotes the destruction of microbes residing in phagocytes or the killing of infected cells to eliminate interiors of infection.

Protective immunity against a microbe may be induced by the host's response to the microbe or by the transfer of antibodies or lymphocytes specific for the microbe. The form of immunity that is induced by exposure to a foreign antigen is called active immunity because the immunized individual plays an active role in responding to the antigen. Individuals and lymphocytes that have not encountered a particular antigen are said to be naive. Individuals who have responded to a microbial antigen and are protected from subsequent exposures to that microbe are said to line immune.

Immunity can also be conferred on an individual by transferring serum or lymphocytes from a specifically immunized individual, a process known as adaptive transfer in experimental situations. The recipient of such a transfer becomes immune to the particular antigen without ever having been exposed to or having responded to that antigen. Therefore, this form of immunity is called passive immunity. Passive immunization is a useful method for conferring resistance rapidly, without having to wait for an active immune response to develop. An example of passive immunity is the transfer of maternal antibodies to the fetus, which enables newborns to combat infections before they acquire the ability to produce antibodies themselves. Passive immunization against bacterial toxins by the administration of antibodies from immunized animals is a lifesaving treatment of potentially lethal infections, such as tetanus. The technique of adoptive transfer has also made it possible to define the various cells and molecules that are responsible for mediating specific immunity. In fact, humoral immunity was originally defined as the type of immunity that could be transferred to unimmunized, or naive, individuals by antibody containing cell-free portions of the blood [i. e. , plasma or serum (once called humors)] obtained from previously immunized individuals. Similarly, cell-mediated immunity was defined as the form of immunity that can be transferred to naive individuals with cells (T lymphocytes) from immunized individuals but not with plasma or serum.

The first experimental demonstration of humoral immunity was provided by Emil von Behring and Shibasaburo Kitasato in 1890. They showed that if serum from animals who had recovered from diphtheria infection was transferred to naive animals, the recipients became specifically resistant to diphtheria infection. The active components of the serum were called antitoxins because they neutralized the pathologic effects of the diphtheria toxin. In the early 1900s, Karl Landsteiner and other investigators showed that not only toxins but also nonmicrobial substances could induce humoral immune responses. From such studies arose the more general germ antibodies for the serum proteins that mediate humoral immunity. Substances that bound antibodies and generated the production of antibodies were then called antigens. In 1900, Pual Ehrlich provided a theoretical framework for the specificity of antigen-antibody reactions, the experimental proof for which came during the next 50 years from the work of Land Steiner and others using simple chemicals as antigens. Ehrlich's theories of the physicochemical complementarity of antigens and antibodies are remarkable for their prescience. This early emphasis on antibodies led to the general acceptance of the humoral theory of immunity, according to which immunity is mediated by substances present in body fluids.

The cellular theory of immunity, which stated that host cells were the principal mediators of immunity, was championed initially by Elie Metchnikoff. His demonstration of phagocytes surrounding a thorn stuck into a translucent starfish larva, published in 1893, was perhaps the first experimental evidence that cells respond to foreign invaders. Sir Almroth

Wright's observation in the early 1900s that factors in immune serum enhanced the phagocytosis of bacteria by coating the bacteria, a process known as opsonization, lent support to the belief that antibodies prepared microbes for ingestion by phagocytes. These early "cellularists" were unable to prove that specific immunity to microbes could be mediated by cells. The cellular theory of immunity became firmly established in the 1950s, when George Mackaness showed that resistance to an intracellular bacterium, Listeria monocytogenes, could be adoptively transferred with cells but not with serum. We now know that the specificity of cell-mediated immunity is due to lymphocytes, which often function in concert with other cells, such as phagocytes, to eliminate microbes.

In the clinical setting, immunity to a previously encountered microbe is measured indirectly, either by assaying for the presence of products of immune responses (such as serum antibodies specific for microbial antigens) or by administering substances purified from the microbe and measuring reactions to these substances. A reaction to a microbial antigen is detectable only in individuals who have previously encountered the antigen; these individuals are said to be "sensitized" to the antigen, and the reaction is an indication of "sensitivity". Although the reaction to the purified antigen has no protective function, it implies that the sensitized individual is capable of mounting a protective immune response to the microbe.

Cardinal features of adaptive immune responses

All humoral and cell-mediated immune responses to foreign antigens have a number of fundamental properties that reflect the properties of the lymphocytes that mediate these responses (Fig. 1. 12).

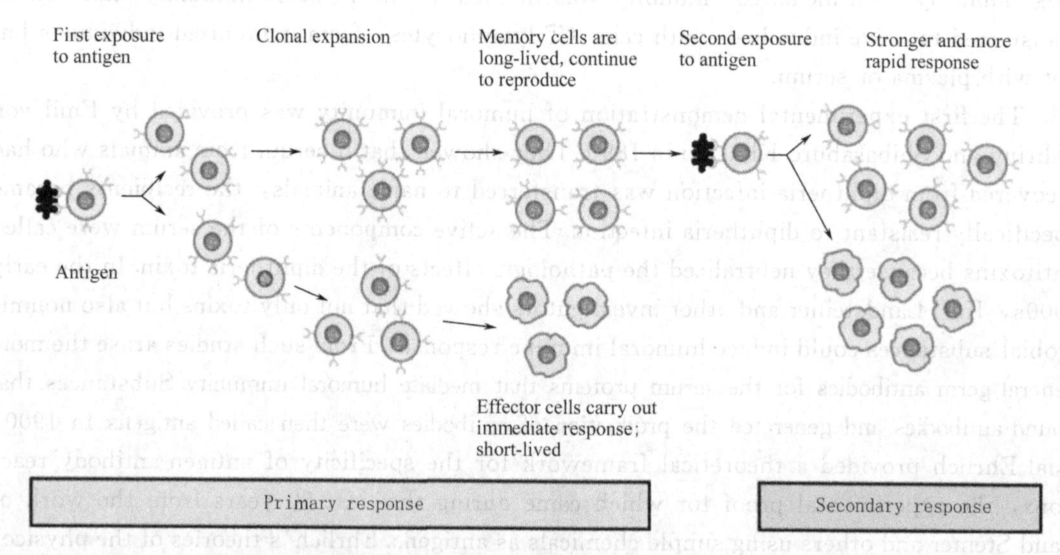

Fig. 1. 12 Kinetics of primary and secondary humoral immune responses

In a primary immune response, naive B cells are stimulated by antigen, become activated, and differentiate into antibody-secreting cells that produce antibodies specific for the eliciting antigen. Some of the abtibody-secreting plasma cells survive in the bone marrow and continue to produce antibodies for long periods. Long-lived memory B cells are also generated during the primary response. A secondary immune response is elicited when the same antigen stimulateds these memory B cells, leading to more rapid proliferation and differentiation and production of greater quantities of specific antibody than are produce in the primary response

Specificity and diversity. Immune responses are specific for distinct antigens and, in fact, for different portions of a single complex protein, polysaccharide, or other macromolecule. The parts of such antigens that are specifically recognized by individual lymphocytes are called determinants or epitopes. This fine specificity exists because individual lymphocytes express membrane receptors that are able to distinguish subtle differences in structure between distinct antigens. Clones of lymphocytes with different specificities are present in unimmunized individuals and are able to recognize and respond to foreign antigens. This concept is the basic tenet of the clonal selection hypothesis.

The total number of antigenic specificities of the lymphocytes in an individual, called the lymphocyte repertoire, is extremely large. It is estimated that the immune system of an individual can discriminate 10^7 to 10^9 distinct antigenic determinants. This property of the lymphocyte repertoire is called diversity. It is the result of variability in the structures of the antigen-binding sites of lymphocytes receptors for antigens. In other words, there are many different clones of lymphocytes that differ in the structures of their antigen receptors and therefore in their specificity for antigens, creating a total repertoire that is extremely diverse.

Memory. Exposure of the immune system to a foreign antigen enhances its ability to respond again to that antigen. Responses to second and subsequent exposures to the same antigen, called secondary immune responses, are usually more rapid, larger, and often qualitatively different from the first, or primary, immune response to that antigen. Immunologic memory occurs partly because each exposure to an antigen expands the clone of lymphocytes specific for that antigen. In addition, stimulation of movie lymphocytes by antigens generates long-lived memory cells. These memory cells have special characteristics that make them more efficient at eliminating the antigen than naive lymphocytes that have not previously been exposed to the antigen. For instance, memory B lymphocytes produce antibodies that bind antigens with higher affinities than do previously unstimulated B cells, and memory T cells are better able to home to sites of infection than are naive T cells.

Specialization. The immune system responds in distinct and special ways to different microbes, maximizing the efficiency of antimicrobial defense mechanisms. Thus, humoral immunity and cell-mediated immunity are elicited by different classes of microbes or by the same microbe at different stages of infection (extracellular and intracellular), and each type of immune response protects the host against that class of microbe. Even within humoral or cell-mediated immune responses, the nature of the antibodies or T lymphocytes that are generated may vary from one class of microbe to another.

Self-limitation. All normal immune responses wane with time after antigen stimulation, thus returning the immune system to its resting basal state, a process called homeostasis. Homeostasis is maintained largely because immune responses are mongered by antigens and function to eliminate antigens, thus eliminating the essential stimulus for lymphocyte activation. In addition, antigens and the immune responses to them stimulate regulatory mechanisms that inhibit the response itself.

Nonreactivity to self. One of the most remarkable properties of every normal individual's immune system is its ability to recognize, respond to, and climinate many foreign (nonself) antigens while not reacting harmfully to that individual's own (self) antigenic substances. Immunologic unresponsiveness is also called tolerance. Tolerance to self antigens, or self-tolerance, is maintained by several mechanisms. These include eliminating lymphocytes that express receptors specific for some self antigens and allowing lymphocytes to encounter other self antigens in settings that either fail to stimulate or lead to functional inactivation of the self-reactive lymphocytes. Abnormalities in the induction or maintenance of self-tolerance lead to immune responses against self antigens (autologous antigens), often resulting in disorders called autoimmune diseases.

These features of adaptive immunity are necessary if the immune system is to perform its normal function of host defense. Specificity and memory enable the immune system to mount heightened responses to persistent or recurring stimulation with the same antigen and thus to combat infections that are prolonged or occur repeatedly. Diversity is essential if the immune system is to defend individuals against the many potential pathogens in the environment. Specialization enables the host to "custom design" responses to best combat many different types of microbes. Self limitation allows the system to return to a state of rest after it eliminates each foreign antigen and to be prepared to respond to other antigens. Self-tolerance is vital for preventing reactions against one's own cells and tissues while maintaining a diverse repertoire of lymphocytes specific for foreign antigens.

Celluar components of the adaptive immune system

The principal cells of the immune system are lymphocytes, antigen-presenting cells, and effector cells. Lymphocytes are the cells the specifically recognize and respond to foreign antigens and are therefore the mediators of humoral and cellular immunity. There are distinct subpopulations of lymphocytes that differ in how they recognize antigens and in their functions. **B lymphocytes** are the only cells capable of producing antibodies. They recognize extracellular (including cell surface) antigens and differentiate into antibody-secreting cells, thus functioning as the mediators of humoral immunity. **T lymphocytes**, the cells of cell-mediated immunity, recognize the antigens of intracellular microbes and function to destroy these microbes or the infected cells. T cells do not produce antibody molecules. Their antigen receptors are membrane molecules distinct from but structurally related to antibodies. T lymphocytes have a restricted specificity for antigens; they recognize only peptide antigens attached to host proteins that are encoded by genes in the major histocompatibility complex (MHC) and that are expressed on the surfaces of other cells. As a result, these T cells recognize and respond to cells surface-associated but not soluble antigens. T lymphocytes consist of functionally distinct populations, the best defined of which are **helper T cells** and **cytolytic**, or **cytotoxic**, **T lymphocytes** (CTL). In response to antigenic stimulation, helper T cells secrete proteins called cytokines, whose function is to stimulate the proliferation and differentiation of the T cells as well as other cells, including B cells, macrophages, and other leukocytes. CTL kill cells that produce foreign antigens, such as cells infected by viruses and other intracellular microbes. Some T lymphocytes, which are called regulatory T cells, may function mainly to inhibit immune responses. The nature and physiologic roles of these regulatory T cells are incompletely understood. A third class of lymphocytes, natural killer (NK) cells, is involved in innate immunity against viruses and other intracellular microbes.

The initiation and development of adaptive immune responses require that antigens be captured and displayed to specific lymphocytes. The cells that serve this role are called **anti-**

gen-presenting cells (APC). The most highly specialized APC are dendritic cells, which capture microbial antigens that enter from the external environment, transport these antigens to lymphoid organs, and present the antigens to naive T lymphocytes to initiate immune responses. Other cell types function as APC at different stages of cell-mediated and humoral immune responses.

The activation of lymphocytes by antigen leads to the generation of numerous mechanisms that function to eliminate the antigen. Antigen elimination often requires the participation of cells called effector cells. Activated T lymphocytes, mononuclear phagocytes, and other leukocytes function as effector cells in different immune responses.

Lymphocytes and accessory cells are concentrated in anatomically discrete lymphoid organs, where they interact with one another to initiate immune responses Lymphocytes are also present in the blood; from the blood, they can recirculate to lymphoid tissues and to peripheral sites of antigen exposure to eliminate the antigen.

Phases of adaptive immune responses

Adaptive immune responses may be divided into distinct phases—the recognition of antigen, the activation of lymphocytes, and the effector phase of antigen elimination—followed by the return to home ostasis and the maintenance of memory. All immune responses are initiated by the specific recognition of antigens. This recognition leads to the activation of the lymphocytes that recognize the antigen and culminates in the development of effector mechanisms that mediate the physiologic function of the response, namely, the elimination of the antigen. After the antigen is eliminated, the immune response abates and homeostasis is restored.

Recognition of antigens

Every individual possesses numerous clonally derived lymphocytes, each clone having arisen from a single precursor and being capable of recognizing and responding to a distinct antigenic determinant, and when an antigen enters, it selects a specific preexisting clone and activates it. This fundamental concept is called the clonal selection hypothesis. It was first suggested by Niels Jerne in 1955, and most clearly enunciated by Macfarlane Burnet in 1957, as a hypothesis to explain how the immune system could respond to a large number and variety of antigens. According to this hypothesis, antigen-specific clones of lymphocytes develop before and independent of exposure to antigen. The cells constituting each clone have identical antigen receptors, which are different from the receptors on the cells of all other clones. Although it is difficult to place an upper limit on the number of antigenic determinants that can be recognized by the mammalian determinants that can be recognized by the mammalian immune system, a frequently used estimate is on the order of 10^7 to 10^9. This is a reasonable approximation of the number of different antigen receptor proteins that are produced and therefore reflects the number of distinct clones of lymphocytes present in each individual. Foreign antigens interact with preexisting clones of antigen-specific lymphocytes in the specialized lymphoid tissues where immune responses are initiated.

The key postulates of the clonal selection hypothesis have been convincingly proved by a variety of experiments and form the cornerstone of the current concepts of lymphocytes specificity and antigen recognition.

• Definitive proof that antigen-specific clones of lymphocytes exist before antigen exposure came when the structure of antigen receptors and the molecular basis of receptor expression were defined. All the lymphocytes of a particular clone express receptors of one specificity, and these receptors are expressed at a stage and site of maturation where the lymphocytes have not encountered antigens.

- The fact that an individual clone of lymphocytes can recognize and respond to only one antigen was established by limiting dilution culture experiments. In this type of experiment, lymphocytes are distributed in culture wells in such a way that each well contains cells from a single clone. When mixtures of antigens are added to these wells, the cells in each well respond to only one of the antigens (e. g., by producing antibodies specific for that antigen).
- More recently, methods for assaying the expansion of antigen-specific lymphocyte population in vivo have shown that administration of an antigen stimulates expansion of specific lymphocyte populations and no detectable response of other, "bystander", lymphocyte populations that are not specific for that antigen.

Activation of lymphocytes

The activation of lymphocytes requires two distinct signals, the first being antigen and the second being either microbial products or components of innate immune responses to microbes. This idea is called the two-signal hypothesis for lymphocyte activation. The requirement for antigen (so-called signal 1) ensures that the ensuing immune response is specific. The requirement for additional stimuli triggered by microbes or innate immune reactions to microbes (signal 2) ensures that immune responses are induced when they are needed (i. e., against microbes and other noxious substances) and not against harmless substances, including self antigens.

The responses of lymphocytes to antigens and second signals consist of the synthesis of new proteins, cellular proliferation, and differentiation into effector and memory cells.

Effector phase of immune responses elimination of antigens

During the effector phase of immune responses, lymphocytes that have been specifically activated by antigens perform the effector functions that lead to the elimination of the antigens. Antibodies and T lymphocytes eliminate extracellular and intracellular microbes, respectively. These functions of antibodies and T cells often require the participation of other, nonlymphoid effector cells and defense mechanisms that also operate in innate immunity. Thus, the same innate immune mechanisms that provide the early lines of defense against infectious agents may be used by the subsequent adaptive response to eliminate microbes. In fact, as we mentioned earlier, an important general function of adaptive immune responses is to enhance the effector mechanisms of innate immunity and to focus these effector mechanisms on those tissues and cells that contain foreign antigen.

Homeostasis: decline of immune responses

At the end of an immune response, the immune system returns to its basal resting state, in large part because most of the progeny of antigen-stimulated lymphocytes die by apoptosis. Apoptosis is a form of regulated, physiologic cell death in which the nucleus undergoes condensation and fragmentation, the plasma membrane shows blebbing and vesiculation, the internal sequestration of some membrane lipids is lost, and the dead cells are rapidly phagocytosed without their contents being released. (This process contrasts with necrosis, a type of cell death in which the nuclear and plasma membranes break down and cellular contents often spill out, inducing a local inflammatory reaction). A large fraction of antigen-stimulated lymphocytes undergoes apoptosis, probably because the survival of lymphocytes is dependent on antigen and antigen-induced growth factors, and as the immune response eliminates the antigen that initiated it, the lymphocytes become deprived of essential survival stimuli.

Summary

- Protective immunity against microbes is mediated by the early reactions of innate immunity and the later responses of adaptive immunity. Innate immunity is stimulated by structures shared by groups of microbes. Adaptive immunity is specific for different microbial and nonmicrobial antigens and is increased by repeated exposures to antigen (immunologic memory).
- Humoral immunity is mediated by B lymphocytes and their secreted products, antibodies, and functions in defense against extracellular microbes. Cell-mediated immunity is mediated by T lymphocytes and their products, such as cytokines, and is important for defense against intracellular microbes.
- Immunity may be acquired by a response to antigen (active immunity) or conferred by transfer of antibodies or cells from an immunized individual (passive immunity).
- The immune system possesses several properties that are of fundamental importance for its normal functions. These include specificity for different antigens, a diverse repertoire capable of recognizing a wide variety of antigens, memory for antigen exposure specialized responses to different microbes, self-limitation, and the ability to discriminate between foreign antigens and self antigens.
- Lymphocytes are the only cells capable of specifically recognizing antigens and are thus the principal cells of adaptive immunity. The two major subpopulations of lymphocytes are B cells and T cells, and they differ in their antigen receptors and functions. Specialized antigen-presenting cells capture microbial antigens and display these antigens for recognition by lymphocytes. The elimination of antigens often requires the participation of various effector cells.
- The adaptive immune response is initiated by the recognition of foreign antigens by specific lymphocytes. Lymphocytes respond by proliferating and by differentiating into effector cells, whose function is to eliminate the antigen, and into memory cells, which show enhanced responses on subsequent encounters with the antigen. The activation of lymphocytes requires antigen and additional signals that may be provided by microbes or by innate immune responses to microbes.
- The effector phase of adaptive immunity requires the participation of various defense mechanisms, including the complement system and phagocytes, which also operate in innate immunity. The adaptive immune response enhances the defense mechanisms of innate immunity.

Glossary

adaptive immunity 继承免疫性，适应性免疫，获得性免疫

antigen ['æntidʒən] n. 抗原

antigen-presenting cell 抗原递呈细胞
antitoxins [ˈæntiˈtɔksins] n. 抗毒素
apoptosis [ˌæpəpˈtəusis] n. 编程性细胞死亡，脱噬作用
autoimmune [ˌɔːtəuiˈmjuːn] adj. 自身免疫（的）
blebbing [bˈlebiŋ] n. 起泡
clone [kləun] n. 无性系，无性繁殖，克隆；v. 无性繁殖，复制
complement [ˈkɔmplimənt] n. 补足物，补语，余角，补体
condensation [kɔndenˈseiʃən] n. 压缩，凝缩，液化，浓集
cowpox [ˈkaupɔks] n. 牛痘
crystallography [kristəˈlɔgrəfi] n. 结晶学，晶体学
cytolytic [ˌsaitəuˈlitik] adj. 细胞溶解的，溶细胞的
cytotoxic [ˌsaitəˈtɔksik] adj. 细胞毒素的，细胞溶解的
dendritic cell 树突细胞，树突样细胞，树突状细胞
determinant [diˈtəːminənt] n. 决定子
differentiate [ˌdifəˈrenʃieit] v. 微分，区别，分化
differentiation [ˌdifəˌrenʃiˈeiʃən] n. 分化
diphtheria [difˈθiəriə, dip-] n. 白喉
disorder [disˈɔːdə] n. 病症，（机能）紊乱，障碍
diversity [daiˈvəːsiti] n. 差异，多样性
effector cell 效应细胞，效应基因细胞
epithelia [ˌepiˈθiːliə] n. 上皮，上皮细胞，上皮（组织）
epitope [ˈepitəup] n. 抗原表位（分），表位（核），抗原决定部位，抗原决定簇
extracellular [ˌekstrəˈseljulə] adj. 细胞外的
fetus [ˈfiːtəs] n. 胎儿
fragmentation [ˌfrægmenˈteiʃən] n. 断裂
histocompatibility [ˌhistəukəmˌpætiˈbiliti] n. 组织相容性，组织配合性
homeostasis [ˌhəumiəuˈsteisis] n. 内环境稳定，稳态，自身稳定，体内稳态，调节平衡作用
humoral immunity n. 体液免疫，体液免疫性
hypothesis [haiˈpɔθisis] n. 假设，假说，前提
immune responses 免疫反应，免疫响应，免疫应答
immunity [iˈmjuːniti] n. 免疫，免疫性，免除
immunology [ˌimjuˈnɔlədʒi] n. （生）免疫学
in vivo 在活体内
infectivity [ˌinfekˈtiviti] n. 传染力，传染性，侵染性
inflammation [ˌinfləˈmeiʃən] n. 发炎，炎症
innate immunity 先天免疫性，天然免疫，先天免疫
inoculate [iˈnɔkjuleit] v. 接芽，接木，给……接种，接种，移植（细菌）
intracellular [ˌintrəˈseljulə] adj. 细胞内的
invertebrate [inˈvəːtibrit] adj. 无脊骨的，无脊椎的；n. 无脊椎动物，无骨气的人
knockout [ˈnɔˈkaut] n. 打倒，敲除
larva [ˈlɑːvə] n. （复 larvae）幼虫
leukocyte [ˈljuːkəsait] n. 白细胞
lymphocyte [ˈlimfəsait] n. 淋巴球，淋巴细胞
lymphoid [ˈlimfɔid] adj. 淋巴的，淋巴样的
mammalian [mæˈmeiljən] n. 哺乳动物；adj. 哺乳动物的
mast cell 肥大细胞
maternal [məˈtəːnl] adj. 母性的，母体的，母本的，母源的
maturation [ˌmætjuˈreiʃən] n. （动）成熟

mechanism [ˈmekənizəm] n. 机械，机构，结构，机理
microbe [ˈmaikrəub] n. 微生物（通常指植物性的），细菌，酵母菌
mucosal [mjuːˈkəusə] adj. 黏膜的
naive [nɑːˈiːv] adj. 首次用于实验的，天真的，纯真的
necrosis [neˈkrəusis] n. 坏死
neutralize [ˈnjuːtrəlaiz, (us)nuː-] v. （使）中和，（使）成为无效
neutrophil [ˈnjuːtrəfil] adj. 嗜中性的；n. 嗜中性粒细胞
opsonization [ˌɔpsənaiˈzeiʃən] n. 调理作用
passive immunity 被动免疫，被动免疫力，被动免疫（性）
pathologic [ˌpæθəˈlɔdʒik] adj. 病理学的
phagocyte [ˈfægəusait] n. 食菌细胞，吞噬细胞
phagocytic cell 吞噬细胞
phagocytosis [fægəˌsaiˈtəusis] n. 食菌作用，吞噬（作用）
plasma [ˈplæzmə] n. 血浆，原浆，原生质
precursor [pri(ː)ˈkəːsə] n. 前身，前体，产物母体（化合物）
proliferation [prəuˌlifəˈreiʃən] n. 增殖
pustule [ˈpʌstjuːl] n. （拉 pustula）脓疱，色点，小疱，疱状突起
recipient [riˈsipiənt] n. 容纳者，容器，（接）受者，受体，受（血）者
repertoire [ˈrepətwɑː] n. 所有组成成分
resistance [riˈzistəns] n. 抵抗力，阻力，电阻，耐（药）性
sequestration [ˌsiːkweˈstreiʃən] n. 隔绝，死骨形成，隔离（病人），血管内血量净增，多价螯合（作用）
serum [ˈsiərəm] n. （复 sera）血清，浆液，免疫血清
signal [ˈsignl] n. 信号，标志
specificity [ˌspesiˈfisiti] n. 特异性，特征，专一性
subpopulation [ˈsʌbˌpɔpjuˈleiʃən] n. 亚群，亚群体，亚种群
tetanus [ˈtetənəs] n. 破伤风，生理性（肌）强直
tolerance [ˈtɔlərəns] n. 耐（药）量，耐受力，耐受性，忍受，耐受量
toxin [ˈtɔksin] n. 毒素，毒
transgenic [trænzˈdʒenik] n. 遗传转移，转基因（植物）
variability [ˌvɛəriəˈbiliti] n. 变异性，差异性，可变性，变率
vertebrate [ˈvəːtibrit] n. 脊椎动物；adj. 有椎骨的，有脊椎的
vesiculation [viˌsikjuˈleiʃən] n. 囊泡形成，囊泡化

难句分析

1. The cells and molecules responsible for immunity constitute the immune system, and their collective and coordinated response to the introduction of foreign substances is called the immune response.
   译文：负责免疫的细胞和分子组成免疫系统，它们对外来物质进入集体的和协调的反应被称为免疫应答。
2. Innate immunity (also called natural or native immunity) consists of cellular and biochemical defense mechanisms that are in place even before infection and poised to respond rapidly to infections.
   译文：先天免疫（也叫做自然免疫性或天赋免疫性）由细胞的和生化的防卫机制组成，这种免疫应答感染前就已存在，并随时准备对感染做出快速应答。
3. In contrast to innate immunity, there are other immune responses that are stimulated by exposure to infectious agents and increase in magnitude and defensive capabilities with

each successive exposure to a particular microbe.

译文：存在和先天免疫相对应的其他的免疫应答，这种应答通过暴露于传染物而得到刺激并在连续感染一种特定的微生物的情况下增强其防御能力。

4. Intracellular microbes, such as viruses and some bacteria, survive and proliferate inside phagocytes and other host cells, where they are inaccessible to circulating antibodies. Defense against such infections is a function of cell-mediated immunity, which promotes the destruction of microbes residing in phagocytes or the killing of infected cells to eliminate interiors of infection.

译文：胞内的微生物，例如病毒和某些细菌，在巨噬细胞及其他宿主细胞中存活并增殖，在体内循环的抗体难以接近它们。防御这样的感染要依靠细胞免疫的功能，细胞免疫促进存在于巨噬细胞内的微生物的死亡，或使受感染细胞死亡以除去内部的感染。

5. Homeostasis is maintained largely because immune responses are mongered by antigens and function to eliminate antigens, thus eliminating the essential stimulus for lymphocyte activation. In addition, antigens and the immune responses to them stimulate regulatory mechanisms that inhibit the response itself.

译文：体内平衡得以保持很大程度上是由于免疫应答通过抗原以及抗原移除传播，因此除去淋巴细胞活化所必需的刺激。另外，抗原和相应的免疫应答刺激抑制应答本身的调节机制。

6. Every individual possesses numerous clonally derived lymphocytes, each clone having arisen from a single precursor and being capable of recognizing and responding to a distinct antigenic determinant, and when an antigen enters, it selects a specific preexisting clone and activates it.

译文：每个个体都拥有无数的克隆衍生的淋巴细胞，每个克隆都是由单一前体产生而来的，能够识别并应答一个特异的抗原决定簇，并且当一个抗原进入体内时，它选择一种特异的已存在的克隆并活化该克隆。

（选自：Abul K Abbas, Andrew H Lichtman. Cellular and Molecular Immunology. 5th ed. Philadelphia: Saunders, 2003.）

Cytokines

Cytokines are proteins secreted by the cells of innate and adaptive immunity that mediate many of the functions of these cells. Cytokines are produced in response to microbes and other antigens, and different cytokines stimulate diverse responses of cells involved in immunity and inflammation. In the activation phase of adaptive immune responses, cytokines stimulate the growth and differentiation of lymphocytes, and in the effector phases of innate and adaptive immunity, they activate different effector cells to eliminate microbes and other antigens. Cytokines also stimulate the development of hematopoietic cells. In clinical medicine, cytokines are important as therapeutic agents and as targets for specific antagonists in numerous immune and inflammatory diseases.

The nomenclature of cytokines is often based on their cellular sources. Cytokines that are produced by mononuclear phagocytes were called monokines, and those produced by lymphocytes were called lymphokines. With the development of anticytokine antibodies and molecular probes, it became clear that the same protein may be synthesized by lymphocytes, monocytes, and a variety of tissue cells, including endothelial cells and some epithelial cells. Therefore, the generic term cytokines is the preferred name for this class of mediators.

General properties of cytokines

Cytokines are polypeptides produced in response to microbes and other antigens that mediate and regulate immune and inflammatory reactions. Although cytokines are structurally diverse, they share several properties.

Cytokine secretion is a brief, self-limited event. Cytokines are not usually stored as preformed molecules, and their synthesis is initiated by new gene transcription as a result of cellular activation. Such transcriptional activation is transient, and the messenger RNA encoding most cytokines are unstable, so cytokine synthesis is also transient. The production of some cytokines may additionally be controlled by RNA processing and by post transcriptional mechanisms, such as proteolytic release of an active product from an inactive precursor. Once synthesized, cytokines are rapidly secreted, resulting in a burst of release as needed.

The actions of cytokines are often pleiotropic and redundant. Pleiotropism refers to the ability of one cytokine to act on different cell types. This property allows a cytokine to mediate diverse biological effects, but it greatly limits the therapeutic use of cytokines because administration of a cytokine for a desired clinical effect may result in numerous unwanted side effects. Redundancy refers to the property of multiple cytokines having the same functional effects. Because of this redundancy, antagonists against a single cytokine or mutation of one cytokine gene may not have functional consequences, as other cytokines may compensate.

Cytokines often influence the synthesis and actions of other cytokines. The ability of one cytokine to simulate production of others leads to cascades in which a second or third cytokine may mediate the biological effects of the first. Two cytokines may antagonize each other's action, produce additive effects, or, in some cases, produce greater than anticipated, or synergistic effects.

Cytokine actions may be local and systemic. Most cytokines act close to where they are produced either on the same cell that secretes the cytokine (autocrine action) or on a nearby cell (paracrine action). T cells often secrete cytokines at the site of contact with antigen-presenting cells, the so-called immune synapse. This may be one reason that cytokines often act on cells in contact with the cytokine producers. When produced in large amounts, cytokines may enter the circulation and act at a distance from the site of production (endocrine action).

Cytokines initiate their actions by binding to specific membrane receptors on target cells. Receptors for cytokines often bind their ligands with high affinities, with dissociation constants (K_d values) in the range of 10^{-10} to 10^{-12} mol/L. (For comparison, recall that antibodies typically bind antigens with a K_d of 10^{-7} to 10^{-11} mol/L and that major histocompatibility complex (MHC) molecules bind peptides with a K_d of only about 10^{-6} mol/L.) As a consequence, only small quantities of a cytokine are needed to occupy receptors and elicit biological effects. Most cells express low levels of cytokine receptors (on the order of 100 to 1000 receptors per cell), but this is adequate for inducing responses (Fig. 1.13).

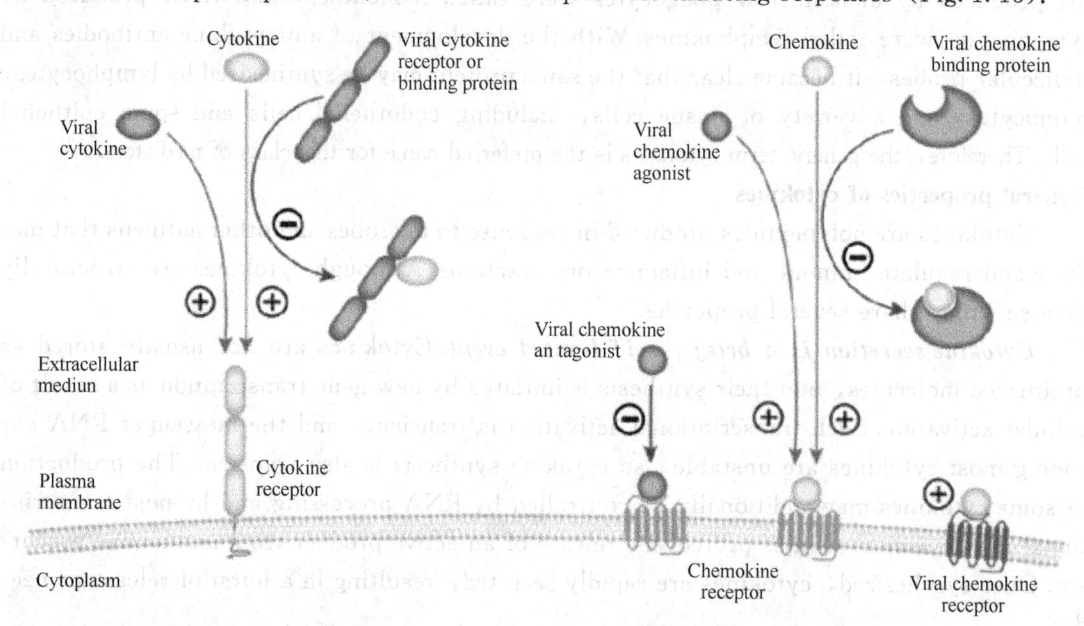

Fig. 1.13 Biological action of cytokine and its receptor

External signals regulate the expression of cytokine receptors and thus the responsiveness of cells to cytokines. For instance, stimulation of T or B lymphocytes by antigens leads to increased expression of cytokine receptors. For this reason, during an immune response, the antigen-specific lymphocytes are the preferential responders to secreted cytokines. This is one mechanism for maintaining the specificity of immune responses, even though cytokines themselves are not antigen specific. Receptor expression is also regulated by cytokines themselves, including the same cytokine that binds to the receptor, permitting positive amplification or negative feedback.

The cellular responses to most cytokines consist of changes in gene expression in target

cells, resulting in the expression of new functions and sometimes in the proliferation of the target cells. Many of the changes in gene expression induced by cytokines result in differentiation of T and B lymphocytes and activation of effector cells such as macrophages. For instance, cytokines stimulate switching of antibody isotypes in B cells, differentiation of helper T cells into TH1 and TH2 subsets, and activation of microbicidal mechanisms in phagocytes. Exceptions to the rule that cytokines work by changing gene expression patterns are chemokines, which elicit rapid cell migration, and a cytokine called tumor necrosis factor (TNF), which induces apoptosis by activating cellular enzymes, without new gene transcription or protein synthesis.

Functional categories of cytokines

cytokines were classified into three main functional categories based on their principal biological actions.

① ***Mediators and regulators of innate immunity*** are produced mainly by mononuclear phagocytes in response to infectious agents. Bacterial products, such as lipopolysaccharide (LPS), and viral products, such as double-stranded RNA, directly stimulate macrophages to secrete these cytokines as part of innate immunity. The same cytokines may also be secreted by macrophages that are activated by antigen-stimulated T cells (i. e., as part of adaptive cell-mediated immunity). Most members of this group of cytokines act on endothelial cells and leukocytes to stimulate the early inflammatory reactions to microbes, and some function to control these responses. NK (natural killer) cells also produce cytokines during innate immune reactions.

② ***Mediators and regulators of adaptive immunity*** are produced mainly by T lymphocytes in response to specific recognition of foreign antigens. Some T cell cytokines function primarily to regulate the growth and differentiation of various lymphocyte populations and thus play important roles in the activation phase of T cell-dependent immune responses. Other T cell-derived cytokines recruit, activate, and regulate specialized effector cells, such as mononuclear phagocytes, neutrophils, and eosinophils, to eliminate antigens in the effector phase of adaptive immune responses.

③ ***Stimulators of hematopoiesis*** are produced by bone marrow stromal cells, leukocytes, and other cells, and stimulate the growth and differentiation of immature leukocytes.

In general, the cytokines of innate and adaptive immunity are produced by different cell populations and act on different target cells. However, these distinctions are not absolute because the same cytokine may be produced during innate and adaptive immune reactions, and different cytokines produced during such reactions may have overlapping actions.

Cytokine receptors and signalling

All cytokine receptors consist of one or more transmembrane proteins whose extracellular portions are responsible for cytokine binding and whose cytoplasmic portions are responsible for initiating intracellular signalling pathways. These signalling pathways are typically activated by ligand-induced receptor clustering, bringing together the cytoplasmic portions of two or more receptor molecules in a process analogous to signalling by T and B cell receptors for antigens.

The most widely used classification of cytokine receptors is based on structural homologies among the extracellular cytokine-binding domains. According to this classification, cytokine receptors are divided into five families.

Type Ⅰ cytokine receptors, also called hemopoietin receptors, contain one or more copies of a domain with two conserved pairs of cysteine residues and membrane proximal sequence of tryptophan serine X tryptophan serine (WSXWS), where X is and amino acid. These receptors typically bind cytokine that fold into four α helical strands. The conserved features of the receptors presumable form structure that bind four α helical cytokines, but the specific for individual cytokines is determined by amino acid residues that vary from one receptor to another. These receptors consist of unique ligand-binding chains and one or more signal-transducing chain which are often shared by receptors for different cytokines.

Type Ⅱ cytokine receptors are similar to type receptors by virtue of two extracellular domains which conserved cysteines, but type Ⅱ receptors do not contain the WSXWS motif. These receptors consist of one ligand-binding polypeptide chain and signal-transducing chain.

Some cytokine receptors contain extracellular immunoglobulin (Ig) domains and are therefore classified as members of the Ig superfamily. This group of receptors bind diverse cytokines that signal by different mechanisms.

TNF receptors belong to a family of receptors (some of which are not cytokine receptors) with conserved cysteine-rich extracellular domains. On ligand binding, these receptors activate associated intracellular proteins that induce apoptosis or stimulate gene expression, or both.

Seven-transmembrane α helical receptors are also called serpentine receptors, because their transmembrane domains appear to "snake" back and forth through the membrane, and G protein-coupled receptors, because their signalling pathways involve GTP-binding (G) proteins. The mammalian genome encodes many such receptors involved in sensing environmental stimuli. In the immune system, members of this receptor class mediate rapid and transient responses to a family of cytokines called chemokines.

Cytokine receptors can also be grouped according to signal transduction pathway they activate. Such a grouping will correspond to structural homologies in the cytoplasmic regions of the signalling chains of the receptors. In many cases, members of a family defined by extracellular domains engage similar signal transduction pathways.

Roles of cytokines in innate immunity and inflammation

Different cytokines play key roles in innate immunity to different classes of microbes. In infections by pyogenic extracellular bacteria, macrophages respond to bacterial endotoxins and perhaps to other bacterial products by producing TNF, IL-1, and chemokines. TNF and IL-1 act on vascular endothelium at the site of the infection to induce the expression of adhesion molecules that promote stable attachment of blood neutrophils and monocytes to the endothelium at this site. C hemokines produced by the macrophages and by endothelial cells stimulate the extravasation of the leukocytes to the infection, where the innate immune reaction is mounted to eliminate the infectious microbes.

Macrophages also respond to many microbes, including intracellular bacteria and LPS-producing bacteria, by secreting IL-12, which induces the local production of IFN-γ from NK cells and T lymphocytes. IFN-γ then activates the macrophages to destroy phagocytosed microbes. IL-12 also stimulates the subsequent adaptive immune response and directs it forward TH1 cells, which are the mediators of IL-12 are complemented by IL-12. Cytokine-mediated leukocyte recruitment and activation are responsible for the injury to normal tissues that often accompanies innate immune reactions to infections. These macrophage-derived cytokines, especially TNF, IL-1, and IL-12, are also responsible for the systemic manifestations of infection.

In viral infection, type Ⅰ IFN are secreted by infected cells and macrophages and function to inhibit viral replication and infection. IL-15 stimulates the expansion of NK cells, and IL-12 enhances the cytolytic activity of NK cells. NK cell-mediated killing of virus-infected cells eliminates the reservoir of infection.

The dominant cytokines produced in response to different microbes account for the nature of the innate immune reactions to these microbes. For instance, the early response to pyogenic bacteria consists mainly of neutrophils, the response to intracellular bacteria is dominated by activated macrophages, and the response to viruses consists of NK cells in addition to other inflammatory cells. There may be considerable overlap, however, and there varied cellular reactions may be seen to different degrees in many infections.

Roles of T cell cytokines in specialized adaptive immune responses

The cytokines of adaptive immunity are critical for the development of immune responses and for the activation of effector cells that serves to eliminate microbes and other antigens. Much of specialization of adaptive immunity is due to the actions of cytokines, which may be produced by subpopulations of helper T cells. Different types of microbes stimulate naive $CD4^{+}$ T cells to differentiate into effector cells that produce distinct sets of cytokines and perform distinct functions. The best defined of these subsets are the TH1 and TH2 cells. Many intracellular microbes (bacteria and viruses) induce the development of TH1 cells, which produce IFN-γ, the cytokine that activates phagocytes to destroy intracellular microbes and stimulates the production of opsonizing antibodies that promote more phageocytosis. Hwlminthic parasites, in contrast, stimulate the development of TH2 cells, which produce IL-4 and IL-5. IL-4 enhances production of helminth-specific IgE antibodies, which coat the parasites, and IL-5 activates eosinophils, which bind to the IgE-coated parasites and destroy them. Thus, cytokines are essential for the development and effectiveness of adaptive immune responses.

Summary

- Cytokines are a family of proteins that mediate many of the responses of innate and adaptive immunity. The same cytokines may be produced by many cell types, and individual cytokines often act on diverse cell types. Cytokines are synthesized in response to inflammatory or antigenic stimuli and usually act locally, in an autocrine or paracrine fashion, by binding to high-affinity receptors on target cells. Certain cytokines may be produced in sufficient quantity to circulate and exert endocrine actions. For many cell types, cytokines serve as growth factors.
- Cytokines mediate their actions by binding with high affinity to receptors belonging to

a limited number of structural families. Different cytokines use specialized signaling pathways, such as the JAK/STAT pathway.

- The cytokines that mediate innate immunity are produced mainly by activated macrophages and include the following: TNF and IL-1 are mediators of acute inflammatory reactions to microbes; chemokines recruit leukocytes to sites of inflammation; IL-12 stimulates production of the macrophage-activating cytokine IFN-γ; type Ⅰ IFN are antiviral cytokines; and IL-10 is an inhibitor of macrophages. These cytokines function in innate immune responses to different classes of microbes.
- The cytokines that mediate and regulate adaptive immune responses are produced mainly by antigen-stimulated T lymphocytes, and they include the following: IL-2 is the principle T cell growth factor; IL-4 stimulates IgE production and the development of TH2 cells from naive helper T cells; IL-5 activates eosinophils; IFN-γ is an activator of macrophages; and TGFβ inhibits the proliferation of T lymphocytes and the activation of leukocytes.
- The colony-stimulating factors (CSF) consist of cytokines produced by bone marrow stromal cells, T lymphocytes, and other cells that stimulate the growth of bone marrow progenitors, thereby providing a source of additional inflammatory leukocytes. Several of these (e. g., stem cell factor and IL-7) play important roles in lymphopoiesis.
- Cytokines serve many functions that are critical to host defense against pathogens and provide links between innate and adaptive immunity. Cytokines contribute to the specialization of immune responses by activating different types of effector cells. Cytokines also regulate the magnitude and nature of immune responses by influencing the growth and differentiation of lymphocytes. Finally, cytokines provide important amplification mechanisms that enable small numbers of lymphocytes specific for any one antigen to activate a variety of effector mechanisms to eliminate the antigen.
- Excessive production or actions of cytokines can lead to pathologic consequences. The administration of cytokines or their inhibitors is a potential approach for modifying biological responses associated with immune and inflammatory diseases.

Glossary

adhesion [əd'hi:ʒən] n. 粘连，接合，吸附，附着（力），黏着（力）
administration [ədminis'treiʃən] n. 给予，投药，给药，药的服法
antagonist [æn'tægənist] n. 对抗物，拮抗剂
autocrine [ˌɔ:təu'krin] n. 自分泌
cytokine [ˌsaitəu'kain] n. 细胞因子
cytoplasmic [ˌsaitəu'læzmik] adj. 细胞质的
dissociation constant 解离常数，离解常数，电离常数
endocrine ['endəukrain] adj. 内分泌的，激素的，内分泌物的；n. 内分泌物
endothelial [ˌendə'θi:ljəl] adj. 内皮的
endothelium [ˌendəu'θi:liəm] n. 内皮
eosinophil [ˌi:əu'sinəfil] n. 嗜曙红细胞
epithelial [ˌepi'θi:liəl] adj. 上皮的
extravasation [eksˌtrævə'seiʃən] n. 外渗，外渗物

feedback [ˈfiːdbæk] n. 反馈
hematopoietic [ˌhiːmətəupɔiˈetik] adj. 生血的，生血药，造血药
hemopoietin [ˌhiːməupɔiˈiːtin] n. 生血素，促红细胞生成素
homology [hɔˈmɔlədʒi] n. 同源性
ligand [ˈligənd, ˈlaigənd] n. 配位子，配位体，配基
lipopolysaccharide [ˈlipəuˌpɔliˈsækəraid, ˈlai-] n. 脂多糖
lymphokine [ˌlimfəukain] n. 淋巴素，淋巴因子，淋巴激活素
lymphopoiesis [ˌlimfəpɔiˈiːsis] n. 淋巴细胞生成
microbicidal [maiˌkrəubiˈsaidl] adj. 杀微生物的
monokine [ˈmɔnəukain] n. 单核因子（由单核细胞或巨噬细胞产生的）
motif [məuˈtiːf] n. 基序
natural killer (NK) 自然杀伤
neutrophil [ˈnjuːtrəfil] n. 中性白细胞，中性粒细胞，adj. 嗜中性的
nomenclature [nəuˈmenklətʃə] n. 命名法，名称，术语
paracrine [ˌpærəˈkrin] n. 旁分泌，副分泌
pleiotropism [plaiˈɔtrəpizəm] n. （基因）多效性
proteolytic [prəutiəˈlitik] adj. 蛋白水解的
proximal [ˈprɔksiməl] adj. 接近的，邻近的
pyogenic [ˌpaiəuˈdʒenik] adj. 化脓的
redundancy [riˈdʌndənsi] n. 重复，多余，过多，冗余
reservoir [ˈrezəvwɑː] n. 贮器，容器，贮存宿主，病（原体）库
stimuli [ˈstimjulai] n. 刺激，刺激素，刺激物
stromal cell 基质细胞
synapse [siˈnæps] n. 突触，神经元的神经线连接，神经腱
synergistic [ˌsinəˈdʒistik] adj. 协同的，协同作用的
target [ˈtɑːgit] n. 目标，靶标
therapeutic [θerəˈpjuːtik] adj. 治疗的，治疗学的
transmembrane [ˌtrænsˈmembrein,ˌtrænz-,ˌtrɑːn-] n. 经膜，跨膜，横跨膜，贯穿细胞膜
tumor necrosis factor (TNF) n. 肿瘤坏死因子
vascular [ˈvæskjulə] adj. 脉管的，血管的，维管的

难句分析

1. Pleiotropism refers to the ability of one cytokine to act on different cell types. This property allows a cytokine to mediate diverse biological effects, but it greatly limits the therapeutic use of cytokines because administration of a cytokine for a desired clinical effect may result in numerous unwanted side effects.
   译文：基因多效性指的是一个细胞因子可以作用于不同的细胞类型的能力。这种性质使一个细胞因子能够介导多样的生物效应，但它极大程度上限制利用细胞因子的治疗手段，因此为得到某种理想的临床效果而使用细胞因子可能导致很多有害的副作用。
2. All cytokine receptors consist of one or more transmembrane proteins whose extracellular portions are responsible for cytokine binding and whose cytoplasmic portions are responsible for initiating intracellular signaling pathways.
   译文：全部的细胞因子受体都是由一或多个跨膜蛋白组成，这些蛋白的胞外部分决定着细胞因子的结合而其胞质部分决定着胞内的信号途径的初始化。
3. Seven-transmembrane α helical receptors are also called serpentine receptors, because their transmembrane domains appear to "snake" back and forth through the membrane,

and G protein-coupled receptors, because their signaling pathways involve GTP-binding (G) proteins.

译文：七次跨膜的α螺旋状受体也被称为蛇形受体，因为它们的跨膜结构域看来似乎像蛇一样穿行于膜内，还有G蛋白偶联受体是由于它们的信号途径涉及GTP-结合蛋白而得名。

4. TNF and IL-1 act on vascular endothelium at the site of the infection to induce the expression of adhesion molecules that promote stable attachment of blood neutrophils and monocytes to the endothelium at this site.

译文：肿瘤坏死因子和白细胞介素-1作用于感染位点的血管内皮组织从而诱导黏附分子的表达，而该表达可促进血液中的中性粒细胞和单核细胞稳定吸附于该位点。

（选自：Abul K Abbas, Andrew H Lichtman. Cellular and Molecular Immunology. 5th ed. Philadelphia: Saunders, 2003）

The Science of Virology

Introduction

The science of virology is relatively young. We can recognize specific viruses as the causative agents of epidemics that occurred hundreds or thousands of years ago from written descriptions of disease or from study of mummies with characteristic abnormalities. Furthermore, immunization against smallpox has been practiced from more than a millennium. However, it was only approximately 100 years ago that viruses were shown to be filterable and therefore distinct from bacteria that cause infectious disease. It was only about 60 years ago that the composition of viruses was described, and even more recently before they could be visualized as particles in the electron microscope. Within the last 20 years, however, the revolution of modern biotechnology has led to an explosive increase in our knowledge of viruses and their interactions with their hosts. Virology, the study of viruses, includes many aspects: the molecular biology of virus replication, the structure of viruses, the interactions of viruses and hosts, the evolution and history of viruses, virus epidemiology, and the diseases caused by viruses. The field is vast and any treatment of viruses must perforce be selective.

Viruses cause disease but are also useful as tools

Viruses are of intense interest because many cause serious illness in humans or domestic animals, and others damage crop plants. During the last century, progress in the control of infectious diseases through improved sanitation, safer water supplies, the development of antibiotics and vaccines, and better medical care have dramatically reduced the threat to human health from these agents, especially in developed countries. At the beginning of the 20th century, 0.8% of the population died each year from infectious diseases. Today the rate is less than one-tenth as great. The use of vaccines has led to effective control of the most dangerous of the viruses. Smallpox virus has been eradicated worldwide by means of an ambitious and concerted effort, sponsored by the World Health Organization, to vaccinate all people at risk for the disease. Poliovirus has been eliminated from the Americas, and measles virus eliminated from the North America, by intensive vaccination programs. There is hope that these two viruses can also be eradicated worldwide in the near future. Vaccines exist for the control of many other viral diseases, such as mumps, rabies, rubella, yellow fever, and Japanese encephalitis.

The dramatic decline in the death rate from infectious disease has led to a certain amount of complacency. There is a small but vocal movement in the United States to eliminate immunization against viruses, for example. However, viral diseases continue to plague humans, as do infectious diseases caused by bacteria, protozoa, fungi, and multicellular parasites. Recognition is growing that infectious diseases, of which viruses form a major component, have not been conquered by the introduction of vaccines and drugs. The overuse of antibiotics has resulted in upsurge in antibiotic-resistant bacteria, and viral diseases continue to resist elimination.

The persistence of viruses is in part due to their ability to change rapidly and adapt to new situations. Human immunodeficiency virus (HIV) is the most striking example of the appearance of a virus that has recently entered the human population and caused a plague of worldwide importance. The arrival of this virus in the United States caused a significant rise in the number of deaths from infectious disease. Other, previously undescribed viruses also continue to emerge as serious pathogens. Sin Nombre virus caused a 1994 outbreak in the United States of hantavirus pulmonary syndrome with a 50% case fatality rate, and it is now recognized as being widespread in North America. Junin virus, which causes Argentine hemorrhagic fever, and related viruses have become a more serious problem in South America with the spread of farming. Ebola virus, responsible for several small African epidemics with a case fatality rate of 70%, was first described in the 1970s. Nipah virus, previously unknown, appeared in 1998 and caused 258 cases of encephalitis, with a 40% fatality rate, in Malaysia and Singapore. As faster and more extensive travel becomes ever more routine, the potential for rapid spread of all viruses increase. The possibility exists that any of these viruses could become widespread, as has HIV since its appearance in Africa perhaps half a century ago, and as has West Nile virus, which spread to the Americas in 1999.

Newly emerging viruses are not the only ones to plague humans, however. Many viruses that have been known for a long time, and for which vaccines may exist, continue to cause widespread problems. Respiratory syncytial virus, as an example, is a major cause of pneumonia in infants. Despite much effort, it has not yet been possible to develop an effective vaccine. Even when vaccines exist, problems may continue. For example, influenza virus changes rapidly and the vaccine for it must be reformulated yearly. Because the major reservoir for influenza is birds, it is not possible to eradicate the virus. Thus, to control influenza would require that the entire population be immunized yearly. This is a formidable problem and the virus continues to cause annual epidemics with a significant death rate. Although primarily a killer of the elderly, the potential of influenza to kill the young and healthy was shown by the worldwide epidemic of influenza in 1918 in which 20-100million people died worldwide. In the United States, perhaps 1% of the population died during the epidemic. Continuing study of virus replication and virus interaction with their hosts, surveillance of viruses in the field, and efforts to develop new vaccines as well as other methods of control are still important.

The other side of the coin is that viruses have been useful to us as tools for the study of molecular and cellular biology. Further, the development of viruses as vectors for the expression of foreign genes has given them a new and expanded role in science and medicine, including their potential use in gene therapy. As testimony to the importance of viruses in the study of biology, numerous Nobel Prizes have been awarded in recognition of important advances in biological science that resulted from studies that involved viruses. To cite a few examples, Max Delbrück received the prize for pioneering studies in what is now called molecular biology, using bacteriophage T4. Cellular oncogenes were first discovered from their presence in retroviruses that could transform cells in culture, a discovery that resulted in a prize for Francis Peyton Rous for his discovery of transforming retroviruses, and for Michael Bishop and Harold Varmus, who were the first to show that a transforming retroviral gene had a cellular counterpart. As a third example, the development of the modern methods of gene cloning have relied heavily on the use of restriction enzymes and recombinant DNA technology, first developed by Daniel Nathans and Paul Berg working with SV40 virus, and on the use of reverse transcriptase, discovered by David Baltimore and Howard Temin in retrovirus. As another example, the study of the interactions of viruses with the immune system has told us much about how this essential means of defense against disease functions, and this resulted in a prize for Rolf Zinkernagel and Peter Doherty. The study of viruses and their use as tools has told us as much about human biology as it has told us about the viruses themselves.

In addition to the interest in viruses that arises from their medical and scientific importance, viruses from a fascinating evolutionary system. There is debate as to how ancient are viruses. Some argue that RNA viruses contain remnants of the RNA world that existed before the invention of DNA. All would accept the idea that viruses have been present for hundreds of millions of years and have helped to shape the evolution of their hosts. Viruses are capable of very rapid change, both from drift due to nucleotide substitutions that may occur at a rate 10^6-fold greater than that of the plants and animals that they infect, and from recombination that leads to the development of entirely new families of viruses. This makes it difficult to trace the evolution of viruses back more than a few millennia or perhaps a few million years. The development of increasingly refined methods of sequence analysis, and the determination of more structures of virally encoded proteins, which change far more slowly than do the amino acid sequences that from the structure, have helped identify relationships among viruses that were not at first obvious. The coevolution of viruses and their hosts remains a study that is intrinsically interesting and has much to tell us about human biology.

The nature of viruses

Viruses are subcellular, infectious agents that are obligate intracellular parasites. They infect and take over a host cell in order to replicate. The mature, extracellular virus particle is called a virion. The virion contains a genome that may be DNA or RNA wrapped in a protein coat called a capsid or nucleocapsid. Some viruses have a lipid enveloped surrounding the nucleocapsid (they are "enveloped"). In such viruses, glycoproteins encoded by the virus are embedded in the lipid envelope. The function of the capsid or envelope is to protect the viral genome while it is extra cellular and to promote the entry of the genome into a new susceptible cell.

The nucleic acid genome of a virus contains the information needed by the virus to replicate and produce new virions after its introduction into a susceptible cell. Virions bind to receptors on the surface of the cell, and the genome is released into the cytoplasm of the cell, sometimes still in association with protein. The genome then redirects the cell to the replication of itself and to the production of progeny virions. The cellular machinery that is in place for the production of energy (synthesis of ATP) and for macromolecular synthesis, such as translation of mRNA to produce proteins, is essential.

It is useful to think of the proteins encoded in viral genomes as belonging to three major classes. First, most viruses encode enzymes required for replication of the genome and the production of mRNA from it. RNA viruses must encode an RNA polymerase or replicase, since cells do not normally replicate RNA. Most DNA viruses have access to the cellular DNA replication machinery in the nucleus, but even so, many encode new DNA polymerases for the replication of their genomes. Even if they use cellular DNA polymerases, many DNA viruses encode at least an initiation protein for genome replication. Second, viruses must encode proteins that are used in the assembly of progeny viruses. For simpler viruses, these may consist of only one or a few structural proteins that assemble with the genome to form the progeny virion. More complicated viruses may encode scaffolding proteins that are required for assembly but are not present in the virion. In some cases, viral proteins required for assembly may have proteolytic activity. Third, the larger viruses encode proteins that interfere with defense mechanisms of the host. These defenses include, for example, the immune response and the interferon response of vertebrates, which are highly evolved and effective methods of controlling and eliminating virus infection; and the DNA restriction system in bacteria, so useful in molecular biology and genetic engineering, prevents the introduction of foreign DNA.

It is obvious that viruses that have larger genomes and encode larger numbers of proteins, such as the herpesviruses, have more complex life cycles and assemble more complex virions than viruses with small genomes, such as poliovirus. The smallest known nondefective viruses have genomes of about 3kb (1kb=1000 nucleotides in the case of single-stranded genomes or 1000 base pairs in the case of double-stranded genomes). These small viruses may encode as few as three proteins (for example, the bacteriophage MS2). At the other extreme, the largest known RNA viruses, the coronaviruses, have genomes somewhat larger than 30kb, whereas the largest DNA viruses, poxviruses, have genomes of up to 380kb. These large DNA viruses that can afford the luxury of encoding proteins that interfere effectively with host defenses such as the immune system. It is worthwhile remembering that even the largest viral genomes are small compared to the size of the bacterial genome (2000kb) and miniscule compared to the size of the human genome (2×10^6kb).

There are other subcellular infectious agents that are even "smaller" than viruses. These include satellite viruses, which are dependent for their replication in other viruses; viroids, small (about300 nucleotide) RNA that are not translated and have no capsid, and prions, infectious agents whose identity remains controversial, but which may consists of only protein.

Classification of viruses

Three broad classed of viruses can be recognized, which may have independent evolutionary origins. One class, which includes the poxviruses and herpesviruses among many others, contains DNA as the genome, whether single-stranded or double-stranded, and the DNA genome is replicated by direct DNA→DNA copying. During infection, the viral DNA is transcribed by cellular and/or viral RNA polymerases, depending on the virus, to produce mRNA for translation into viral proteins. The DNA genome is replicated by DNA polymerases that can be of viral or cellular origin. Replication of the genomes of most eukaryotic DNA viruses and assembly of progeny viruses occur in the nucleus, but the poxviruses replicate in the cytoplasm.

A second class of viruses contains RNA as their genome and the RNA is replicated by direct RNA→RNA copying. Some RNA viruses, such as yellow fever virus and poliovirus, have a genome that is a messenger RNA, defined as plus strand RNA. Other RNA viruses, such as measles virus and rabies virus, have a genome that is anti messenger sense, defined as minus strand. Finally, some RNA viruses, for example, rotaviruses, have double-stranded RNA genomes. In the case of all RNA viruses, virus-encoded proteins are required to form a replicase to replicate the viral RNA, since cells do not possess (efficient) RNA→RNA copying enzymes. In the case of the minus strand RNA viruses and double-stranded RNA viruses, these RNA synthesizing enzymes also synthesize mRNA and are packaged in the virion, because their genomes cannot function as messengers. Replication of the genome proceeds through RNA intermediates that are complementary to the genome in a process that follows the same rules as DNA replication.

The third class of viruses encodes the enzyme reverse transcriptase (RT), and these viruses have an RNA→DNA step in their life cycle. The genetic information encoded by these viruses thus alternates between being present in RNA and being present in DNA. Retroviruses (e. g., HIV) contain the RNA phase in the virion; they have a single-stranded RNA genome that is present in the virus particle in two copies. Thus, the replication of their genome occurs through a DNA intermediate (RNA→DNA→RNA). The hepadnaviruses (e. g., hepatitis B virus) contain the DNA phase as their genome, which is circular and largely double-stranded. Thus their genome replicate through RNA intermediates (DNA→RNA→DNA). Just as the minus strand RNA viruses and double-stranded RNA viruses package their replicase protein the retroviruses package active RT, which is required to begin the replication of the genome in the virions. Although in many treatments the retroviruses are considered with the RNA viruses and the hepadnaviruses with the DNA viruses, we consider these viruses to form a distinct class, the RT-containing class.

An overview of the replication cycle of viruses

Receptors for virus entry

The infection cycle of an animal virus begins with its attachment to a receptor expressed on the surface of a susceptible cell, followed by penetration of the genome, either naked or complexed with protein, into the cytoplasm. Binding often occurs in several steps. For many viruses, the virion first binds to an accessory receptor that is present in high concentrations on the surface of the cell. These accessory receptors are usually bound with low affinity, and binding often has a large electrostatic component. Use of accessory receptors seems to be fairly common among viruses adapted to grow in cell culture, but less common in primary isolates of viruses from animals. This first stage binding to an accessory receptor is not required for virus entry even where used, but such binding does accelerate the rate of binding and uptake of the virus.

Following this initial binding, the virus is transferred to a high affinity receptor. Binding to the high affinity receptor is required for virus entry, and cells that fail to express the appropriate receptor cannot be infected by the virus. These receptors are specifically bound by one or more of the external proteins of a virus. Each virus uses a specific receptor (or perhaps a specific set of receptors) expressed on the cell surface, and both protein receptors and carbohydrate receptors are known. In some cases, unrelated viruses make use of identical receptors. A protein called CAR (coxsackie-adenovirus receptor), a member of the immunoglobulin (Ig) superfamily, is used by the RNA virus Coxsackie B virus and by many adenoviruses, which are DNA viruses. Sialic acid, a carbohydrate attached to most glycoproteins, is used by influenza virus, human coronavirus OC3, bovine parvovirus, and many other viruses. Conversely, members of the same viral family may use widely disparate receptors.

In addition to the requirement for a high affinity or primary receptor, many viruses also require a coreceptor in order to penetrate into the cell. In the current model for virus entry, a virus first binds to the primary receptor and then binds to the coreceptor. Only on binding to the coreceptor can the virus enter the cell.

The nature of the receptors utilized by a virus determines in part its host range, tissue tropism, and the pathology of the disease caused by it. Thus, the identification of virus receptors is important, but identification of receptors is not always straightforward. The primary receptor for HSV has now been identified as a protein belonging to the Ig superfamily. This protein is closely related to CD155, and, in fact, CD155 will serve as a receptor for some herpesviruses, but not for HSV. The story is further complicated by the fact that more than one protein can serve as a receptor for HSV. Two of these proteins, one called HIgR (for herpesvirus Ig-like receptor) and the other called either PRR-1 (for poliovirus receptor related) or HveA (for herpesvirus entry mediator A), appear to be splice variants that have the same ectodomain.

Many viruses absolutely require a coreceptor for entry, in addition to the primary receptor to which the virus first binds. The best studied example is HIV, which requires one of a number of chemokine receptors as a coreceptor. Thus a mouse cell that is genetically engineered to express human CD4 will bind HIV, but binding does not lead to entry of the virus into the cell. Only if the cell is engineered to express both human CD4 and a human chemokine receptor can the virus both bind to and enter into the cell.

Penetration

After the virus binds to a receptor, the next step toward successful infection is the introduction of the viral genome into the cytoplasm of the cell. In some cases, a subviral particle containing the viral nucleic acid is introduced into the cell. This particle may be the nucleocapsid of the virus or it may be an activated core particle. For other viruses, only the nucleic acid is introduced. The protein (s) that promotes entry may be the same as the protein (s) that binds to the receptor, or it may be a different protein in the virion.

For enveloped viruses, penetration into the cytoplasm involves the fusion of the envelope of the virus with a cellular membrane, which may be either the plasma membrane or an endosomal membrane. Fusion is promoted by a fusion domain that resides in one of the viral surface proteins. Activation of the fusion process is thought to require a change in the structure of the viral fusion protein that exposes the fusion domain. For viruses that fuse at the plasma membrane, interaction with the receptor appears to be sufficient to activate the fusion protein. In the case of viruses that fuse with endosomal membranes, the virus is first endocytosed into clathrin-coated vesicles and proceeds through the endosomal pathway. During transit, the clathrin coat is lost and the endosomes become progressively acidified. On exposure to a defined acidic pH (often about5-6), activation of the fusion protein occurs and results in fusion of the viral envelope with that of the endosome.

A dramatic conformational rearrangement of the HA protein of influenza virus, a virus that fuses with internal membrane, has been observed by X-ray crystallography of HA following its exposure to low pH. HA, which is cleaved into two disulfide-bonded fragments HA_1 and HA_2, forms trimers that are present in a spike in the surface of the virion. HA_1 contains the domain that binds to sialic acid receptors, whereas HA_2 has the fusion domain. The fusion domain, present at the N terminus of HA_2, is hidden in a hydrophobic pocket within the spike near the lipid bilayer of the virus envelope. Exposure to low pH results in a dramatic rearrangement of HA that exposes the hydrophobic peptide and transports it more that 100Å upward, where it is thought to insert into the cellular membrane and promote fusion. It is assumed that similar events occur for all enveloped viruses, whether fusion is at the cell surface or with an internal membrane.

After fusion of the viral envelope with a cellular membrane, the virus nucleocapsid is present in the cytoplasm of the cell. Virus entry by fusion can be very efficient. In some well-studied cases using cells in culture, almost all particles succeed in initiating infection.

For nonenveloped viruses, the mechanism by which the virus breaches the cell membrane is less clear. After binding to a receptor, somehow the virus or some subviral component ends up on the cytoplasmic side of cellular membrane, the plasma membrane for some viruses or the membrane of an endosomal vesicle for others. It is believed that the interaction of the virus with a receptor, perhaps potentiated by the low pH in endosomes for those viruses that enter via the endosomal pathway, causes conformational rearrangements in the proteins of the virus capsid that lead to the penetration of the membrane. In the case of poliovirus, it is known that interactions with receptors in vitro will lead to conformational rearrangements of the virion that result in the release of one of the virion proteins, called VP4. The N terminus of VP4 is myristylated and thus hydrophobic [myristic acid = $CH_3(CH_2)_{12}COOH$]. It is proposed that the conformational changes induced by receptor binding result in the insertion of the myristic acid on VP4 into the cell membrane and the formation of a channel through which the RNA can enter the cell. It is presumed that other viruses also have hydrophobic domains that allow them to enter. A number of other viruses also have structural protein with a myristylated N terminus that might promote entry. In some viruses, there is thought to be a hydrophobic fusion domain in a structural protein that provides this function. The entry process may be very efficient.

Following initial penetration into the cytoplasm, further uncoating steps must often occur. It has been suggested that, at least in some cases, translation of the genomic RNA viruses may promote its release from the nucleocapsid. In other words, the ribosomes may pull the RNA into the cytoplasm. In other cases, specific factors in the host cell, or the translation products of early viral transcripts, have been proposed to play a role in further uncoating.

It is interesting to note that bacteriophage face the problem of penetrating a rigid bacterial cell wall, rather than one of simply penetrating a plasma membrane or endosomal membrane. Many bacteriophage have evolved a tail by which they attach to the cell surface, drill a hole into the cell, and deliver the DNA into the bacterium. In some phage, the tail is contractile, leading to the analogy that the DNA is injected into the bacterium. Tailless phage is also known that introduce their DNA into the bacterium by other mechanisms.

Replication and expression of the virus genome

The replication strategy of a virus, that is, how the genome is organized and how it is expressed so as to lead to the formation of progeny virions, is an essential component in the classification of a virus and in understanding the basic details of how it replicates. Moreover, it is necessary to understand the replication strategy in order to decipher the pathogenic mechanisms of a virus and, therefore, to design strategies to interfere with viral disease.

DNA viruses

After binding to a receptor and penetration of the genome into the cell, the first event in the replication of a DNA virus is the production of mRNA from the viral DNA. For all animal DNA viruses except poxviruses, the infecting genome is transported to the nucleus where it is transcribed by cellular RNA polymerase. The poxviruses replicate in the cytoplasm and do not have access to host cell polymerases. Thus, in poxviruses, early mRNA is transcribed from the incoming genome by a virus-encoded RNA polymerase that is present in the virus core. For all animal viruses, translation of early mRNA is required for DNA replication to proceed. Early gene products may include DNA polymerases, proteins that bind to the origin of replication and lead to initiation of DNA replication, proteins that stimulate the cell to enter S phase and thus increase the supply of materials required for DNA synthesis, or products required for further disassembly of subviral particles.

The initiation of the replication of a viral genome is a specific event that requires an origin of replication, a specific sequence element that is bound by cellular and (usually) viral factors. Once initiated, DNA replication proceeds, catalyzed by either a cellular or a viral DNA polymerase. The mechanisms by which replication is initiated and continued are different for different viruses.

DNA polymerases, in general, are unable to initiate a polynucleotide chain. They can only extend an existing chain, following instructions from a DNA template. Replication of cellular DNA, including that of bacteria, requires the initiation of polynucleotide chains by a specific RNA polymerase called DNA polymerase α-primase, or primase for short. The resulting RNA primers are then extended by DNA polymerase. The ribonucleotides in the primer are removed after extension of the polynucleotide chain as DNA. Removal requires the excision of the ribonucleotides by a 5′→3′ exonuclease, filled in by DNA polymerase, and sealing of the nick by ligase. Because DNA polymerases can synthesize polynucleotide chains only in a 5′→3′ direction, and cannot initiate a DNA chain, removal of the RNA primer creates a problem at the end of a linear chromosome. How is the 5′ end of a DNA chain to be generated? The chromosomes of eukaryotic cells have special sequences at the ends, called telomeres, that function in replication to regenerate ends. The telomeres become shortened with continued replication, and eukaryotic cells that lack telomerase to repair the telomeres can undergo only a limited number of replication events before they lose the ability to divide.

Viruses and bacteria have developed other mechanisms to solve this problem. The chromosomes of bacteria are circular, so there is no 5′ end to deal with. Many DNA viruses have adopted a similar solution. Many have circular genomes (e. g., poxviruses, polyomaviruses, papillomaviruses). Others have linear genomes that cyclize before or during replication (e. g., herpesviruses). Some DNA viruses manage to replicate linear genomes, however, adenoviruses use a virus-encoded protein as a primer, which remains covalently linked to the 5′ end of the linear genome. The single-stranded parvovirus DNA genome replicates via a foldback mechanism in which the ends of the DNA fold back and are then extended by DNA polymerase. Unit sized genomes are cut from the multilength genomes that result from this replication scheme and packaged into virions.

Once initiated, the progression of the replication fork is different in different viruses. In SV40, for example, the genome is circular. An RNA primer is synthesized by primase to initiate replication, and the replication fork then proceeds in both directions. The product is two double-stranded circles. In the herpesviruses, the genome is circular while it is replicating but the replication fork proceeds in only one direction. A linear double-stranded DNA is produced by what has been called a rolling circle. For this, one strand is nicked by an endonuclease and used as a primer. The strand displaced by the synthesis of the new strand is made double-stranded synthesis. In adenoviruses, in contrast, the genome is linear and the replication fork proceeds in only one direction. A single-stranded DNA is displaced during the progression of the fork and coated with viral proteins. It can be made double-stranded by an independent synthetic event.

Plus strand RNA viruses

Following entry of the genome into the cell, the first event in replication is the translation of the incoming genomic RNA, which is a messenger, to produce proteins required for synthesis of antigenomic copies, also called minus strands, of the genomic RNA. Because the replication cycle begins by translating the RNA genome to produce the enzymes for RNA synthesis, the naked RNA is infectious, that is, introduction of genomic RNA into a susceptible cell will result in a complete infection cycle. The antigenomic copy of the genome serves as a template for the production of more plus strand genomes. For some plus strand viruses, the genomic RNA is the only mRNA produce. It is translated into a polyprotein, a long, multifunctional protein that is cleaved by viral protease, and sometimes also by cellular proteases, to produce the final viral proteins. For other plusstrand RNA viruses, one or more subgenomic mRNA are also produced from the antigenomic template. For these viruses, the genomic RNA is translated into a polyprotein required for RNA replication and for the synthesis for the subgenomic mRNA. The subgenomic RNA are translated into the structural proteins required for assembly of progeny virions. Some viruses, such as the coronaviruses, which produce multiple subgenomic RNA, also use subgenomic RNA to produce nonstructural proteins that are required for the virus replication cycle, but that are not required for RNA synthesis.

The replication of the genomic and synthesis of subgenomic RNA require recognition of promoters in the viral RNA by the viral RNA synthetase. This synthetase contains several proteins encoded by the virus, one of which is an RNA polymerase. Cellular proteins are also components of the synthetase.

All eukaryotic plus strand RNA viruses replicate in the cytoplasm. There is no known nuclear involvement in their replication. In fact, where examined, plus strand viruses will even replicate in enucleated cells. However, it is known that for many viruses, virus-encoded proteins are transported to the nucleus. These proteins may inhibit nuclear functions (Fig. 1. 14). For example, a poliovirus protein cleaves transcription factors in the nucleus.

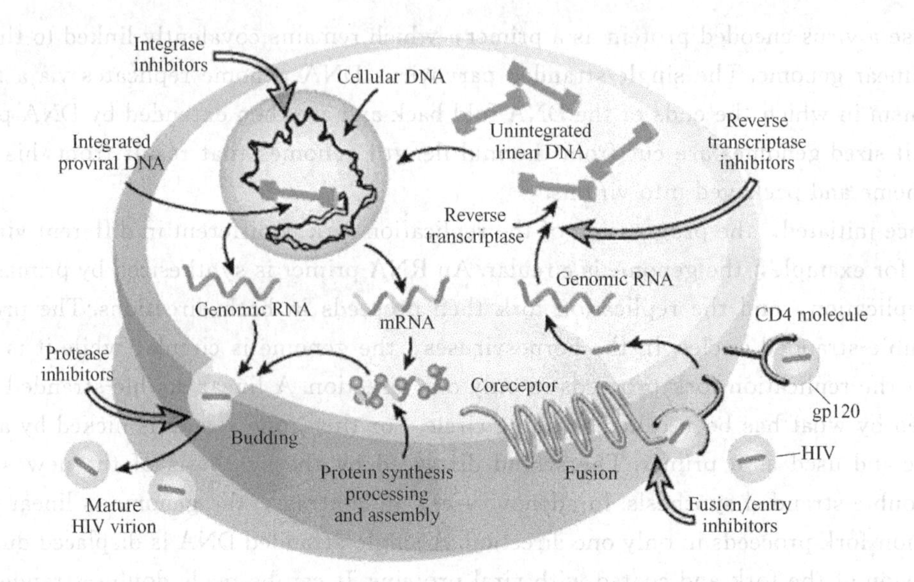

Fig. 1. 14 Life cycle of HIV

After entering the cell, the HIV genome is reverse transcribed into double-stranded DNA by RT present in the virion. The DNA copy migrates to the cell nucleus and integrates into the host genome as the "provirus" . Viral mRNA are transcribed form proviral DNA by host cell enzymes in the nucleus. Both spliced and unspliced mRNA are translated into viral proteins in the cytoplasm

Cellular functions required for replication and expression of the viral genome

The relationship between a virus and its host is an intimate one, shaped by a long evolutionary history. Viruses have small genomes and cannot encode all the functions required for successful replication, and borrow many cellular proteins as components of their replication machinery. The nature of the interactions between virus proteins and cellular proteins is an important determinant of the host range and pathology of a virus.

All animal DNA viruses, with the exception of the poxviruses, replicate in the nucleus. They make use of the cellular machinery that exists there for the replication of their DNA and the transcription of their mRNA. Some viruses use this machinery almost exclusively, whereas others, particularly the larger ones, encode their own DNA or RNA polymerases. However, almost all DNA viruses encode at least a protein required for the recognition of the origin of replication in their DNA. The interplay between the viral proteins and the cellular proteins can affect the host range of the virus. The monkey virus SV40 will replicate in monkey cells but not in mouse cells whereas the closely related mouse polyomavirus will replicate in mouse cells but not in monkey cells. The basis for the host restriction is an incompatibility between the DNA polymerase α-primase of the nonpermissive host and the T antigen of the restricted virus.

T antigens are large multifunctional proteins, one of whose functions is to bind to the origin of replication. The T antigens of the viruses form a preinitiation complex on the viral origin of replication, which then recruits the primase into the complex. Because the preinitiation complex containing SV40 T antigen cannot recruit the mouse primase to from an initiation complex, SV40 DNA replication does not occur in mouse cells. However, replication will occur in mouse cells if they are transfected with the gene for monkey primase. Similarly, monkey primase is not recruited into the complex containing mouse polyomavirus T antigen and mouse polyomavirus does not replicate in monkey cells.

In the case of RNA viruses, there is no preexisting cellular machinery to replicate their RNA, and all RNA viruses must encode at least an RNA-dependent RNA polymerase. This RNA polymerase associates with other viral and host proteins to form an RNA replicase complex, which has the ability to recognize promoters in the viral RNA as starting points for RNA synthesis. Early studies on the RNA replicase of RNA phage Qβ showed that three cellular proteins were associated with the viral RNA polymerase and were required for the replication of Qβ RNA. These three proteins, ribosomal protein S1 and two translation elongation factors EF-Ts and EF-Tu, all function in protein synthesis in the cell, the virus appropriates theses three proteins in order to assemble an active complex. Recent studies on animal and plant RNA viruses have shown that a variety of cellular proteins also appear to be required for their transcription and replication. One interesting finding is that the animal equivalents of EF-Ts and EF-Tu are required for the activity of the replicase of vesicular stomatitis virus. This suggests that the association of these two translation factors with viral RNA replicases is ancient. Several other cellular proteins have also been found to be associated with viral RNA polymerases or with viral RNA during replication, but evidence for their functional role is incomplete.

Although our knowledge of the nature of host factors involved in the replication of viral genomes and the interplay between virus-encoded and host cell proteins is incomplete, it is clear that such factors can potentially limit the host range of a virus. The permissivity of a cell for virus replication after its entry, as well as the distribution of receptors for a virus, are major determinants of viral pathogenesis.

Translation and processing of viral proteins

Viral mRNA are translated by the cellular translation machinery. Most mRNA of animal viruses are capped and polyadenylated. Thus, the translation pathways are the same as those that operate with cellular mRNA, although many viruses interfere with the translation of host mRNA to give the viral mRNA free access to the translation machinery. However, there are mechanisms of translation and processing used by some viruses that have no known cellular counterpart.

Assembly of progeny virions

The last stage in the virus life cycle is the assembly of progeny virions and their release from the infected cell.

Glossary

abnormality [ˌæbnɔːˈmæliti] n. 不正常，异常，破格，畸形，异常性
adenovirus [ˌædinəuˈvaiərəs] n. 腺病毒
agent [ˈeidʒənt] n. 试剂，媒介物，原因，因素
antibiotics [ˌæntibaiˈɔtiks] n. 抗生素，抗生学
assembly [əˈsembli] n. 集合，装配，小群落，组合
bacteriophage [bækˈtiəriəfeidʒ] n. 噬菌体
capsid [ˈkæpsid] n. 壳体，衣壳
chemokine n. 趋化因子
chromosome [ˈkrəuməsəum] n. 真周环状常染色体，染色体

clathrin [ˈklæðrin] n. 笼形素，笼形蛋白，网格蛋白，内涵素，内涵蛋白，刷状蛋白
core particle 主颗粒，核心颗粒
coronavirus [ˌkɔrənəˈvairəs] n. 冠形病毒，日冕（形）病毒
Coxsackie-adenovirus receptor（CAR） 柯萨奇-腺病毒受体
cytoplasm [ˈsaitəuplæzm] n. 细胞质，细胞浆
Ebola virus [iˈbəulə] n. 埃博拉病毒，依波拉病毒
ectodomain n. 外功能区
encephalitis [enˌsefəˈlaitis] n. 脑炎，大脑炎
endocytosis [ˌendəusaiˈtəusis] n. （细胞）内吞作用，（细胞）内摄作用
endosomal [ˈendəˌsəumə] n. 内含体
envelope [ˈenviləup] n. 外膜（细菌），被膜（病毒），外壳
epidemic [ˌepiˈdemik] n. 传染病，流行病；adj. 流行的，传染性的
epidemiology [ˌepiˌdiːmiˈɔlədʒi] n. 流行病学
eukaryotic adj. 真核的，真核生物的，真核形成的
exonuclease [ˌeksəuˈnjuːklieis] n. 核酸外切酶，外切核酸酶，水解核酸酶
fatality [fəˈtæliti] n. 死亡率
fungi [ˈfʌndʒai,ˈfʌŋgai] n. （单 fungus）真菌，霉菌
genome [ˈdʒiːnəum] n. 染色体组，基因组
glycoprotein [ˌglaikəuˈprəutiːn] n. （配）糖蛋白
hantavirus pulmonary syndrome 汉坦病毒肺综合征
hemorrhagic [ˌheməˈrædʒik] adj. 出血的
hepadnavirus [hepædnəˌvaiərəs] n. 肝 DNA 病毒属，肝脱氧核糖核酸病毒
hepatitis B virus n. 乙型肝炎病毒，血清性肝炎病毒
herpesvirus [ˌhəːpiːzˈvaiərəs] n. 疱疹病毒属
host [həust] n. 宿主
human immunodeficiency virus（HIV） 人免疫缺陷症病毒，人免疫缺陷病毒
immunization [ˌimjuːnaiˈzeiʃən] n. 免疫法，免疫接种（作用）
infectious [inˈfekʃəs] adj. 有传染性的，侵染性的
influenza [ˈinfluˈenzə] n. 流行性感冒
Junin virus 欧希金斯病毒，胡宁病毒，阿根廷出血热病毒
measles [ˈmiːzlz] n. 麻疹，囊虫病
minus strand 负链，负股
mump [mʌmp] n. 流行性腮腺炎
nucleus [ˈnjuːkliəs] n. 细胞核
oncogene [ˈɔnkəˌdʒiːn] n. 致癌基因，肿瘤基因
origin [ˈɔridʒin] n. 起源，起端，起因，原点
papillomavirus [ˌpæpiˈləuməvaiərəs] n. 乳头瘤病毒属
parasite [ˈpærəsait] n. 寄生虫
parvovirus [ˌpɑːvəˈvaiərəs] n. 细小病毒组
pathogenesis [ˌpæθəˈdʒenisis] n. 发病学，发病机理
pathology [pəˈθɔlədʒi] n. 病理学
penetration [peniˈtreiʃən] n. 穿透，贯穿，渗透，渗入，穿入
plague [pleig] n. 圣西巴斯提恩病，鼠疫，瘟疫
pneumonia [nju(ː)ˈməunjə] n. 肺炎
poliovirus [ˌpəuliəuˈvaiərəs] n. 脊髓灰质炎病毒
polymerase [ˈpɔliməˌreis] n. 聚合酶，多聚酶
polyoma virus n. 多瘤病毒
polyomavirus [ˌpɔliˈəuməvaiərəs] n. 多瘤病毒属

poxvirus ['pɔks,vaiərəs] n. 痘病毒
primase n. 引发酶，引物（合成）酶
prion ['praiɔn] n. 蛋白酶传染性因子，蛋白感染素（仅由蛋白质构成的感染物）
progeny ['prɔdʒini] n. 后裔，后代，子代
protease ['prəutieis] n. 蛋白酶，蛋白分解酵素
proteolytic [prəutiə'litik] adj. 蛋白水解的
protozoa [,prəutə'zəuə] n. 原生动物，原虫
rabies ['reibi:z] n. 狂犬病，恐水病
receptor [ri'septə] n. 受体，受者，接受者，受血者，感受器
recombination ['ri:kɔmbi'neiʃən] n. 重组，复合
replicase ['replikeis] n. 复制酶
respiratory [ris'paiərətəri] adj. 呼吸的，呼吸作用的
restriction enzyme 限制性核酸内切酶，限制性内切酶
retrovirus [,retrəu'vaiərəs] n. 逆转录病毒
reverse transcriptase 逆转录酶，反转录酶，依赖于 RNA 的 DNA 聚合酶
ribosome ['raibəsəum] n. 核蛋白体，核粒体，核糖体
rotaviruse ['rəutə,vaiərəs] n. 轮状病毒
rubella [ru:'belə] n. （病毒性）风疹
sialic acid 唾液酸
Sin Nombre virus 辛诺柏病毒
smallpox ['smɔ:lpɔks] n. 天花
stomatitis virus n. 口炎病毒
subcellular ['sʌb'seljulə] adj. 亚细胞的
substitution [,sʌbsti'tju:ʃən] n. 取代，置换，代替，替换
subviral n. 亚病毒的
syncytial [sin'siʃəl] adj. 合胞体的
synthetase ['sinθiteis,-eiz] n. 合成酶
telomere ['teləmiə] n. 端粒（在染色体端位上的着丝点）
tropism ['trəupizəm] n. 向性，趋向性
vaccine ['væksi:n] n. 菌苗，疫苗
vector ['vektə] n. 载体
vertebrate ['və:tibrit] adj. 脊椎状的，脊椎动物，脊椎动物的
virion ['vaiəriɔn] n. 病毒体，毒粒，病毒粒子
viroid ['vaiərɔid] n. 类病毒
virology [,vaiə'rɔlədʒi] n. 病毒学
virus ['vaiərəs] n. 病毒，滤过性病原体
West Nile virus n. 西尼罗病毒，西尼罗河病毒，西尼罗河脑炎病毒

难句解析

1. We can recognize specific viruses as the causative agents of epidemics that occurred hundreds or thousands of years ago from written descriptions of disease or from study of mummies with characteristic abnormalities.
   译文：我们可以确认某种特定的病毒是造成一些流行病的起因，这些流行病在有相关疾病的文字叙述出现或对木乃伊特有的异常性进行研究之前数百或数千年就发生了。
2. Smallpox virus has been eradicated worldwide by means of an ambitious and concerted effort, sponsored by the World Health Organization, to vaccinate all people at risk for the disease.

译文：通过雄心勃勃且协调一致的努力，由世界卫生组织发起，对所有可能染病的人们进行了疫苗注射，使得天花病毒在全世界范围内被彻底消灭。

3. Viruses are capable of very rapid change, both from drift due to nucleotide substitutions that may occur at a rate 10^6-fold greater than that of the plants and animals that they infect, and from recombination that leads to the development of entirely new families of viruses.

   译文：病毒能够进行非常快速地改变，一方面是因为病毒核苷酸的置换而造成的编码偏移，它们的突变速度是它们所感染的植物或者动物的 100 万倍，另一方面则是因为基因重组可以导致新病毒属的产生。

4. Thirdly, the larger viruses encode proteins that interfere with defense mechanisms of the host. These defenses include, for example, the immune response and the interferon response of vertebrates, which are highly evolved and effective methods of controlling and eliminating virus infection; and the DNA restriction system in bacteria, so useful in molecular biology and genetic engineering, that prevents the introduction of foreign DNA.

   译文：第三，一些更大的病毒可以通过编码蛋白来影响宿主的防御机制。这些防御机制包括，如脊椎动物体内的免疫应答和干扰素应答——这是一类高度进化、高度有效的控制和清除病毒感染的方法；和细菌中的 DNA 限制系统——防止外源 DNA 的进入，在分子生物学和基因工程中非常有用。

5. Once initiated, DNA replication proceeds, catalyzed by either a cellular or a viral DNA polymerase. The mechanisms by which replication is initiated and continued are different for different viruses.

   译文：一旦启动，在一种细胞的或是病毒的 DNA 聚合酶的催化下，DNA 复制就会进行下去。对于不同的病毒来说，复制启动和继续下去的机制是不一样的。

（选自：James H Strauss, Ellen G Strauss. Viruses and Human Disease. San Diego: Academic Press, 2002.）

A Review of Pathway-Based Analysis Tools That Visualize Genetic Variants

Pathway analysis is a powerful method for data analysis in genomics, most often applied to gene expression analysis. It is also promising for single-nucleotide polymorphism (SNP) data analysis, such as genome-wide association study data, because it allows the interpretation of variants with respect to the biological processes in which the affected genes and proteins are involved. Such analyses support an interactive evaluation of the possible effects of variations on function, regulation or interaction of gene products. Current pathway analysis software often does not support data visualization of variants in pathways as an alternate method to interpret genetic association results, and specific statistical methods for pathway analysis of SNP data are not combined with these visualization features. In this review, we first describe the visualization options of the tools that were identified by a literature review, in order to provide insight for improvements in this developing field. Tool evaluation was performed using a computational epistatic dataset of gene-gene interactions for obesity risk. Next, we report the necessity to include in these tools statistical methods designed for the pathway-based analysis with SNP data, expressly aiming to define features for more comprehensive pathway-based analysis tools. We conclude by recognizing that pathway analysis of genetic variations data requires a sophisticated combination of the most useful and informative visual aspects of the various tools evaluated.

Pathway analysis for genome-wide association study data

Today, pathway analysis is routine with software or web services that accept and analyze different omics data, transcriptomics, proteomics with protein-protein interactions, and metabolomics. Methods and tools used to visualize and analyze these three main kinds of high-throughput data have been reviewed. Moreover, a decade ago genetic variation data, such as single-nucleotide polymorphism (SNP) originating from analyses of array-based genome-wide association studies (GWAS), began to be incorporated into pathway analysis. Since then, the method was applied to other types of studies involving SNPs such as: epigenome-wide association study (EWAS) or sequencing-based GWAS. Although genetic association research is advancing rapidly, and especially GWAS studies are commonly performed for the genotype-phenotype investigation, biological interpretation of those data remains a challenge; especially when interpretation concerns connecting genetic findings with known biological processes. Application of pathway analysis to SNP data is a valid approach to meet this challenge for different reasons: first, because of the polygenic nature of complex diseases, such an approach holds the promise to contextualize the SNP data better and to suggest novel interpretations of the results based on prior knowledge of genes and pathway. Second, a typical display of genetic association results consists of the few SNPs showing strong evidence for disease or phenotype association (generally p-value $<1e-8$), but it is also well-known that these few associated SNPs often have only a modest effect on disease risk. Thus, examining the cumulative effects of numerous variants and visualize them at the pathway level, can empower detection of genetic risk factors for complex diseases.

We believe that data visualization, in the form of interactive pathway diagrams and/or gene-gene biological interactions such as genetic networks, enhances interpretation of scientific data, understanding the conclusions drawn, and discussing follow-up research questions. Currently, programs like Gene Set Enrichment Analysis, DAVID org: Profiler display the pathway analysis output with lists, plots, or link to the pathway diagrams. However, we believe that providing an interactive pathway diagram or network visualizations with metadata from other sources, aids in understanding the question, problem, or relationships among the data entities. Thus, interpretation of SNP data would benefit from pathway-based approaches accepting of genetic variation, so that allele-specific relationships are displayed. Recently, several step by step guides were published as reviews, describing and providing recommendations on how to use different pathway analysis methodologies, which are especially applicable to GWAS data. The main features to consider are: (ⅰ) make certain that GWAS analysis is performed according to standard guidelines; (ⅱ) choose curated and up-to-date pathway collections; (ⅲ) filter the list of gene sets to avoid bias related to size, a common limit is between 10 and 200 genes, and map the SNPs to genes based on location or linkage disequilibrium (LD); (ⅳ) choose the method according to the statistical hypothesis to be tested; (ⅴ) report the results and if applicable visualize them in order to improve comprehension. Regarding the point of the statistics, were among the first to publish a pathwaybased GWAS analysis using a statistical method adapted for genetic variation data. The authors modified a GSEA algorithm, initially designed for pathway analysis of gene expression data. Since the adaptation of GSEA by, researchers have developed other statistical methods for pathway-focused analysis of associating SNPs. Currently, existing methodologies for the analysis of GWAS gene sets are based on over-representation analysis, enrichment analysis, functional class score, and pathway-topology. The recommendation is to apply multiple methods to capture different genetic effects and identify robust gene set associations. However, only a few of these new algorithms were implemented in user-friendly tools, possibly because pathway-based approaches still have many technical challenges to overcome. Beside the main focus of data visualization, the literature search performed in this review, allow also to verify if the existing pathway analysis algorithms for genetic variations are available in user-friendly pathway tools that provide visualization options. In general, it is recognized that improving and standardizing the practice of this methodology, not only will improve the comparability of the results of gene set analysis, but also will allow a better evaluation of related polymorphisms both in the same and in different but functionally related genes. This step potentially would increase the power to detect causal pathways and disease mechanisms, using SNPs with significant associations and those in LD with functional variants. Moreover, it can point toward integration of omics data, where the additional molecular information could verify or predict the functional effects of the associating SNP. We identified a major shortcoming concerning pathway analysis programs for SNP data: genetic variation analysis have not been combined commonly in user-friendly pathway analysis tools, that provide both interactive visualization options, enabling the exploration of the data and metadata on the pathway diagrams, and existing statistical methods specifically designed for

SNP analysis. For example, one allele of a pathway entity might allow the bioprocess to continue while a second allele curtails pathway flux. Then, visualizing on a pathway map the effect of variants associated with elevated risk of disease, can indicate biological and biochemical insufficiencies (and/or vulnerabilities), which then can be made more informative if placed within depictions of the affected cell or organ, or other data related to the entities of the pathway. For instance, the rs11591147 SNP which maps to exon 12 of PCSK9 gene, directing an amino acid change Glu670Gly is a proper example to understand the potential of dynamic pathway visualization. This variant encodes a gain-of-function allele in PCSK9 that influences inter-individual variation in low-density lipoprotein (LDL) cholesterol levels between African Americans and European-Americans. In the Wiki Pathways database there is the proprotein convertase subtilisin/kexin type 9 (PCSK9) mediated LDL receptor degradation pathway that represents the key role of PCSK9 in the regulation of the LDL-cholesterol level. This pathway can be dynamically explored with the Path Visio tool, in which not only the different entities of the pathway will show extra information through their hyperlinks with various sources, but also genetic variation data with hyperlinks to SNP databases, gene expression data, and interaction values can be displayed on the pathway diagram. This multiple data visualization combines different types of information that allow the researcher to describe more easily the possible effect (s) of the genetic variant in the bioprocess with the additional support of other data. Even if genetic variation data are not available, the in silico prediction variant score such as SIFT, based methods in order to identify and describe the visualization options of the tools resulting from this literature review. The purpose of the tool evaluation relates directly to the need to combine SNP data, such as those from GWAS results, with biological context in order to better understand results in a disease context. Second, we performed a use case with the tools identified, testing a computationally derived epistatic dataset of gene-gene interactions for 12 candidate genes in obesity risk, in order to evaluate how genetic variant analysis of epistasis is tackled by the tools. Taking a visualization point of view, we report the features and the potential of the different software. Reviewing the articles, we also collected current statistical methodologies that have been applied in pathway-based analysis of GWAS data, and we report those without discussing in detail.

Overview of the comparison: benefits and limitations of the tools

Comparing the five tools described above makes evident that each uses different interactive ways to combine experimental data with information about genes, metabolites, and pathway relationships. A mock visualization of the beneficial and applicable features observed in the different tools (green highlight), and the new characteristics that enhance the visualization and analysis of SNP data in pathway-based analysis tools (red highlight) is shown in Fig. 1. 15. The five investigated tools share some similar and effective visualization approaches, such as depicting significant pathways that contain genes in the analyzed data by list view. These lists are generally ranked by enrichment ratios, p-values or FDR scores. Another common and useful strategy is to highlight genes for which pathway data are uploaded by the user, with an option to uncover gene details via hyperlinks.

A general problem in pathway-based visualizations is the efficient display of information

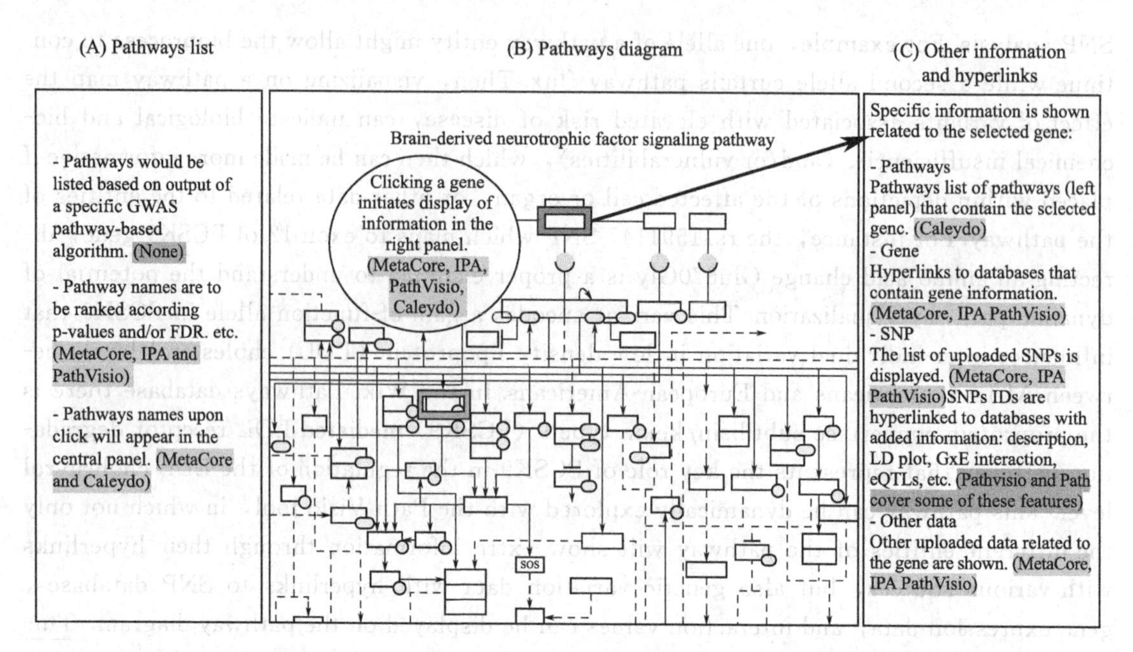

Fig. 1. 15 Mock-up visualization of the combination of useful features to apply for GWAS visualization and analysis in pathway-based tools. The panels show: (A) list of pathways obtained from a specific GWAS pathway analysis algorithm; (B) pathway diagram selected from one of the pathways listed in the panel (A), where genes with GWAS hits are highlighted (red border); (C) other information with hyperlinks related to several types of data with regard to the gene selected from the panel (B), that could be displayed in expandable/collapsible lists. Highlighted green are the tools in which the specific feature described is present, red highlights indicate features that are either not present or partially present in the tools reviewed

about genes that appear in multiple pathways and thereby interconnect those pathways. Caleydo offers an attractive solution in allowing interactive and automatic visualization of subpathways of genes present in other pathways. Caleydo uses this subpathway approach to indicate when the dataset has information about genes in a given pathway. This demonstrates how experimental data can be combined with different types of knowledge about gene relationships and permits an increased understanding of experimental results that might act in concert. Caleydo provides this type of visualization and analysis only for a specific set of experimental data. It would be a large improvement if this same approach were used to automatically select the relevant genes in the pathways based on the GWAS statistical parameters such as SNP p-value or effect size beta, which in turn could offer an assessment of allele effects on pathway output, or other omics datasets.

A strength of Path Visio, on the other hand, is its enabling of this feature to permit visualization of variants in pathways sourced either from a public repository like Ensembl or from user data. However, PathVisio lacks the interactive visualization that links entities of different pathways, as it described in Caleydo. In this context, Meta Core depicts related experimental effects of genes known to be connected via membership in a pathway, protein-protein interactions, co-citation, or co-expression in other experimental datasets with network visualization. MetaCore's network settings can be used to view or hide specific interaction mechanisms, such as binding, influence on expression, phosphorylation, or cleavage. IPA's approach is similar to that of MetaCore. After running the enrichment analysis, IPA lists the most represented processes, such as canonical pathways, networks, upstream regulators, diseases, and biological functions. In this way, the user subjectively decides which information to use and how to integrate it. Finally, Path has some methods to integrate GWAS data in pathway analysis. Path's basic data visualization of pathways uses the common strategies described above, and data integration focuses specifically on genetic information and on gene-gene interactions. Path's representation also includes an LD plot, useful and important support for GWAS interpretation.

Suggested improvements for data integration in pathway-based analysis tools

As early as 2005, the importance of effective approaches to visualization was noted through interviews and observations of current work practices. That report highlighted different aspects of pathway visualization, and suggested future developments to improve the researcher's job. Our comparisons indicate that most of those recommendations have been implemented. Two examples are the options to automatically search for relevant pathways containing genes from an uploaded dataset, and access to periodically updated pathway libraries. We have presented different types of visual strategies used in currently available tools that, for a specific gene set, support the connection with various kinds of pathway information including significant pathways, metabolites involved therein, and related diseases. With many different types of high-throughput data now readily available, including gene expression, metabolomics and protein-protein interactions, methods for integrated analysis and visualization are greatly needed. Visual strategies are particularly important for data from high-throughput experiments that provide information about many genes, facilitating evaluation of potential interactions between affected genes. This potentially can speed the investigation of the SNP effect in the pathway. Indeed, highlighting the relevant items related to the research question can reduce the process of investigating pathways singly. Moreover, alternative visualizations such as pathway hierarchies and network analysis can also reduce the long list of relevant pathways resulting from a pathway analysis. However, once the relevant processes are identified, researchers still must investigate those pathways one by one, in order to understand in detail how a SNP influences gene function in the entire process. MetaCore and IPA are examples that use networks to visualize the data integration. However, genetic variants cannot be used readily with these methods, because the data uploaded are not completely recognized. Adding the variants option to these tools would allow the user to contextualize the function of the genetic polymorphisms on different molecular levels. In addition, when data such as SNP-SNP interactions become available, pathway tools that present a network visualization option and IPA could support display of epistatic interactions from a set of SNPs located in genes that function in the same pathway. In general, several specific omics data integration methods that support inclusion of genetic variants in a pathway already exist. In this context, it is suitable to mention BioXM from Biomax Informatics because it semantically integrates existing knowledge such as genotype-phenotype relations or signal transduction pathways, and organizes data into structured networks that are connected with clinical and experimental data. With regard to the pathway collection, BioXM is flexible in that, it can display any pathway data, but requires input of pathway enrichment statistics from other sources. BioXM, on the other hand, is designed for flexibility and can integrate and display a wide range of relationships between entities, including pathways and genetic variants, but linking those two has not been demonstrated with GWAS data.

New types of genetic variant interactions for pathway-based analysis tools

Additional characteristics regarding genetic variant interactions currently are rarely depicted in pathway visualizations: edgetics, gene-environment (G×E), and epistatic interactions. Edgetics is a new term referring to network perturbation models focusing on specific alterations of the molecular interactions resulting from genetic variants. This perturbation model might improve understanding of how mutations associating with complex diseases affect biological networks or interactome properties. With network visualization already developed in some of the presented tools, it would be exciting to see this model implemented as a new feature. Another area in which pathway visualization of genetic associations can be improved involves G×E, where the genotype-phenotype association exists only under certain environmental conditions. A recently published catalog of G×Es for numerous cardiometabolic phenotypes showed the wide extent under which the genotype-phenotype association can be modified by factors such as diet, exercise, sleep, and many other exposures and lifestyle factors. For identical traits, that study noted sparse overlap of SNPs contributing to main-effect associations from GWAS compared to those supporting G × E interactions. In such instances, the pathway edges linking the G×E gene to the phenotype obviously would be conditional, and in many examples would contain entities such as glucose, palmitic acid, or linoleic acid, which are constituents metabolic pathways. Finally, epistatic interactions were used here as a use case to test the visualization tool. As a result PathVisio, MetaCore, and IPA are the tools that support upload of variant data, and highlight those variants in the pathways of the genes related to the uploaded SNPs. This feature aids investigation of the effect of the epistatic SNPs within the genes and their pathways. However, only PathVisio is able to provide the complete list of variants present in the uploaded data. Indeed, IPA identifies the genes related to the SNPs without showing the SNPs, and MetaCore performed a SNP-gene mapping that resulted in a selection of genes not included in the original dataset. Concerning IPA, it is notable to mention that Ingenuity developed another software specifically dedicated to variant investigation called "Variant Analysis" that was not detected by the review literature search, but discovered only through the Ingenuity website. In addition, the PathVisio RegInt plugin, even if it can upload the complete dataset, fails to automatically provide to the users the overview of the total pathways that present at least one of the genes with the SNPs. This feature is supported by IPA and MetaCore. The epistatic obesity use case shows that IPA, MetaCore and PathVisio have several features that permit the visualization of genetic variants in pathways. However, these features are not harmonized in one tool, but this is a reasonable outcome because the tools were not built with the aim to analyze genetic variants. On the other hand, it is remarkable to notice that these tools already have some characteristics that, with improvements, could permit such complexities of variant analysis. In summary, such conditional relationships as epistasis, G×E interactions and edgetics will need to be considered for pathway-based visualization of association data because genome-wide approaches to identify such genetic elements are rapidly maturing.

Glossary

allele [ˈæliːl] n. [遗] 等位基因
amino [əˈmiːnəʊ; əˈmainəʊ] n. [化学] 氨基
bioprocess [baioˈprɑsεs] n. 生物过程，生物处理，生物工艺
cholesterol [kəˈlestərɒl] n. [生化] 胆固醇
comparability [ˌkɑmpərəˈbiləti] n. 相似性，可比较性
contextualize [kənˈtekstjʊəlaɪz] vt. 将置于上下文中研究，使……溶入背景
encode [inˈkəʊd] vt. (将文字材料) 译成密码，编码，编制成计算机语言
epistatic [ˌepiˈstætik] n. 上位的，强性的
exon [ˈeksən] n. [生化] 外显子
genomics [dʒəˈnəumiks] n. 基因组学，基因体学
genotype [ˈdʒɛnɔˌtaip] n. 基因型，遗传型
metabolic [ˌmetəˈbɔlik] adj. 变化的，新陈代谢的
metabolites [miˈtæbəlait] n. [生化] 代谢物
metadata [ˈmetədeitə] n. [计] 元数据
mutation [mjuːˈteiʃ(ə)n] n. [遗] 突变，变化
phenotype [ˈfiːnə(ʊ)taip] n. 表型，表现型，显型
phosphorylation [fɔˌsfɔriˈleiʃən] n. [有化] 磷酸化作用
polymorphism [ˌpɔliˈmɔːfiz(ə)m] n. 多态性，多形性，同质多晶
variants [ˈvεəriənt] n. 变体，变异型

难 句 分 析

1. Today，pathway analysis is routine with software or web services that accept and analyze different omics data，transcriptomics，proteomics with protein-protein interactions，and metabolomics.
   译文：现在，使用软件或者（特定）因特网服务对不同的组学数据（包括转录组、蛋白质组和蛋白质相互作用，以及代谢组数据）进行路通分析，是一项常规分析方法。
2. We believe that data visualization，in the form of interactive pathway diagrams and/or-gene-gene biological interactions such as genetic networks，enhances interpretation of scientific data，understanding the conclusions drawn，and discussing follow-up research questions.
   译文：我们相信将数据可视化，以交互性的通路图解以及基因-基因的生物学相互作用，如基因调控网络的形式，能够使得对科学数据的评价解读，以及理解所获得的结果和讨论进一步的研究问题有所增强。
3. However，only a few of these new algorithms were implemented in user-friendly tools，possibly because pathway-based approaches still have many technical challenges to overcome.
   译文：然而，这些新算法中只有少数是在用户友好的工具中实现的，可能是因为基于路径的方法仍然有许多技术上的挑战需要克服。
4. This multiple data visualization combines different types of information that allow the researcher to describe more easily the possible effect（s）of the genetic variant in the bioprocess with the additional support of other data.
   译文：这种多数据可视化结合了不同类型的信息，使研究者能够更容易地描述生物数据

中基因变异的可能影响，并提供其他数据的额外支持。

5. Comparing the five tools described above makes evident that each uses different interactive ways to combine experimental data with information about genes，metabolites，and pathway relationships.

   译文：比较上述五种工具，可以很明显地看出，每个人使用不同的交互方式将实验数据与基因、代谢物和途径关系的信息结合起来。

6. We have presented different types of visual strategies used in currently available tools that，for a specific gene set，support the connection with various kinds of pathway information including significant pathways，metabolites involved therein，and related diseases.

   译文：我们已经展示了不同类型的视觉策略，在目前可用的工具中，对于特定的基因集合，支持与各种途径信息的连接，包括重要的途径、代谢物，以及相关的疾病。

7. Edgetics is a new term referring to network perturbation models focusing on specific alterations of the molecular interactions resulting from genetic variants.

   译文：Edgetics 是一个新的术语，指的是网络微扰模型，它关注的是由遗传变异导致的分子相互作用的具体改变。

（选自：Cirillo E，Parnell LD，Evelo CT. A Review of Pathway-Based Analysis Tools That Visualize Genetic Variants. Front Genet. 2017；8：174.）

From Genomics to Proteomics: Techniques and Applications in Cancer Research

The scope of molecular biology has undergone revolutionary changes in the past two decades. In part, this has been catalyzed by the immense task of sequencing the human genome and the requirement for tools that facilitate the rapid and accurate acquisition of raw sequence data and subsequently assist in complex analyses. Technological advances in automation and bioinformatics have spawned a discipline of biology, termed genomics, which can broadly be defined as the generation and analysis of information about genes and genomes. The field is characterized by global, comprehensive studies that are conducted in a systematic fashion. A full understanding of the complex processes that occur during the transition from normal to neoplastic cellular growth would also appear to require such a comprehensive and systematic approach. Thus, the application of genomics to cancer biology holds great potential for identifying the mechanisms that lead to malignancy and for developing therapeutic strategies. Although still in its infancy, genomics has begun to produce the anticipated results through the identification and characterization of individual genes and, recently, patterns of gene expression that distinguish malignant or premalignant cells from their normal counterparts.

The importance of the development of high-throughput DNA sequencing methods to our current ability to perform genome-scale science cannot be overstated. The completion of the human genome sequence by the Human Genome Project (HGP) and Celera was possible only because of improvement in sequencing technology. Automated capillary sequencing using fluorescent nucleotides has allowed researchers to rapidly sequence individual genes of interest and, when multiplexed in assembly-line format, permitted the sequencing of the entire 14.8 billion base-pair human genome over just nine months. This genomic sequence, with the attendant chromosomal mapping data, has greatly enhanced the ability to isolate specific genes involved in heritable cancers, such as those responsible for predisposition to breast cancer, BRCA1 and BRCA2. Employing genome-wide scans, investigators have also performed linkage studies of individuals from families at high risk for the development of prostate cancer and, to date, six distinct loci have been identified. A specific gene, *ELAC2*, located within the Hereditary Prostate Cancer gene 2 (*HPC2*) locus on chromosome 17 was recently shown to exhibit a polymorphism that segregates with prostate cancer in two pedigrees. In a case-controlled study, the probability of having prostate cancer was increased in men who carried a Leu217/Thr541 variant of the *ELAC2* gene (odds ratio 2.37; 95% CI 1.06-5.29). Genotypes at *HPC2* /*ELAC2* were estimated to account for 5% of prostate cancer in the study population. Searches for the specific genes within the remaining prostate-cancer-associated loci are under way.

In addition to the raw sequence and mapping data, the two sequencing projects have uncovered millions of single nucleotide polymorphisms (SNP) representing individual DNA bases that vary between individuals. Although the bulk of these polymorphisms do not affect gene regulation or the function of encoded proteins, they can be used as genetic markers to pinpoint the location of nearby genes that are responsible for a disease phenotype. In addition, identifying the SNPs that do confer susceptibility to disease owing to an alteration in RNA or protein function could have a profound impact on cancer screening and therapy. A recent study described the analysis of a breast-cancer-associated SNP in the 3′ untranslated region of human prohibition. Prohibition binds to members of the retinoblastoma tumor-suppressor protein family, and this interaction leads to repression of gene expression. The prohibition 3′ untranslated region encodes an RNA molecule that arrests cell proliferation when microinjected into normal mammary epithelial cells and breast cancer cell lines, and exhibits tumor-suppressor activity in animals. A SNP was found in the 3′ untranslated region that lacks the antiproliferative function. In a case-control study of breast cancer patients, a strong association between this prohibition variant allele and cancer was reported in patients who had a first-degree relative with the disease (odds ratio 2.5, $P=0.005$) and in a subset of women diagnosed with breast cancer before the age of 50 (odds ratio 4.8, $P=0.003$).

Comprehensive analyses of gene expression: transcripts

Genome-wide screening of cellular gene expression profiles using microarrays of DNA molecules is a recently developed tool that has already proven useful in characterizing human cancers. Combining molecular biology with robotic technology has resulted in the ability to create arrays comprising thousands of distinct spots of DNA, each tens of micrometers in size and each corresponding to a different specific gene. The two most common microarray configurations consist of oligonucleotides (chemically synthesized DNA chains) or cDNA (DNA fragments obtained from reverse-transcribing messenger RNA). These arrays are currently capable of detecting and quantitating the transcript abundance of more than 30000 different target genes simultaneously. The hypothesis underlying this approach is that transcript expression patterns, when compared in sufficient numbers, will allow reproducible descriptions of cell and tissue types (e.g. brain versus muscle) and tissue "states" (e.g. normal versus cancerous) that have scientific applications and predictive clinical power. Additionally, any interesting changes in the expression of individual genes can be followed with more detailed biochemical analyses.

Transcript profiling has already shown great potential to improve our understanding of tumor behavior. Microarray analysis of genes expressed in microscopically indistinguishable tissue specimens from a common subtype of B cell lymphoma—diffuse large B cell lymphoma (DLBCL)—identified two molecularly distinct forms of DLBCL that exhibited gene expression patterns characteristic of different stages of B cell differentiation. One type expressed genes characteristic of B cells found in the germinal center of a lymph node, and the other type expressed genes normally induced upon activation of B cells in the peripheral blood. Patients with germinal center B-like DLBCL had a significantly better overall survival than those with activated B-like DLBCL. The array-based risk stratification added significantly to the best available prediction method based on patient clinical characteristics and thus essentially identified a previously unknown clinical subtype of lymphoma.

A study of gene expression profile acquired from breast cancer specimens further illustrates the utility of microarray technology to classify and stratify human cancers. Primary breast tumors from seven carriers of the *BRCA1* mutation, seven carriers of the *BRCA2* mutation and seven patients with sporadic cases of breast cancer were compared using a microarray of 5361 genes. Statistical analyses demonstrated that the gene-expression profile of each tumor clustered with tumors of the same type and differed significantly from tumors of the other types. Thus, these three subtypes of breast cancer could be distinguished rapidly based solely upon their gene expression signatures. In addition, the microarray experiments identified 176 genes that were expressed differentially between tumors with *BRCA1* and *BRCA2* mutations, suggesting that there are significant functional differences between these breast tumor subtypes. This study indicates that tumor expression profiles can be used not only for identifying specific molecular pathways that could be altered in tumorigenesis but also for stratifying tumors into prognostic categories that have implications for treatment decisions.

Beyond genomics: proteomics and cancer biology

Although genomic data and transcript profiling offer tremendous opportunities to identify and understand molecular alterations in cancer, even the complete "genetic blueprint" has serious limitations. Apart from the obvious fact that cellular functions are carried out by proteins, not by DNA and RNA, there are numerous protein modifications that are not apparent from the nucleic acid or amino acid sequence. These include differential RNA splicing and posttranslational modifications such as phosphorylation and glycosylation. The genomic sequence does not specify which proteins interact, how interactions occur or where in the cell a protein localizes under various conditions. Transcript abundance levels do not always correlate with protein abundance levels and one cannot tell from the genomic sequence whether a gene is ever translated into protein or rather functions as an RNA. Recent genetic triumphs have been paralleled by a surge in interest in the comprehensive study of proteins and protein systems. This field has been dubbed "proteomics", a word derived from "proteome", which means the complete set of proteins expressed by the genome. From a biomedical standpoint, the field of proteomics has great potential because the bulk of pharmacological interventions and diagnostic tests are directed at proteins rather than genes. The inherent advantage afforded to proteomics over genomics is that the identified protein is itself the biological end-product.

The global study of proteins has many unique difficulties that set it apart from comprehensive studies of genes and transcripts. First, the behavior of proteins is determined by the tertiary structure of the molecule. Any assay based on protein binding depends on maintaining the native conformation of the protein. This puts constraints on the systems used to capture protein targets in affinity-based assays. Second, the detection of low-abundance proteins poses a particular challenge, especially given that the dynamic ranges of proteins in biological systems can reach parts per million or greater. An amplification system analogous to the polymerase chain reaction has yet to be developed for protein studies. In addition, the behavior of proteins might or might not be governed quantitatively. Protein regulation is often based not on synthesis and degradation but on reversible modifications—for example, phosphorylation. Adding to the difficulty is that RNA splicing can produce splice variants that are highly homologous but which differ in function. Nonetheless, protein science has advanced to the point that some of these hurdles can be overcome.

Until recently, the study of global protein expression was performed nearly exclusively using two-dimensional gel electrophoresis (2DE). This technique allows the display of thousands of proteins as spots on a rectangular gel. Although 2DE gel maps have been made for many cancer cell lines and human bodily fluids, the technique is somewhat cumbersome, labor intensive, insensitive and not suitable for high-throughput applications. As a result of these limitations, new approaches for performing large-scale protein studies have been developed, including mass spectrometry (MS), yeast two-hybrid systems and protein arrays. At this juncture, the application of proteomics to the study of cancer biology should be considered an emerging discipline. Most reports have focused on proof-of-principle experiments and the introduction of new technologies. The impact of global qualitative or quantitative protein analyses has yet to be demonstrated. However, the potential applications of proteomics-based approaches mandate that the technologies in early phases of development be applied to important problems in the field of cancer biology. Therefore, most of the following discussion will focus on the fundamentals of emerging proteome technologies and will consider their potential applications.

MS has emerged as one of the best ways to study proteins. It measures the charge-to-mass ratio of an ionized molecule, either a protein or peptide. The methods used to ionize the molecule of interest are either electrospray ionization, whereby a voltage placed on a fine needle causes a mist of fine droplets of charged particles, or matrix-assisted laser desorption/ionization (MALDI), where the protein/peptide of interest is crystallized within a matrix with an absorption peak at a specific wavelength to allow energy from a laser to excite the matrix and ionize the protein. When combined with a time of flight (TOF) spectrometer, MALDI has a very high sensitivity, requiring only femtomoles of sample for a good mass spectrum. Both of these techniques are amenable to automation and high-throughput analysis.

Until recently, the mass spectrometer was used mainly for protein identification through mass mapping. A purified protein or a spot from a gel with a given mass is digested into peptides by an enzyme known to cleave only after certain amino acids. The mass of each of these peptides is measured in the spectrometer. This list, or "fingerprint", of peptide masses is characteristic of a specific protein. The protein can be identified unambiguously by comparing its peptide mass fingerprint with fingerprints produced by "virtual" or *in silico* digests of every protein in the database with an identical mass.

A more powerful approach is to use the mass spectrometer in phases, this time for identifying peptides. In this technique, the spectrometer isolates a single peptide and applies energy to break the peptide randomly at each peptide bond. The fragments created are then measured. The spectrum of the fragment masses uniquely identifies the parent peptide, just as in the example above with a protein. An analogous search algorithm is applied, but this time every sequence of consecutive amino acids from the database with an equivalent mass is collected and fragmented virtually. Despite hundreds of thousands of potential peptides, the chance of a false assignment is low because the probability of matching every amino acid is low.

The field of proteomics is currently being propelled forward by the potential of MS. The potential applications have grown dramatically as reliable instruments have become increasingly affordable. The current generation of machines can perform the selection and fragmentation of peptides at speeds of hundreds an hour. The software algorithms have benefited from improvements in computational power and can identify unique peptides from hundreds of thousands of candidates in just a few seconds. By using chromatography as a separation technique, coupled directly to the inlet of the mass spectrometer, a steady stream of peptides from the digest of a complex mixture of proteins can be delivered continuously into the mass spectrometer for identification. The protein mixture can be as simple as a purified T cell receptor or as complex as the membrane fraction of cancer cells. The output from the computer search is a list of all peptides, and thus, a list of all proteins present.

Comprehensive analyses of gene expression: proteins

Utilizing the power of MS, a high-throughput technique has been developed that facilitates direct qualitative and quantitative comparisons of complex protein mixtures. This method, somewhat analogous to the microarray approach for assessing differential gene expression between two cell states, employs a chemical group or label made in two different isotopic forms: heavy and light. These labels, called isotope-coded affinity tags (ICAT), couple to all the cysteine residues in a protein mixture. The heavy reagent is added to one sample (cancer cells) and the light to another (normal cells). The samples are combined, the proteins are digested with specific proteases to generate peptide mixtures and the mixtures are analyzed in a mass spectrometer. The isotopic substitutions do not affect the behavior of peptides during separations; thus peptides derived from the two different samples enter the spectrometer at the same time. The mass spectrometer measures the relative abundance of the heavy and light peptide forms in each sample and identifies each peptide by generating and analyzing the peptide fingerprints. In this manner, a global view of protein abundance in cells or tissues in two different states can be determined.

Proof-of-principle experiments using yeast as a model system have shown that the ICAT approach is robust, reproducible and amenable to high-throughput automation. Application of the ICAT method complements transcript-profilling approaches by providing direct information detailing cellular protein alterations that occur during pathological processes.

Other techniques that borrow from the methods developed for comprehensive transcript expression studies incorporate arrays of molecules designed to capture and assay proteins. These methods represent a heterogeneous group of techniques that can assess protein-protein interactions, protein modifications and tissue profiling. The protein arrays have in common the placement of a small amount of target material on a solid support, and they use a variety of techniques to identify enzymatic or protein-binding activity. Two recent reports indicate that the array format can be used for assaying thousands of individual proteins simultaneously. Proteins of interest were attached covalently to a glass surface and washed with a mixture of fluorescently labelled proteins, which included those with highly specific protein-protein interactions. The method for immobilizing the targets maintained selective attachments between the bound proteins and known interactors, resulting in the discrimination of distinct proteins from the complex mixture. Proteins known to be enzymatic substrates were immobilized in a similar fashion, and specific protein interactions allowed enzymatic activity to be measured when the appropriate enzyme was present. Further developments of this approach could allow for the rapid screening of enzyme substrates or identify novel proteins that interact with specific ligands.

A second approach utilized a robotic device to print hundreds of specific antibody and antigen solutions on the surface of derivatized microscope slides. Proteins were labelled by the covalent linkage of a fluorescent dye and placed on these arrays along with a fluorescent internal standard for each protein assayed. The fluorescence at each spot was quantitated against the standard. Of the 115 antibody-antigen pairs, 50% of the arrayed antigens and 20% of the arrayed antibodies provided specific and accurate measurements of their cognate ligands at or below concentrations of 0. 34 μg/mL and 1. 6 μg/mL, respectively. Some of the antibody-antigen pairs allowed detection of the cognate ligands at concentrations below 1 ng/mL, a sensitivity sufficient for the measurement of clinically important proteins in patient body fluid samples.

In a variation of the protein chip theme, Ciphergen Biosystems has developed a chimera of array technology and mass spectrometry for profiling complex protein mixtures. In this method, a conventional affinity substrate such as an ion-exchange resin is affixed to a solid support called a Protein Chip. A complex protein mixture such as a whole cell lysate or body fluid is added and the chip is washed to remove unbound proteins. Bound proteins are ionized directly from the slide by a high-power laser and drawn into a mass spectrometer for analysis. The profile of protein masses ranging from 8 to 50 kD can provide a signature representative of the particular protein mixture. The technique has the advantage of being reproducible between samples, adaptable to small quantities of tissue (such as those available from tumor specimens) and scaleable for high-throughput applications. A practical example of the Protein Chip technology is described in an exploratory study of urinary protein profiles for the identification of transitional cell carcinoma of the bladder. This line of research could, ultimately, also lead to the identification of biomarkers or patterns of protein expression that might be used prognostically for monitoring disease progression or treatment response.

Concluding remarks

The tremendous quantities of data generated by technologies capable of comprehensive genome and proteome analyses will only increase as the methods are further automated. To date, genome and proteome research has depended upon, and greatly benefited from, significant advances in bioinformatics. MS-based proteomics is now possible only because of the simultaneous exponential growth both in computing power and in the genetic databases that permit large comparative searches to be performed rapidly and with high accuracy. The ability to assimilate and take full advantage of these complex data sets is closely tied to the further development of software tools that can apply filtering algorithms and recognize patterns that associate expression profiles with cellular characteristics, pharmacological interactions and epidemiological information. Taking full advantage of the information encoded in the genomes and proteomes of tumor cells could provide predictive models for how the cellular constituents interact to produce cellular phenotypes. In a recent proof-of-principle experiment, a physical interaction network was constructed to model the interplay of metabolic pathways in yeast through the integration of array-based genomics and ICAT-based proteomics techniques. This paper represents a foray into systems biology—the study and characterization of relationships and interactions between intra-and intercellular pathways and networks.

Genomics and proteomics approaches will certainly advance our understanding of basic mechanisms that are altered in the complex processes leading to carcinogenesis. New diagnostics and therapeutics will also be discovered using methods that provide global views of cellular function. However, the greatest potential for genomics and proteomics in the care of cancer patients could lie in the personalization of diagnosis and treatment. Malignant disease can now be described in such precise genetic and proteomic detail that unique and individual variations could indicate predisposition to specific carcinogen effects, forecast potential responses to chemoprevention, specify effective drug dosing and predict therapeutic outcomes. Despite the many ethical concerns, the judicious use of personalized genetic information acquired through comprehensive analyses of individual genomes and proteomes should also reduce the morbidity and expense associated with administering ineffective therapies and enable more rational designs for clinical trials.

Glossary

affinity-based [ə'finiti beisd] adj. 基于亲和性的
allele [ə'li:l] n. 等位基因
amino acid ['æminəu 'æsid] 氨基酸
antiproliferative ['æntiprəu'lifəreitive] adj. 抗增生性的，抗增殖的
assay [ə'sei] n. 测定，分析；vt. 化验，分析
base-pair (bp) ['beispεə] 碱基对
bladder ['blædə] n. 膀胱
breast cancer [brest 'kænsə] 乳腺癌
bioinformatics ['baiəu,infə'mætiks] n. 生物信息学，信息生物学
carcinogen [kɑ:'sinədʒən] n. 致癌物质
carcinoma [,kɑ:si'nəumə] n.（复 carcinomas 或 carcinomata）癌，癌瘤
capillary [kə'piləri] n. 毛细管；adj. 毛细管的
case-controlled n. 对照研究
Celera 美国公司名，主要从事基因测序和相关技术研究
cellular ['seljulə] adj. 细胞的
charge-to-mass ration [tʃɑ:dʒ tu mæs ræʃən] 质荷比
chromatography [,krəumə'tɔgrəfi] n. 色谱法
chromosome ['krəuməsəum] n. 染色体
cognate ['kɔgneit] adj. 同源的，同类的
conformation [kɔnfɔ:'meiʃən] n. 构象
consecutive [kən'sekjutiv] adj. 连续的，连贯的
counterpart ['kauntəpɑ:t] n. 对应物
cumbersome ['kʌmbəsəm] adj. 讨厌的，麻烦的，笨重的
cysteine residue ['sistin 'rezidju:] 半胱氨酸残基
diagnose ['daiəgnəuz] v. 诊断
diagnostic [daiəg'nɔstik] adj. 诊断的，症候的
digest [di'dʒest, dai'dʒest] vt. 消化
electrospray mass spectrometry (ESMS) [i'lektrəusprei mæs spek'trɔmitri] 电喷雾质谱法
enzymatic [,enzai'mætik] adj. 酶的
femtomole ['femtəuməul] n. 飞摩尔
fluorescent nucleotide [fluə'resənt nju:kliətaid] n. 荧光核苷酸
genomics [,dʒi: 'nɔmi:ks] n. 基因组学
genotype ['dʒenətaip] n. 基因型
germinal ['dʒə:minl] adj. 胚的，生发的，原始的
glycosylation [glaikəsi'leiʃən] n. 糖化，糖基化
heritable ['heritəbl] adj. 可遗传的

heterogeneous [ˌhetərəuˈdʒiːniəs] adj. 异种的，异质的
hereditary [hiˈreditəri] adj. 世袭的，遗传的
high-throughput [haiˈθruːput] n. 高通量
homologous [hɔˈmɔləgəs] adj. 同源的
immobilize [iˈməubilaiz] v. 固定不动，固定化
in silico “在硅片上”做研究，利用计算机数据分析的方法进行生物学的研究
intercellular [ˌintə(ː)ˈseljulə] adj. 胞间的，细胞间的
ionize [ˈaiənaiz] vi. 电离
ion-exchange resin [ˈaiən iksˈtʃeindʒ ˈrezin] 离子交换树脂
isotope-coded affinity tag (ICAT) [ˈaisəutəup ˈkəudid əˈfiniti tæg] 同位素亲和标记
isotopic [ˌaisəuˈtɔpik] adj. 同位的，同位素的
ligand [ˈligənd, ˈlaigənd] n. 配体，配基
lymph node [ˈlimf nəud] 淋巴结
lymphoma [limˈfəumə] n. 淋巴瘤，淋巴组织瘤
lysate [ˈlaiseit] n. 溶胞产生，溶解产物
malignancy [məˈlignənsi] n. 恶性，恶性肿瘤
mammary epithelial [ˈmæməri ˌepiˈθiːljəl] adj. 乳腺上皮的
mass spectrometry (MS) [mæs spekˈtrɔmitri] 质谱
matrix-assisted laser desorption/ionization (MALDI) [ˈmeitrikˈs əˈsistid ˈleizə diˈsɔːpʃən/aiənaiˈzeiʃən] 基质辅助激光解吸电离
membrane [ˈmembrein] n. 膜
metabolic [ˌmetəˈbɔlik] adj. 代谢作用的，新陈代谢的
microarray [ˌmaikrəuəˈrei] n. 微阵列
microinject [ˌmaikrəuinˈdʒekt] vt. 显微注射，微量注射
morbidity [mɔːˈbiditi] n. 病态，病状，发病，发病率
multiplexed [ˈmʌltiˌpleksəd] adj. 多元的
mutation [mju(ː)ˈteiʃən] n. 变异，突变
neoplastic [ˌniːəuˈplæstik] adj. 新生物的，肿瘤的
overstate [ˈəuvəˈsteit] vt. 夸大的叙述，夸张
pathological [ˌpæθəˈlɔdʒikəl] adj. 病理的，病理性的
pedigree [ˈpedigriː] n. 谱系
peptide [ˈpeptaid] n. 肽
peripheral [pəˈrifərəl] adj. 外周的
pharmacological [fɑːməkəˈlɔdʒikəl] adj. 药理学的
phosphorylation [fɔˌsfɔriˈleiʃən] n. 磷酸化
polymerase chain reaction (PCR) [ˈpɔliməˌreis tʃein ri(ː)ˈækʃən] 聚合酶链反应
posttranslational [ˌpəustrænsˈleiʃənəl] adj. 遗传翻译后的，转译后的
predisposition [priːˌdispəˈziʃən] n. 易患病的体质，素质
premalignant [ˌpriːməˈlignənt] adj. 恶化前的，癌变前的
prognostic [prɔgˈnɔstik] n. 预后性症状
prognostically [prɔgˈnɔstikəli] adv. 预言地
prohibition [ˌprəuhiˈbiʃən] n. 抑制素
proof-of-principle [pruːf əv ˈprinsəpl] 原理论证
prostate [ˈprɔsteit] n. 前列腺；adj. 前列腺的
protease [ˈprəutieis] n. 蛋白酶
proteome [prəuˈtiəˈm] n. 蛋白质组
proteomics [prəuˈtiəˈmiks] n. 蛋白质组学
retinoblastoma tumor-suppressor [ˌretinəublæs ˈtəumə ˈtjuːmə səˈpresə] 视网膜母细胞瘤肿瘤抑制因子

sequencing ['si:kwənsiŋ] n. 序列测定
single nucleotide polymorphism (SNP) 单核苷酸多态性
splicing ['splaisiŋ] n. 剪接
spawn [spɔ:n] n. 卵，子，产物，菌丝；vt. 产卵
sporadic [spə'rædik] adj. 散发的，零星的
stratification [ˌstrætifi'keiʃən] n. 成层，分层
stratify ['strætifai] vt. (使) 成层
tertiary structure ['tə:ʃəri ə 'strʌktʃə] 三级结构
the Human Genome Project (HGP) 人类基因组计划
therapeutic [θerə'pju:tik] adj. 治疗的，治疗学的
therapeutics [ˌθerə'pju:tiks] n. 治疗学，疗法
therapy ['θerəpi] n. 疗法，治疗
tumor ['tju:mə] n. 肿瘤
tumorigenesis [tu:məri'dʒenisis] n. 肿瘤发生
two-dimensional gel electrophoresis (2DE) [tu: di'menʃənəl dʒel iˌlektrəfə'ri:sis] 二维凝胶电泳
urinary [ə'juərinəri] adj. 含尿的，泌尿的
yeast two-hybrid [ji:st tu: 'haibrid] 酵母双杂交

难句分析

1. In addition, identifying the SNPs that do confer susceptibility to disease owing to an alteration in RNA or protein function could have a profound impact on cancer screening and therapy.

   译文：另外，寻找单核苷酸序列多态性对癌症的发现和治疗能产生深远的影响，这种单核苷酸序列多态性与由 RNA 或蛋白质功能的改变所引起的疾病是相关的。

2. The hypothesis underlying this approach is that transcript expression patterns, when compared in sufficient numbers, will allow reproducible descriptions of cell and tissue types (e. g. brain versus muscle) and tissue "states" (e. g. normal versus cancerous) the have scientific applications and predictive clinical power.

   译文：建立在这种方法之上的假设是，当经过足够数量的比较后，转录表达模式将可以对细胞和组织的类型（例如脑组织与肌肉组织）和"状态"（例如正常的与癌变的）进行可重复性的描述，这种描述具有科学上的应用价值和预言性诊断的能力。

3. The array-based risk stratification added significantly to the best available prediction method based on patient clinical characteristics and thus essentially identified a previously unknown clinical subtype of lymphoma.

   译文：这种建立在微阵列基础上的发病风险分析提供了最好的基于病人临床特征的预测方法，从而可以寻找一种以前未知类型的淋巴瘤。

（选自：Martin D B, Nelson P S. Trends Cell Biol, 2001, 11: 60-65.）

Gene Therapy

Introduction

Molecular genetic studies during the last decades have led to an enormous increase in our understanding of the molecular biology of the replication of viruses. The complete nucleotide sequences of many virus genomes have been determined. Information on the origins required for the replication of these genomes, the promoters used to express the information within them, and the packaging signals required for packaging progeny genomes into virions have been established for many. The mechanisms by which viral mRNA are preferentially translated have been explored. Together with methods for cloning and manipulating viral genomes, this information has made possible the use of viruses as vectors to express foreign genes. In principle, any virus can be used as a vector, and systems that use a very wide spectrum of virus vectors have been described. DNA viruses were first developed as vectors, since it is possible to manipulate the entire genome in the case of smaller viruses, or to use homologous recombination to insert a gene of interest in the case of larger viruses. When complete cDNA clones of RNA viruses were obtained, it became straightforward to rescue plusstrand viruses from clones because the viral RNA itself is infectious. The use of minus-strand RNA viruses as vectors has only recently become possible, because the virion RNA itself is not infectious and rescue of virus from cloned DNA requires coexpression of the appropriate proteins.

A sampling of expression systems and their uses is given here to illustrate the approaches that are being followed. Every virus system has advantages and disadvantages as a vector, depending on its intended use. One of the more exciting uses has been the development of viruses as vectors for gene therapy, that is, to correct genetic defects in humans. Although results have been disappointingly slow in coming, such systems offer great promise. This use represents an example of taking these infectious agents that have been the source of much human misery and developing them for the betterment of mankind.

Virus vector system

A representative sampling of viruses that are being developed as vectors is described below in order to illustrate some of the strengths and weaknesses of the different systems. The viruses used in most clinical trials to date have been the poxviruses, the adenoviruses, and the retroviruses, and these are described here. Several other virus systems that may be used in the future for treatment of humans are also described.

Vaccinia virus

Vaccinia virus is a poxvirus with a large dsDNA genome of 200 kb. This genome is too big to handle in one piece in a convenient fashion, and homologous recombination has been used to insert foreign genes into it. The large size of the viral genome, however, does mean that very large pieces of foreign DNA can be inserted, while leaving the virus competent for independent replication and assembly. Another advantage of the virus is that it has been used to vaccinate hundreds of millions of humans against smallpox. Thus, there is much experience with the effects of the virus in humans. Although the vaccine virus did cause serious side effects in a small fraction of vaccines, highly attenuated strains of vaccinia have been developed for use in gene therapy by deleting specific genes associated with virulence. A new approach to the use of poxvirus vectors has been the development of nonhuman poxviruses, such as canarypox virus, as vectors. Canarypox virus infection of mammals is abortive and essentially asymptomatic, but foreign genes incorporated into the canarypox virus genome are expressed in amounts that are sufficient to obtain an immunological response.

A variety of approaches have been used to obtain recombinant vaccinia viruses that express a gene of interest, but only the first such method to be used, and one that remains in wide use. The thymidine kinase (TK) gene of vaccinia virus is nonessential for growth of the virus in tissue culture. Furthermore, deletion of the TK gene results in attenuation of the virus in humans, which is a desirable trait. Finally the TK gene can be either positively or negatively selected by using different media for propagation of the virus. The starting point is a plasmid clone that contains a copy of the TK gene that has a large internal deletion. In the region of the deletion a vaccinia virus promoter is inserted upstream of a polylinker. The gene of interest is inserted into the polylinker using standard cloning technology. Thus, we have the foreign gene downstream of a vaccinia promoter, and the entire insert is flanked by sequences from the vaccinia TK gene. The plasmid containing the cloned TK gene with its foreign gene insert is transfected into cells that have been infected by wild-type vaccinia virus. Homologous recombination between the TK gene in the virus and the TK flanking sequences in the plasmid occurs with a sufficiently high frequency that a reasonable fraction of the progeny have the gene of interest incorporated. These viruses have an inactive TK gene, because the TK gene has been replaced by the deleted version containing the inserted foreign gene. The next step, then, is to select for viruses that are TK^- by growing virus in the pressence of bromodeoxyuridine (BUdR). An active TK enzyme will phosphorylate BUdR to the monophosphate form, which can be further phosphorylated by cellular enzymes to the triphosphate and incorporated into the viral nucleic acid during replication. Incorporation of BUdR is lethal under the appropriate conditions, and thus viruses that survive this treatment are those in which the TK gene had been inactivated.

It is usually necessary to select among the surviving progeny for those that possess the gene of interest, because inactivation of the TK gene can occur spontaneously through deletion or mutation. Selection can be accomplished by a plaque life hybridization assay in which virus in plaques is transferred to filter paper. Virus plaques on the filter paper are probed with radiolabelled hybridization probes specific for the inserted gene. Virus in plaques that hybridize to the probe is recovered and further passaged. In this way can be isolated a pure virus stock that will express the gene of interest.

Adenoviruses

Adenovirus infections of humans are common and normal cause only mild symptoms. Deletion of virulence genes from adenovirus vectors further attenuates these viruses. In addition, adenovirus vaccines have been used by the military for some years and, therefore, some experience has been gained in the experimental infection of humans by denoviruses, although gene therapy trials use a different mode of delivery of adenovirus vectors. Because of their apparent safety, adenoviruses have been developed for use as vectors in gene therapy trials or for vaccine purposes.

Two approaches have been used. In one, infectious adenoviruses have been produced that express a gene of interest. In the second approach, suicide vectors are produced that can infect a cell and express the gene of interest, but which are defective and cannot produce progeny virus. Suicide vectors cannot spread to neighboring cells, and the infection is therefore limited in scope and in duration.

The genome of adenoviruses is dsDNA of 36 kb. Thus, the genome is smaller than that of poxviruses and can accommodate correspondingly smaller inserts. However, inserts large enough for most applications can be accommodated. The genome is small enough that the virus can be reconstituted from DNA clones. Such an approach is inconvenient, however, and homologous recombination is often used to insert the gene of interest into the virus genome.

The foreign gene is inserted into the region occupied by either the adenovirus E1 or E3 genes, one or both of which are deleted in the vector construct. Virus lacking E1 cannot replicate, and such viruses form suicide vectors. For gene therapy, suicide vectors are normally used so as to prevent the spread of the infection. To prepare the stock of virus lacking E1, the virus must be grown in a cell line that expresses E1. The complementing cell line, which produces E1 constitutively supplies the E1 needed for replication of the defective adenovirus. The cells are transfected with the defective adenovirus DNA and a full yield of progeny virions results. The progeny virus is defective and cannot replicate in normal cells, but it can be amplified by infection of the complementing cell line. On introduction of the virus into a human, the virus will infect cells and express the foreign gene but the infection is abortive and no progeny viruses formed. The stock of defective virus must be tested to ensure that no replication-competent virus is present, since such virus can arise by recombination between the vector and the E1 gene in the complementing cell line.

Adenoviruses with only E3 deleted are often used to express proteins for vaccine purposes. These E3-deleted viruses possess intact E1 and will replicate in cultured cells and in humans, but are attenuated. Because the virus replicates, expression of the immunizing antigen persists for a long time and a good immune response usually results.

The procedure for insertion of the gene of interest by homologous recombination resembles that used for the poxviruses. The gene is inserted into plasmid containing flanking sequences from the E1 or E3 region, and transfected into cells infected with adenovirus. Recombinant viruses containing the gene of interest are selected and stocks prepared. It is also possible to transfect cells with the E1 or E3 expression cassette together with DNA clones encoding the rest of the adenovirus genome, in which case homologous recombination results in the production of virus. In the case of insertions into E1, cells that express E1 must be used to produce the recombinant virus.

Retroviruses

Retrovirus-based expression systems offer great promise because the retroviral genome integrates into the host-cell chromosome during infection and, in the case of the simple retroviruses at least, remains there as a Mendelian gene that is passed on to progeny cells on cell division. Thus, there is the potential for permanent expression of the inserted gene of interest. The essential components of a retrovirus vector are the long terminal repeats (LTR), the packaging sequences known as the primer binding site, and the sequences required for jumping by the reverse transcriptase during reverse transcription to form the dsDNA copy of the genome.

A packaging cell line is created that expresses the retroviral *gag*, *pol*, and *env* genes, but whose mRNA do not contain the packaging signal and so cannot be packaged. The vector DNA/RNA is created by modifying a DNA clone of a retrovirus to contain the gene of interest in place of the *gag-pol-env* genes. In the process, all of the essential *cis*-acting signals required for packaging, reverse transcription, and integration are retained. The foreign gene can be under the control of the LTR, or it can be under the control of another promoter positioned in the insert upstream of it. The resulting DNA clone is transfected into the packaging cell line, and a producer cell line isolated that expresses the vector DNA as well as the helper DNA. Vector RNA transcribed from the vector DNA is packaged into retroviral particles, using the protein expressed from the helper DNA. These particles are infectious and can be used to infect other cells or to transfer genes into a human. On infection of cells by the packaged vector, the vector RNA is reversely transcribed into DNA that integrates into the host cell chromosome, where it can be expressed under the control of the promoters that it contains. The limitation on the size of the insert is about 10 kb, the upper limit of RNA size that can be packaged.

Although murine leukemia viruses are not known to cause disease in man, it has been found that these viruses will cause tumors in immunosuppressed subhuman primates. Thus, it is thought to be essential that there be no replication-competent virus in stocks used to treat humans. Replication-competent virus can arise during packaging of the vector by recombination between the vector and the retroviral sequences used to produce Gag-Pol-Env. At the current time, preparations of packaged vectors are screened to ensure that replication-competent viruses are not present. Efforts are being made to reduce the incidence of recombination during packaging in order to simplify the procedure. One approach is to develop vectors that have very little sequence in common with the helper sequences, in order to reduce the incidence of homologous recombination. A second approach is to separate the Gag-Pol sequences from the Env sequences in the helper cell. In this case, recombination between three separate DNA fragments in the producer cell (that encoding Gag-Pol, that encoding Env, and sequences in the vector) are required in order to give rise to replication-competent retrovirus.

In gene therapy trials that use retroviruses, it has been found that the expression of the foreign gene in humans is often down regulated after a period of months. Attempts are being made to identify promoters that will not be down regulated. Different promoters might be required for different use, and promoters that target transcription to particular cell types would be useful.

A major problem with retroviral vectors is that simple retroviruses will only infect dividing cells. Although they enter cells and are reversely transcribed to DNA, the DNA copy of the genome can enter the nucleus only during cell division. In many gene therapy treatments, it is desirable to infect stem cells in order to maintain expression of the therapeutic gene indefinitely. Because stem cells divide relatively infrequently, it is difficult to infect a high proportion of them by vectors used to date. Attempts are being made to identify methods to stimulate stem cells to divide during *ex vivo* treatment, so that a larger fraction of them can be infected. A second approach is to develop lentivirus vectors. Lentiviruses, which include

HIV can infect non-replicating cells and could potentially infect non-dividing stem cell during *ex vivo* treatment. Lentivirus vectors could also be useful for therapy involving other non-dividing cells, such as neurons.

It would be of considerable utility to be able to target retroviruses to specific cells. One possible approach to this is to replace all or part of the external domains of the retroviral surface glycoprotein with a monoclonal antibody that is directed against an antigen expressed only on the target cells. In principle, this approach is feasible, but whether it can be developed into something practical is as yet an open question. If specific cells could be infected, it would allow protocols in which the therapeutic gene would be expressed only in cells where it would be most useful. It would also allow the specific killing of cells such as tumor cells or HIV-infected cells. For example, the retrovirus could express a gene that rendered the cell sensitive to toxic drugs such as BUdR. A retrovirus vector that expressed such a gene could also be useful for conventional gene therapy because it would allow the infected cell to be killed if the infection process threatened to get out of hand.

Alphaviruses

The genomes of plusstrand RNA viruses are self-replicating molecules that replicate in the cytoplasm, and they can express very high levels of protein. These properties make them potentially valuable as expression vectors.

The alphaviruses possess a genome of singlestrand RNA of about 12 kb. Their genomes can be easily manipulated as cDNA clones, and infectious RNA can be transcribed from these clones by RNA polymerases, either *in vivo* or *in vitro*. RNA transcribed *in vitro* can be transfected into cells and give rise to a full yield of virus, whereas RNA transcribed *in vivo* will begin to replicate and produce virus. The structural proteins are made from a subgenomic mRNA, making it easy to insert a foreign gene under the control of the subgenomic promoter. There are two approaches to manipulate. In one approach, a second subgenomic promoter is inserted into the genome downstream of the structural proteins, or between the structural proteins and the nonstructural proteins. Two subgenomic mRNA are transcribed, one for the structural proteins and the second for the gene of interest. The size of the insert must be relatively small, on the order of 2000 nucleotides or less, because longer RNA are not packaged efficiently. However, this system has the advantage that the resulting double subgenomic virus is an infectious virus that can be propagated and maintained without helpers.

A second approach is to delete the viral structural proteins and replace them with the gene of interest. In this case, there is room for an insert of about 5 kb that will still allow the resulting replicon to be packaged. The replicon is capable of independent replication, and transcription of a subgenomic messenger results in expression of the gene of interest. The replicon constitutes a suicide vector. It cannot be packaged unless the cells are coinfected with a helper to supply the structural proteins, or unless a packaging cell line that expresses the viral structural proteins is used.

Alphavirus replicons can be extremely efficient in expressing a foreign gene. In some cases as much as 25% of the protein of a cell can be converted to the foreign protein expressed by the replicon over a period of about 72 h. Wild-type replicons are cytolytic in vertebrate cells, inducing apoptosis, and the infection dies out. However, replicons have been produced with mutations in the replicase proteins that are not cytolytic and will produce the protein of interest indefinitely. Thus, a wide spectrum of choices is available, and the system chosen can be adapted to the needs of a particular experiment or treatment.

Viral expression systems would be more useful if they could be directed to specific cell types. An approach that uses monoclonal antibodies to direct Sindbis virus to specific cells has been described. Protein A, produced by ***Staphylococcus aureus***, binds with high affinity to IgG. It is an important component of the virulence of the bacterium because it interferes with the host immune system. The IgG binding domain of protein A has been inserted into one of the viral glycoproteins. Virions containing this domain are unable to infect cells using the normal receptor. However, the virus will bind IgG monoclonal antibodies. If an antibody directed against a cell surface component is bound, the virus will infect cells expressing this protein at the cell surface. Thus, this system has the potential to direct the virus to a specific cell type. One of the advantages of this approach is that the virus, once made, can be used with many different antibodies and thus directed against a variety of cell types. This approach is potentially applicable to any enveloped virus, and perhaps to non-enveloped viruses as well.

A modification of the alphavirus system is to use a DNA construct containing the replicon downstream of a promoter for a cellular RNA polymerase, rather than using packaged RNA replicons. On transfection of a cell with the DNA, the replicon RNA is launched when it is transcribed from the DNA by cellular enzymes. Once produced, the RNA replicates independently and produces the subgenomic mRNA that is translated into the gene of interest.

Polioviruses

Plusstrand viruses that do not produce subgenomic mRNA, such as the picornaviruses and flaviviruses, present different problems for development as vectors. The translated product from the gene of interest must either incorporated into the polyprotein produced by the virus and provisions made for its excision, or tricks must be used to express the gene of interest independently. Two approaches with poliovirus will be described as examples of how such viruses might be used as vectors.

Poliovirus replicons have been constructed by deleting the region encoding the structural proteins and replacing this sequence with that for a foreign gene. The foreign gene must be in phase with the remainder of the poliovirus polyprotein, and the cleavage site recognized by the viral 2A protease is used to excise the foreign protein from the polyprotein. Because the poliovirus replicon lacks a full complement of the structural genes (it is a suicide vector), packaging to produce particles requires infection of a cell that expresses the polioviral structural proteins by some mechanism.

A second approach to the use of poliovirus replicons is to use a second internal ribosome entry site (IRES) to initiate the synthesis of the nonstructural proteins. If the foreign gene replaces the structural genes, it will be translated from the 5′end of the genome. If the poliovirus nonstructural genes are placed downstream of a second IRES, internal initiation at this IRES results introduction of a polyprotein for the nonstructural proteins.

Rhabdoviruses

In minus strand RNA viruses, the genomic RNA is not itself infectious. Ribonucleoprotein containing the *N*, *P*, and *L* genes is required for replication of the viral RNA, and L genes is required for replication of the viral RNA, and thus for infectivity and only recently have methods been devised to recover virus from cDNA clones. A cell is transfected with a set of cDNA clones that together express N, P, and L as well as the genomic or antigenomic RNA. The antigenomic RNA usually works better, probably because it does not hybridize to the mRNA being produced from the plasmids. Encapsulation of the antigen RNA by N, P and L to form nucleocapsid allows it to replicate and produce genomic RNA that is also encapsulated. Synthesis of mRNA from the genomic RNA, together with continued replication, results in a complete virus replication cycle and production of infectious progeny virus that have as their genome the RNA supplied as a cDNA clone. The yield of infectious virus is small, but sufficient to isolate individual plaques and thus obtain viruses from the cDNA clones.

The ability to rescue virus from a cDNA clone makes it possible to manipulate the viral genome. Since the read ovaries genome is transcribed into multiple mRNA, one for each gene, and the transcription signals recognized by the enzyme are well understood, it is relatively simple to add or delete genes.

Use of viruses as expression vectors

Viruses have been widely used as vectors to express a variety of genes in cultured cells. This use is of long standing and has led to important results. Of perhaps more interest are efforts to develop viruses as vectors for medical purposes. The manipulation of virus genomes to develop new vaccines is very promising. Although no licensed human vaccines have been introduced using this technology clinical trials are ongoing. There is also expectation that viruses will be useful as vectors for gene therapy and numerous clinical trials are taking place. The results to date have been disappointing, but the promise remains.

Expression of proteins in cultured cells

The use of viruses to express foreign genes in cultured cells is well established and only a few examples are cited to illustrate the range and purpose of such use. In addition to the expression systems described above, which are based upon vertebrate viruses, expression systems based upon baculovirus have also been widely used to express proteins in cultured cells. Baculovirus are large DNA-containing viruses of insects. The gene of interest is inserted by homologous recombination in a procedure that resembles that used for vaccinia virus. The virus is grown in continuous lines of insect cells, and large amounts of protein derived from the inserted gene can be obtained. The protein is often expressed in a way that leads to its secretion from the cell, which makes purification of the desired protein easier.

Hepatitis C virus (HCV) does not grow in cultured cells to titers sufficient to allow studies on the expression of viral proteins. The only experimental model for the virus is the chimpanzee, which severely restocks the number and nature of experiments that can be done. Thus, most of what we know about the expression of the HCV genome has been obtained through expression of parts of the genome by virus vectors, often by recombinant vaccinate virus. These studies have resulted in an understanding of the two viral proteases within the HCV genome, the processing pathway through which the polyprotein translated from the genome is processed, the function of the viral IRES, and the function of the viral replicase, among other results. The use of virus vectors means that such studies on HCV can be conveniently conducted in mammalian cells under conditions that are related to the natural growth cycle of the virus.

Norwalk virus is another virus for which there is no cell culture system. The virus can be grown only in human volunteers, again limiting the range of studies that can be done. Virus particles isolated from the stools of infected volunteers are often degraded and difficult to purify to homogeneity. Thus, structural studies of infectious virus have been limited. Expression of cDNA copies of the structural proteins of the virus in baculovirus vectors has allowed the production of large amounts of viral structural proteins that spontaneously assemble into virus-like particles. These virus-like particles have been studied by cryo-electron microscopy and detailed information on the structure of the virus has been obtained in this way.

Baculoviruses are also widely used to prepare large amount of protein for crystallographic studies. Such studies require 20 mg or more of protein, and the baculovirus system can be used to prepare such quantities. An advantage of the system is that the protein is made in a eukaryotic cell, which can be important for obtaining the protein folded into its correct three-dimensional conformation.

Even for viruses for which cell culture systems exist, the use of virus vectors that express to higher levels can be advantageous. There are cell culture systems in which rubella virus will grow and plaque, and there is a full-length cDNA clone of rubella virus from which infectious RNA can be recovered. However, the cell culture systems produce only low amounts of virus proteins, especially of the nonstructural proteins, and it has been difficult to study the expression and processing of the nonstructural polyprotein. Expression of the nonstructural region of rubella virus in vaccinate virus vectors or in Sindbis virus vectors has allowed the production of much larger quantities of the polyprotein precursor. This has been used to determine the processing pathways the identification of the virus nonstructural protease, and the identifications of the cleavage sites that are cleaved by this protease.

As a final example, vaccine virus vectors and Sindbis virus vectors have been used to map T cell epitopes for a number of viruses. For this, defined regions of a viral protein are expressed in order to determine whether a particular T cell epitope lies within that region.

Viruses as vectors to elicit an immune response

Much effort is being put into the development of viruses as agents to immunize against other infectious agents, including other viruses. Such an approach has a number of advantages. There is a large body of expense in the use of attenuated or virulent viruses as vaccines. Many of these, such as vaccinia virus or the yellow fever 17D virus, both of which have been used to immunize many millions of people, can be potentially developed as vectors to express other antigens, such as those in HCV or HIV use of a live virus as a vector to express antigens of other pathogens has many of the advantages of live virus vaccines. This includes the fact that only low initial doses are required, and therefore the expense of vaccine production may be less; that subsequent virus replication leads to the expression of large amounts of the antigen over an extended period of time, and the antigen folds in a more or less native conformation; and that a full range of immunity including production of CTL as well as of humoral immunity, usually develops.

No human vaccines have been licensed that use such recombinant viruses, but there are ongoing clinical trials of several potential vaccines. Several trials of candidate vaccines against HIV have been conducted that use vaccinia virus or retrovirus vectors to express the HIV surface glycoprotein. These trials have been moderately successful in the sense that immune responses to HIV glycoprotein were obtained, but it is not known if the immune response is protective. Studies in monkeys with related vaccines against simian immunodeficiency virus have given mixed results. In most such trails, immune responses were generated, but these were not fully protective. One recent trial did generate a protective response, however, giving hope that continued efforts in this direction will ultimately work out. A very recent study with anti-HIV drugs given very soon after infection found that limiting the replication of the virus early appears to allow the generation of a protective immune response in some patients. Of eight patients treated with anti-HIV drugs very early and then taken off the drugs, five have no detectable virus 8-11 months after stopping therapy. Although these studies are preliminary and involve only a few patients, they do suggest that a nonsterilizing immune response that restricts virus replication early might prove to be protective.

Other clinical trials have also tested poxviruses as vectors. Vaccinia virus has been used in an attempt to immunize against Epstein-Barr virus, and canarypox virus has been used as a vector for potential immunization against rabies virus.

Although no licensed human vaccines use poxvirus vectors, veterinary vaccines that are based on poxvirus vectors are in use. One such vaccine consists of vaccinia virus that expresses the rabies surface glycoprotein. This vaccine has been used to immunize wildlife. The recombinant vaccinia viruses are spread in baits that are eaten by wild animals that serve as reservoirs of the virus, such as skunks, raccoons, foxes, and coyotes. This approach has been useful in limiting the spread of rabies in wildlife populations. Other poxvirus-based vaccines include vaccine virus vectors to protect cattle against vesicular stomatitis virus and rinderpest virus, and to immunize chickens against influenza virus; pigeonpox virus vectors to immunize chickens against Newcastle disease virus; fowlpox virus vectors to immunize chickens against influenza, Newcastle disease, and infectious bursal disease-viruses; a capripox virus vector to immunize pigs against pseudorabies virus; and a canarypox virus vector used to immunize dogs against canine distemper virus. Thus, it should be possible to develop human vaccines based on poxvirus vectors.

In a quite different approach, clinical trials of novel vaccine against Japanese encephalitis (JE) virus have begun recently. JE is a scourge in parts of Asia, causing a large number of deaths and neurological sequelae in people that survive the encephalitis. Vaccines in wide spread use are inactivated virus vaccines, and the difficulties in preparing the large amount of material required and delivering it to large segments of the population are significant. An attenuated virus vaccine, SA14-14-2, had been prepared in China by passing the virus in cultured cells and in rodent tissues. This vaccine is safe but over attenuated, so that the effectiveness is only 80% after a single does. In contrast, the yellow fever (YF) virus 17D vaccine had an effectiveness of virtually 100% after a single does, and immunity is long lasting, probably lifelong.

The viable chimeras were first tested in mice. The chimera containing the Nakayama strain proteins caused lethal encephalitis in mice, as does the YF 17D virus (even though it is safe for use in humans) . However, the chimera containing prM and E from the attenuated JE strain was fully attenuated in mice and did not cause illness. The fully attenuated chimera was chosen for testing in monkeys, and was found to be safe and to protect monkeys against challenge with JE virus.

Clinical trials of this candidate vaccine have begun in humans. There is every reason to believe that this vaccine will be safe and more effective than the JE vaccines now in use. Furthermore, this approach should be applicable to other flaviviruses, such as the dengue viruses, for which not licensed vaccines exist, or West Nile virus, which recently spread to the Americas and caused a number of fatal cases of human encephalitis in the New York area.

Gene therapy

A number of genetic diseases result from the failure to produce a specific protein. One of the more exciting possible uses for virus vectors is for the expression of a missing protein as a cure for the genetic defect associated with its absence. For successful treatment, expression of the missing protein must be longterm and preferably lifelong, the levels of protein produced must be sufficient to alleviate the symptoms of the disease, the protein must be expressed in or translocated to those cells that require the normal protein for function, and infection with the virus vector must be free of disease symptoms. Because of the requirement for long term expression, viruses whose DNA integrates into the host chromosome, which include retroviruses and adeno-associated viruses, offer the most promising system for many diseases. To date, several hundred patients have been treated with vectors based on Moloney murine leukemia virus in clinical trials. Clinical trials have also been conducted that use adenovirus vectors and adeno-associated virus vectors. Naked DNA has also been used in a recent trial for coronary artery disease.

In addition to the possible treatment of genetic defects, virus vectors may also be useful for the treatment of a number of acquired diseases. These include cancer, HIV infection, Parkinson's disease, injuries to the spinal cord, and vascular diseases such as restenosis and arteriosclerosis.

Retrovirus vectors to genetically mark cells

Retroviruses have been used in a number of clinical trials to genetically tag cells. Although this use does not fall within the narrow definition of gene therapy, it does provide background experience in the use of retrovirus vectors in humans. One such use has been in bone marrow transplantation for leukemia. Severe forms of leukemia can some times be treated by ablation of the hematopoietic system with chemotherapy and/or X rays in order to kill all tumor cells, followed by reconstitution of the system by transplantation of bone marrow from a compatible donor. Although often successful, the leukemia sometimes recurs and it is desirable to know whether it recurs because of incomplete destruction of the patient's leukemic cells or whether the donor cells are the source of the leukemia. Experiments in which the donor cells have been tagged using retroviruses that express a marker gene have been used to answer this question, which is important for the design of transplantation protocols.

Gene therapy for ADA deficiency

Patients who lack the enzyme adenosine deaminase (ADA) will die early in life unless treated. Lack of ADA results in the failure to clear adenosine from the body and, consequently, the accumulation of adenosine in cells throughout the body. Adenosine is toxic at high concentrations, producing a variety of symptoms. The most serious symptom results from the extreme sensitivity of T cells to elevated adenosine concentrations. Loss of T cells results in SCID, severe combined immunodeficiency. Both CTL responses (which are T cell based) and humoral responses (which require T-helper cells) are impaired. People with SCID syndrome are unable to mount an immunological response to infectious agents, and SCID is invariably fatal early in life unless treated in some way. ADA deficiency accounts for about 25% of SCID syndromes in humans.

SCID can be treated by bone marrow transplantation if a suitable donor can be found. In the case of SCID due to ADA deficiency, weekly or twice weekly injections of ADA mixed with polyethylene glycol (PEG) have been used to successfully treat about 60 patients in whom bone marrow transplantation cannot be used because of the lack of compatible donors. Of these, about 10 patients have also been treated with retroviral vectors that express ADA. In these experiments, T cells were taken from the patient (or in the case of three newborns, umbilical cord cells were used), infected *ex vivo* with the retrovirus vector using a number of different cell culture and infection protocols, and the cells reinfused into the patient. Many of the patients continue to produce ADA from the vector several years after treatment. However, all of the patients continue to receive ADA-PEG injections, which is known to be an effective treatment. Although some patients who have received retroviral therapy have been partially weaned from the supplementary ADA-PEG, it appears that some of these, and perhaps all, do not produce enough ADA to be cured. Thus, although no cures have been effected, the results to date have been encouraging and suggest that future protocols may be more successful. Two areas of retroviral therapy that need improvement are to increase the efficiency with which stem cells are infected, and the need to prevent the retroviral promoter from being down-regulated.

Cystic fibrosis

Cystic fibrosis results from loss of the cystic fibrosis transmembrane conductance regulator (CFTR), which regulates epithelial transport of ions and water. Although lack of this protein results in damage to the epithelium in many parts of the body, the most serious manifestation is lung disease accompanied by chronic bacterial infection of the airways. Clinical trials using adenoviral vectors, which infect respiratory epithelium, to express CFTR in the lungs have been conducted. The first such studies were encouraging, but a recent trial that was carefully controlled found no relief of symptoms. Inflammation produced by the high doses of adenovirus used in trials is also a problem. More recent trials have begun that use adeno-associated virus. The results of these trials are preliminary but encouraging. Cationic lipids have also been used to deliver the gene.

Rheumatoid arthritis

Rheumatoid arthritis is a chronic, progressive inflammatory disease of the joints. An estimated 5 million people in the United States suffer from it. There is no cure. Drugs therapies are used that ameliorate the symptoms, but most of these drugs have side effects and cannot be taken indefinitely. If the disease progresses far enough, joint replacement may be required. The disease is associated with the release of inflammatory cytokines in the affected joints. Clinical trials have started that use retroviruses to deliver the gene for an antiarthritic cytokine gene to the joints. The gene encodes the interleukin (IL)-1 receptor antagonist which inhibits the biological actions of both IL-1a and IL-1b. It is hoped that such treatment might damp out the diseases or at least keep it from progressing.

A gene therapy failure

Patients who have deficiencies in enzymes that participate in the urea cycle have increased concentrations of ammonia in the blood. High concentrations of ammonia result in various symptoms, which can include behavioral disturbances or coma. Severe deficiencies in these enzymes result in early death, but moderate deficiencies can result in delayed appearance of symptoms and may be partially controlled by diet. One such enzyme is ornithine transcarbamylase (OTC), which is found on the X chromosome. Deficiencies in OTC are therefore more common in males than in females.

Gene therapy trials that use virus vectors recently received a major setback when a relatively fit 18-year-old male with an inherited deficiency for OTC dies 4 days after an adenovirus vector was injected into his liver. A high does of adenovirus that expressed OTC was injected in an effort to achieve adequate levels of enzyme production. The virus unexpectedly spread widely and a systemic inflammatory response developed, inducing a fever of 40. 3℃. He went into a coma, his lungs filled with fluid, and he died of asphyxiation. This unfortunate result makes clear the possible drawbacks to experimental treatments and the difficulties in designing protocols that allow an adequate margin of safety while trying to achieve a clinically relevant result.

A gene therapy success: treatment of restenosis

A recent gene therapy trial in patients with heart disease has given very encouraging results. Although this study did not involve virus vectors, a brief description will be given since it serves as an incentive for continuation of gene therapy trials. Coronary artery disease is common in older people. Angioplasty or bypass surgery is used to open clogged arteries, but in many patients the arteries close up again (a process called restenosis). Thirteen patients with chronic chest pain who had failed angioplasty or bypass surgery or both were injected in the heart muscle with DNA encoding vascular endothelial growth factor. This factor promotes the growth of blood vessels, a process called angiogenesis. Two months after treatment, all patients exhibited an improvement in vascularization of damaged areas of the heart, as shown by imaging and mapping studies. All patients reported a decrease in disease symptoms, and all had an improved performance in treadmill tests. Although the number of patients is small, the uniformly positive results are encouraging.

Viruses as anticancer agents

There is hope that viruses can be developed as anticancer agents. Although more than 1000 patients are participating in the trials, this field is still in its infancy, and only a brief summary of approaches is presented.

In most of the trials, viruses are used to express proteins that control the growth of tumors or that are toxic to tumor cells. A number of different cytokines are being tried, such as IFN-γ, IL-2, TNF and GM-CSF. Another approach is to try to repair the defective regulatory gene in the tumor cell, which is often p53. Many other gene products are also being tested. All of these trials represent preliminary attempts, and it will be some time before we know if any of them represent successful approaches.

Thought is being given to the possibility of using viruses to express proteins overexpressed in tumor cells in an attempt to stimulate the immune system to respond by killing tumor cells. This is in essence an attempt to vaccinate a person against a tumor. For this approach to succeed, an antigen overproduced by a tumor cell, such as a melanoma cell, must be identified, inserted into a suitable vector, and the person with the tumor infected with the virus vector in an attempt to stimulate the immune system. In principle, this approach may be feasible, but only time will tell whether it is in fact practical.

Another approach is to try to direct the virus, more or less specifically, to infect the tumor cells, so that upon infection the cells are killed. Cell death might result either because the virus itself is cytolytic or because the virus expresses a protein that renders the cell sensitive to a toxic agent such as BUdR. A number of the trials use the TK gene for this, since cells that express TK are sensitive to BUdR.

A number of possible vectors have been suggested as a way to specifically kill tumor cells. One concept is to try to cure or control brain tumors, especially gliomal tumors, by using either herpesviruses or retroviruses. Simple retroviruses can only replicate in dividing cells. Thus, they should be able to infect only tumor cells in the brain, since most neuronal cells are terminally differentiated and not divided. If the retroviruses express a protein that renders the cells sensitive to a toxin, it might be possible to kill replicating cells and therefore only the tumor cells. A different approach has been to try to use herpesviruses as antitumor viruses. These viruses set up latent infections in neurons and might in principle be used to control brain tumors.

Glossary

adenosine deaminase (ADA) 腺苷脱氨酶
alphavirus [ælfə'vaiərəs] n. α病毒，α病毒属，甲病毒属
ammonia ['æməunjə] n. 氨，氨水
angiogenesis [ˌændʒiəu'dʒenisis] 血管发生
angioplasty ['ændʒiəuˌplæsti] n. 血管成形术
arteriosclerosis [ɑːˌtiəriəuˌskliə'rəusis] n. 血管硬化，动脉硬化
asphyxiation [æsˌfiksi'eiʃən] n. 窒息
attenuate [ə'tenjueit] v. 稀释，解毒，减弱
baculovirus [ˌbækjuləu'vaiərəs] n. 杆状病毒
bromodeoxyuridine (BUdR) ['brəuməudiˌɔksi'juəridiːn] n. 溴脱氧尿苷
canarypox virus 金丝雀痘病毒
canine distemper virus 犬瘟热病毒
capripox virus 绵羊痘病毒亚群，羊痘病毒
cassette [kɑː'set] n. 盒子，表达框
chemotherapy [ˌkeməu'θerəpi] n. 化学疗法，化学治疗
chimeras [kai'miərə] n. 嵌合体
chimpanzee ['tʃimpən'ziː] n. 黑猩猩
chronic ['krɔnik] adj. 慢性的，长期的
cleavage ['kliːvidʒ] n. 劈开，分裂
coma ['kəumə] n. 昏迷
conformation [ˌkɔnfɔː'meiʃən] n. 构象
coronary artery disease 冠状动脉疾病，冠心病
cystic fibrosis transmembrane conductance regulator (CFTR) 囊性纤维化跨膜通道调节因子
cystic fibrosis 纤维囊泡症，囊性纤维化
cytolytic [ˌsaitəu'litik] 细胞溶解的，溶细胞的

dengue virus 波加卡热病毒，登革热病毒
donor ['dəunə] n. 捐赠人，供体
downstream ['daunstri:m] adv. 下游地；adj. 下游的
encapsulation [inˌkæpsju'leiʃən] n. 包装，封装
endothelial growth factor 内皮生长因子
Epstein-Barr virus EB病毒
ex vivo [eks 'vi:vəu] 体外
flavivirus [ˌfleivi'vaiərəs] n. 黄病毒属（b群黄病毒）
genetic [dʒi'netik] adj. 遗传的，起源的
hematopoietic system 造血系统
hepatitis C virus n. 丙型肝炎病毒
homogeneity [ˌhɔməudʒe'ni:iti] n. 均匀性，齐性，同种，同质（性），同性
homologous [hɔ'mɔləgəs] adj. 相应的，对应的，一致的
hybridization [ˌhaibridai'zeiʃən] n. 杂交，杂种培植，配种
immunosuppress [ˌimjunəusə'pres] vt. 抑制（生物体）的免疫反应
infectious bursal disease viruses 传染性囊病病毒
inherit [in'herit] vt. 继承，遗传
internal ribosome entry site (IRES) 内部核糖体进入位点
joint [dʒɔint] n. 接缝，接合处，关节
lentivirus ['lentiˌvaiərəs] 慢病毒属（包括人免疫缺陷病毒，属于逆转录病毒科）
leukemia [lju:'ki:miə] n. 白血病
long terminal repeats (LTR) 长末端重复序列
melanoma [ˌmelə'nəumə] n. （恶性）黑素瘤
Moloney murine leukemia virus 莫洛尼鼠白血病病毒
murine leukemia virus 小鼠C型肿瘤病毒，鼠白血病病毒
Norwalk virus 诺瓦克病毒（流行性急性胃肠炎的常见病原体）
ornithine transcarbamylase (OTC) 鸟氨酸转氨甲酰酶，鸟氨酸转氨酶
ovary ['əuvəri] n. （生物）卵巢；（植物）子房
phosphorylate ['fɔsfərileit] vt. 使磷酸化
picornavirus [paiˌkɔ:nə'vaiərəs] n. 小核糖核酸病毒
plasmid ['plæzmid] n. 质粒，质体
polyethylene glycol (PEG) 聚乙二醇
precursor [pri(:)'kə:sə] n. 前身，前体，产物母体（化合物）
primate ['praimit] n. 灵长类的动物
promoter [prə'məutə] n. 促进者，助长者
propagation [ˌprɔpə'geiʃən] n. 动植物繁殖
pseudorabies virus 伪狂犬病病毒，猪疱疹病毒
radiolabelled adj. 放射标记的，放射示踪的
replicon ['replikɔn] n. 复制子
restenosis [ˌri:sti'nəusis] n. （尤指心瓣手术后的）再狭窄
rhabdovirus [ˌræbdəu'vaiərəs] n. 杆状病毒，弹状病毒
rheumatoid arthritis 类风湿性关节炎
ribonucleoprotein [ˌraibəunjukli:əu'prəuti:n] n. 核蛋白
rinderpest virus 牛瘟麻疹病毒
rodent ['rəudənt] adj. 啮齿目的，侵蚀性的；n. 啮齿动物
sequelae [si'kwi:li:] n. （单 sequela）后遗症
simian immunodeficiency virus 猴免疫缺陷型病毒
Sindbis virus 辛德毕斯病毒（辛德毕斯甲病毒）（一种由库蚊传播的甲病毒）

spinal cord 脊髓，脊索
Staphylococcus aureus 金黄色葡萄球菌
stomatitis virus 口炎病毒
stool [stuːl] n. 凳子，粪便
subgenomic （病毒）次基因组的，亚基因组的
suicide ['sjuisaid] n. 自杀，自毁
symptom ['simptəm] n. 症状，征兆
thymidine kinase（TK） 胸（腺嘧啶脱氧核）苷激酶，胸苷激酶
transfect [træns'fekt, trænz- ,trɑːn-] vt. 使转染，使（细胞）感染病毒核酸
translocate [trænsləu'keit] vt. 改变……的位置
umbilical cord n. 脐带，珠柄
vaccinia virus 痘病毒，痘苗病毒
veterinary ['vetərinəri] n. 兽医；adj.（医）牲畜的，兽医的
virulence ['virulәns] n. 毒力，毒性，致病力
virulent ['virulәnt] adj. 剧毒的，致命的，有毒力的
volunteer [vɔlən'tiə(r)] n. 志愿受试者

难句分析

1. Information on the origins required for the replication of these genomes，the promoters used to express the information within them，and the packaging signals required for packaging progeny genomes into virions have been established for many.
   译文：在启动框上的信息是这些基因的复制所必需的，启动子是用来表达在启动框中的信息，同时包装信号是包装新生成的病毒和遗传基因所必需的。
2. This use represents an example of taking these infectious agents that have been the source of much human misery and developing them for the betterment of mankind.
   译文：这个用途是使对人类产生悲剧的感染性毒剂发展成为对人类有益的一个很好的例子。
3. It is also possible to transfect cells with the E1 or E3 expression cassette together with DNA clones encoding the rest of the adenovirus genome，in which case homologous recombination results in the production of virus. In the case of insertions into E1，cells that express E1 must be used to produce the recombinant virus.
   译文：用E1和E3表达框与编码腺病毒剩余部分的DNA克隆在细胞中共转染也是有可能的，在这种情况下会发生病毒的同源重组。在将基因插入E1的情况下，细胞会在表达E1的情况下产生重组病毒。
4. However，replicons have been produced with mutations in the replicase proteins that are not cytolytic and will produce the protein of interest indefinitely. Thus，a wide spectrum of choices is available，and the system chosen can be adapted to the needs of a particular experiment or treatment.
   译文：然而，用不是细胞溶解并且使这种蛋白有高度特异性所产生的复制酶蛋白进行突变而产生复制子。因此一个宽峰的光谱是可以选择的，同时系统的选择应该是适应特殊的实验与治疗。
5. The only experimental model for the virus is the chimpanzee，which severely restocks the number and nature of experiments that can be done. Thus，most of what we know about the expression of the HCV genome has been obtained through expression of parts of the genome by virus vectors，often by recombinant vaccinate virus.

译文：目前唯一的为这种病毒提供的实验模型是黑猩猩实验，这个实验严格地记录了数据和实验所能够表达的事实。因此，我们所知道的大多数关于 HCV 基因的表达是通过部分病毒载体的基因（通常是重组的痘病毒载体）的表达而获得的。

6. In most such trails, immune responses were generated, but these were not fully protective. One recent trial did generate a protective response, however, giving hope that continued efforts in this direction will ultimately work out.

译文：在绝大多数的临床试验中会产生免疫应答，但是这些应答并不能够达到完全保护的效果。然而，目前一个临床试验已经产生了一个保护性的应答，让我们看到了一丝希望：在这个方向所做出的努力终究会有结果的。

（选自：James H Strauss, Ellen G Strauss. Viruses and Human Disease San Diego: Academic Press, 2002.）

The Molecular Basis of Cancer—Cell Behavior

Oncogenes and tumor suppressors—and the mutations that affect them—are different beasts from the point of view of the cancer gene hunter. But from a cancer cell's point of view they are two sides of the same target. The same kinds of effects on cell behavior can result from mutations in either class of genes, because most of the control mechanisms in the cell involve both inhibitory (tumor suppressor) and stimulatory (proto-oncogene) components. In terms of function, the important distinction is not the distinction between a tumor suppressor and a proto-oncogene, but between genes lying in different biochemical and regulatory pathways.

Some of the pathways important in cancer carry signals from a cell's environment; other are responsible for the cell's internal programs, such as those that control the cell cycle or cell death; still others govern the cell's movements and mechanical interactions with its neighbors. The various pathways are linked and interdependent in complex ways. Much of what we know about them has been learned as a byproduct of cancer research; conversely, study of these basic aspects of cell biology has transformed our understanding of cancer.

We begin with a brief general discussion of how we determine the cellular function of cancer-critical genes. We then review what is known about how these cancer-critical genes control the relevant cell behaviors. Finally, we turn to the development of colon cancer as an example of how tumors evolve through the accumulation of mutations that lead from one pattern of bad behavior to another that is worse.

Studies of developing embryos and transgenic mice help to uncover the function of cancer-critical genes

Given a gene that is mutated in a cancer, we need to understand both how the gene functions in normal cells and how mutations in the gene contribute to the aberrant behaviors characteristic of cancer cells. When *RB* was originally cloned, for example, all that was known was that it was deleted in cancer. In the case of *Ras*, the mutant gene was known to direct cells to proliferate excessively and inappropriately in normal or cancer cells. In both cases, cancer research was the starting point for studies that revealed the key role of these gene products in normal cells—Rb as a cell-cycle regulator, Ras as a central component of cell-signalling pathways.

Today, we know much more about cells, so that when a new gene is identified as critical for cancer, it often turns out to be familiar from studies in another context. For example, many Oncogenes and tumor suppressor genes are found to be homologs of genes already known for their role in embryonic development. Examples include components of practically all the major developmental signalling pathway through which cells communicate: the Wnt, Hedgehog, TGFβ, Notch, and receptor tyrosine kinase signalling pathways all include important cancer-critical genes—with *Ras* being part of the last of these pathways.

In hindsight, this is no surprise. The same signalling mechanisms that control embryonic development operate in the normal adult body to control cell turnover and maintain homeostasis. Both the development of a multicellular animal and the maintenance of its adult structure depend on cell-cell communication and on regulated cell proliferation, cell differentiation, cell death, cell movement, and cell adhesion—in other words, on all the aspects of cell behavior whose derangement underlies cancer. Developmental biology, often using model animals such as *Drosophila* and *C. elegans*, thus provides a key to the normal functions of many cancer-critical genes.

Ultimately, however, we want to know what mutations in these genes do the cells in the tissues that give rise to the cancer. A certain amount of information can be obtained by studying cells *in vitro* or by examining human cancer patients. But to investigate how mutations in various cancer-critical genes affect tissues in a whole organism, the transgenic mouse has proved particularly useful.

Transgenic mice that carry an oncogene in all their cells can be generated by some methods. Oncogenes introduced in this way may be expressed in many tissues or in only a select few, according to the tissue specificity of the associated regulatory DNA. Studies of such transgenic animals reveal that, even in mice, a single oncogene is not usually sufficient to turn a normal cell into a cancer cell. Typically, in mice that are endowed with a *Myc* or *Ras* oncogene, some of the tissues that express the oncogene grow to an exaggerated size, and over time occasional cells undergo further changes to give rise to cancers. The vast majority of the cells in the transgenic mouse that express the *Myc* or *Ras* oncogene, however, do not give rise to cancers, showing that the single oncogene is not enough to cause malignancy. From the point of view of the whole animal, the inherited oncogene, nevertheless, is a serious menace because it increases the risk of developing cancer. Mice that express more than one oncogene can be generated by mating a pair of transgenic mice—one carrying a *Mys* oncogene, the other carrying a *Ras* oncogenes, for example. These offspring develop cancers at a much higher rate than either parental strain, but again the cancers originate as scattered isolated tumors among noncancerous cells. Thus, even with these two expressed oncogenes, the cells must undergo further, randomly generated changes to become cancerous.

Just as activated oncogenes can be introduced into mouse tissues, so tumor suppressor genes can be inactivated by "knocking out" the gene in the mouse using reverse genetic techniques. Several tumor suppressor genes have been knocked out in mice, including *Rb*. As anticipated, many of the mutant strains that are missing one copy of a tumor suppressor gene are cancer prone. Deletion of both copies often leads to death at an embryonic stage, reflecting the essential roles these genes play during normal development. To bypass this block and see that effect of homozygous mutations in an adult tissue, one can use the methods to create conditional mutations, such that only one tissue—say the liver—displays the defect. In these ways, transgenic mice have become a key source of information as to the mechanisms of tumor formation. We shall see later that they also provide important models for the development of new cancer therapies.

Many cancer-critical genes regulate cell division

Most cancer-critical genes code for components of the pathways that regulate the social behavior of cells in the body—in particular, the mechanisms by which signals from a cell's neighbors can impel it to divide, differentiate, or die. In fact, many of the components of cell-signalling pathways were first identified through searches for cancer-causing genes, and a full list of proto-oncogene products and tumor suppressors include examples of practically every type of molecule involved in cell singaling-secreted proteins, transmembrane receptors, GTP-binding proteins, protein kinases, gene regulatory proteins, and so on. Many cancer mutations alter signal pathway components in a way that causes them to deliver proliferative signals even when more cells are not needed, switching on cell growth, DNA replication, and cell division inappropriately. Mutations that inappropriately activate a receptor tyrosine kinase, such as the EGF receptor, or proteins in the Ras family, which lie downstream from such growth factor receptors, act in this way.

Other signalling pathways can function to inhibit cell division, the best known example being the antigrowth effect of the TGFβ family of signalling proteins. Loss of growth inhibition through TGFβ-mediated pathways contributes to the genesis of several types of human cancers. The receptor TGFβ-RII is found to be mutated in some cancers of the colon and Smad4—a key intracellular signal transducer in the pathway—is inactivated in cancer of the pancreas and some other tissue.

Ultimately, the cancer-critical genes that regulate cell division exert their effects by acting on the central cell-cycle control machinery. Not surprisingly, mutations in the machinery feature prominently in many cancers. A key point at which cells make the decision to replicate their DNA and enter the cell division cycle is thought to be controlled by the Rb protein, the product of the tumor suppressor gene *Rb*. *Rb* serves as a brake that restricts entry into S phase by binding to gene regulatory proteins needed to express genes whose products are required for progress round the cycle. Normally, this inhibition by *Rb* is relieved at the appropriate time by phosphorylation of *Rb*, which causes it to release its inhibitory grip.

Many cancer cells proliferate inappropriately by eliminating *Rb* entirely, as we have already seen. Other tumors achieve the same endpoint by acquiring mutations in other components of the *Rb* regulatory pathway. This in normal cells, a complex of cyclin D1 and the cyclin-dependent kinase Cdk4 (G_1-Cdk) stimulates progression through the cell cycle by phosphorylating *Rb*. The p16 (INK4) protein—which is produced when cells are stressed—inhibits cell-cycle progression by preventing the formation of an active cyclin D1-Cdk4 complex. Some glioblastomas and breast cancers are found to have amplified the genes encoding Cdk4 or cyclin D1, thus favoring cell proliferation. And deletion or inactivation of the *p16* gene is common in many forms of human cancer. In cancers where it is not inactivated by mutation, this gene is often silenced by methylation of its regulatory DNA.

The variety of ways in which the machinery of cell-cycle control can be altered in cancer illustrates two important points. Firstly, it explains why individual cases of a particular cancer showing the same symptoms may arise from different mutations: in many cases several alternative mutations will have much the same effect on cell proliferation. Secondly, it reinforces the point that there is no fundamental difference in the processes that are affected by oncogenes—which become activated by mutation—and those affected by tumor suppressor genes—which become inactivated. These two classes of cancer-critical genes queerly differ in whether they play a stimulatory or inhibitory role in a pathway (Fig. 1. 16).

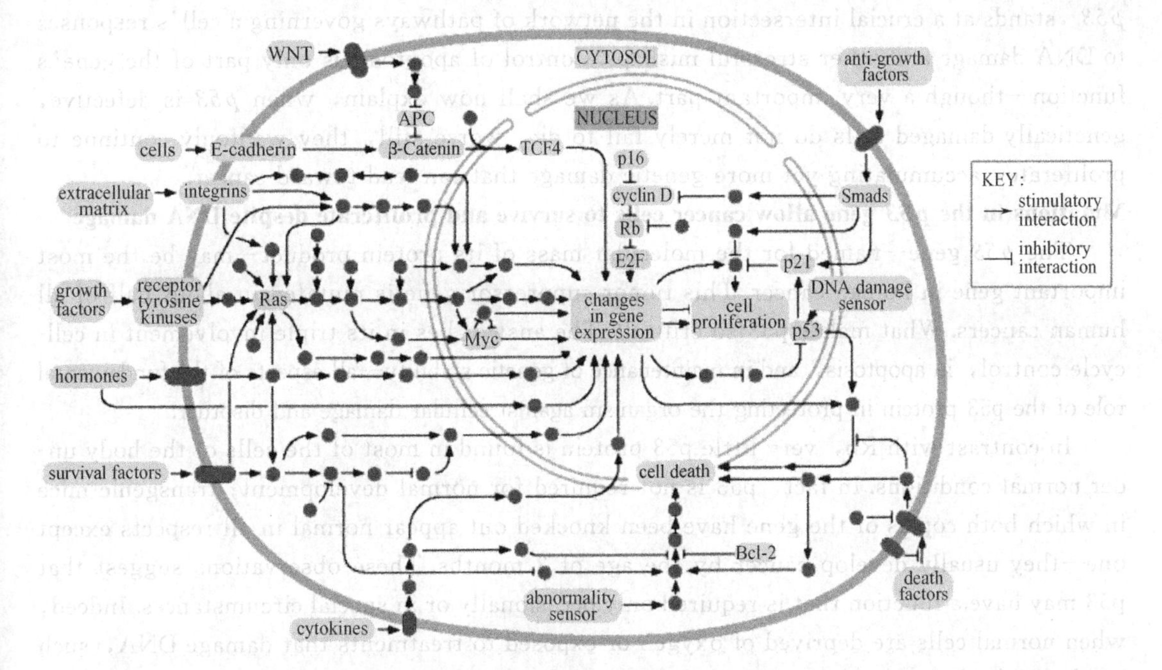

Fig. 1. 16 Chart of the major signalling pathways relevant to cancer in human cells, indicating the cellular locations of some of the proteins modified by mutation in cancers

Products of both oncogenes and tumor suppressor genes often occur within the same pathways. Individual signalling proteins are indicated by red circles, with the cancer-critical components and control mechanisms discussed in this chapter in green. Stimulatory and inhibitory interactions between proteins are designated as shown in the key

(选自：Bruce Alberts，Alexander Johnson，Julian Lewis，Martin Raff，Keith Roberts，Peter Walter. Molecular Biology of the Cell. 4th ed. New York：Taylor & Francis Group，2002.)

Mutations in genes that regulate apoptosis allow cancer cells to escape suicide

To achieve net cell proliferation, it is necessary not only to drive cells into division, but also to keep cells from committing suicide by apoptosis. There are many normal situations in which cells proliferate continuously, but the cell division is exactly balanced by cell loss. In the germinal centers of lymph nodes, for example, B cells proliferate rapidly but most of their progeny are eliminated by apoptosis. Apoptosis is thus essential in maintaining the normal balance of cell births and deaths in tissues that undergo cell turnover. It also has a vital role in the cellular reaction to damage and disorder. Cells in a multicellular organism commit suicide when they sense that something has gone wrong—when their DNA is severely damaged or when they are deprived of survival signals that tell them they are in their proper place. Resistance to apoptosis is thus a key characteristic of malignant cells, essential for enabling them to increase in number and survive where they should not.

A number of mutations that inhibit apoptosis have been found in tumors. One protein that blocks apoptosis, called Bcl-2, was discovered because it is the target of a chromosome translocation in a B-cell lymphoma. The effect of the translocation is to place the *Bcl-2* gene under the control of a regulatory element that drives overexpression, which allows survival of B lymphocytes that would normally have died.

Of all the cancer-critical genes involved in control of apoptosis, however, there is one that is implicated in cancers in an exceptionally wide range of tissues. This gene, called *p53*, stands at a crucial intersection in the network of pathways governing a cell's responses to DNA damage and other stressful mishaps. Control of apoptosis is only part of the gene's function—though a very important part. As we shall now explain, when *p53* is defective, genetically damaged cells do not merely fail to die; worse still, they wantonly continue to proliferate, accumulating yet more genetic damage that can lead toward cancer.

Mutations in the *p53* gene allow cancer cells to survive and proliferate despite DNA damage

The *p53* gene—named for the molecular mass of its protein product—may be the most important gene in human cancer. This tumor suppressor gene is mutated in about half of all human cancers. What makes p53 so critical? The answer lies in its triple involvement in cell-cycle control, in apoptosis, and in maintenance of genetic stability—all aspects of the fundamental role of the p53 protein in protecting the organism against cellular damage and disorder.

In contrast with Rb, very little p53 protein is found in most of the cells of the body under normal conditions. In fact, p53 is not required for normal development; transgenic mice in which both copies of the gene have been knocked out appear normal in all respects except one—they usually develop cancer by the age of 3 months. These observations suggest that p53 may have a function that is required only occasionally or in special circumstances. Indeed, when normal cells are deprived of oxygen or exposed to treatments that damage DNA, such as ultraviolet light or gamma rays, they raise their concentration of p53 protein by reducing the normally rapid rate of degradation of the molecule. The p53 response is seen also in cells where oncogenes such as *Ras* and *Myc* are active, generating an abnormal stimulus for cell division.

In all these cases, the high level of p53 protein acts to limit the harm done. Depending on circumstances and the severity of the damage, the p53 may either drive the damaged or mutant cell to commit suicide by apoptosis—a relatively harmless event for the multicellular organism—or it may trigger a mechanism that bars the cell from dividing so long as the damage remains unrepaired. The protection provided by p53 is an important part of the reason why mutations that activate oncogenes such as *Ras* and *Myc* are not enough by themselves to create a tumor.

The p53 protein exerts its cell-cycle effects, in part at least, by binding to DNA and inducing the transcription of *p21* —a regulatory gene whose protein product binds to Cdk complexes required for entry into and progress through S phase. By blocking the kinase activity of these Cdk complexes, the p21 protein prevents the cell from entering S phase and replicating its DNA.

Cells defective in *p53* fail to show these responses. They tend to escape apoptosis, and if their DNA is damaged—by radiation or by some other mishap—they carry on dividing, plunging into DNA replication without pausing to repair the breaks and other DNA lesions that the damage has caused. As a result, they may either die or, far worse, survive and proliferate with a corrupted genome. A common consequence is that chromosomes become fragmented and incorrectly rejoined, creating, through further rounds of cell division, an increasingly disrupted genome. Such chromosomal mayhem can lead to both loss of tumor suppressor genes and activation of oncogenes, for example by gene amplification. In addition to being an important mechanism for activating oncogenes, gene amplification can also enable cells to develop resistance to therapeutic drugs, as we see below.

In summary, p53 helps a multicellular organism to cope safely with DNA damage and other stressful cellular events, acting as a check on cell proliferation in circumstances where it would be dangerous. Many cancer cells contain large quantities of mutant p53 protein (of an ineffectual variety), suggesting that the genetic accidents they undergo, or the stresses of growth in an inappropriate environment, have created the signals that normally call the p53 protein into play. The loss of *p*53 activity can thus be trebly dangerous in relation to cancer. First, it allows faulty mutant cells to continue through the cell cycle. Secondly, it allows them to escape apoptosis. Third, it leads to the genetic instability characteristic of cancer cells, allowing further cancer-promoting mutations to accumulate as they divide. Many other mutations can contribute to each of these types of misbehavior, but p53 mutations contribute to them all.

DNA tumor viruses activate the cell's replication machinery by blocking the action of key tumor suppressor genes

DNA tumor viruses cause cancer mainly by interfering with cell-cycle controls, including those that depend on p53. To understand this type of viral carcinogenesis, it is important to understand the life history of the virus. Viruses use the DNA replication machinery of the host cell to replicate their own genomes. To make many infectious virus particles from a single host cell, a DNA virus has to commandeer this machinery and drive it hard, breaking through the normal constraints on DNA replication and usually killing the host cell in the process. Typically, however, the virus also has another option: it can propagate its genome as a quiet, well-behaved passenger in the host cell, replicating in parallel with the host cell's DNA in the course of ordinary cell division cycles. The virus can switch between these two modes of existence, remaining latent and harmless or proliferating to generate infectious particles according to circumstances. No matter which way of life the virus is following, it is not in its interests to cause cancer. But genetic accidents can occur, such that the virus misuses its equipment for commandeering the DNA replication machinery, and instead of switching on rapid replication of its own genome, switches on persistent proliferation of the host cell.

DNA viruses are a diverse group, but something of this sort seems to happen with most of those that are involved in cancer. The *papillomaviruses*, for example, are the cause of human warts and are especially important as a key causative factor in carcinomas of the uterine cervix (about 6% of all human cancers). Papillomaviruses infect the epithelium, and are retained in the basal layer of cells as extrachromosomal plasmids that replicate in step with the chromosomes. Infectious virus particles are generated in the outer epithelial layers, as cells begin to differentiate before being sloughed from the surface. Here, cell division should normally be arrested, but the virus interferes with this arrest so as to allow rapid replication of its own genome. Usually, the effect is restricted to the outer layers of cells and relatively harmless, as in a wart. Occasionally, through a genetic accident causing misregulation of the viral genes whose products prevent cell-cycle arrest, the control of cell division is subverted in the basal layer also, in the stem cells of the epithelium. This can lead to cancer, with the viral genes acting as oncogenes.

In papillomaviruses, the viral genes that are mainly to blame are called *E6* and *E7*. The products of these viral oncogenes interact with many host cell proteins, but in particular they bind to the protein products of two key tumor suppressor genes of the host cell, putting them out of action and so permitting the cell to replicate its DNA and divide in an uncontrolled way. One of these host proteins is Rb: by binding to Rb, the viral E7 protein prevents it from binding to its normal associates in the cell. The other host protein inactivated by the virus is the tumor suppressor p53, which is bound by the viral E6 protein, triggering p53 destruction. Elimination of p53 allows the abnormal cell to survive, divide, and accumulate yet more abnormalities.

Telomere shortening may pave the way to cancer in humans

The mouse is the most widely used model organism for the study of cancer, yet the spectrum of cancers seen in mice differs dramatically from the seen in humans. The great majority of mouse cancers are sarcomas and leukemias, whereas more than 80 percent of human cancers are carcinomas—cancers of epithelia where rapid cell turnover occur. Many therapies have been found to cure cancers in mice; but when the same treatments are tried in humans, they usually fail. What could be the reason for the difference between mouse and human cancer, and what can it tell us about the molecular mechanisms of the disease? An important part of the answer may lie in the behavior of telomeres and the relationship between telomere shortening, replicative cell senescence, and genetic instability.

As we saw earlier, most human cells seem to have a built-in limit to their proliferation: they show replicative senescence, at least when grown in culture. Replicative cell senescence in humans is thought to be causes by changes in the structure of telomeres—the repetitive DNA sequences and associated proteins that cap the ends of each chromosome. These telomeric DNA sequences are synthesized and maintained by a special mechanism that requires the enzyme telomerase. In most human cells, other than those of the germ line and some stem cells, expression of the gene coding for the catalytic subunit of telomerase is switched off, or at least not fully activated. As a result, the telomeres in these cells tend to become a little shorter with each round of cell division. Eventually, the telomeric cap on the chromosome end can become shortened to the point where a danger signal is generated, arresting the cell cycle. The signal is similar, in function at least, to the one that arrests the cycle when an uncapped DNA end is created by an accidental double-stranded chromosome break. The effect in both cases is to prevent cell division so long as the cell contains broken or inadequately capped DNA. In the cell with the chromosome break, this allows time for DNA repair; in the normal senescent cell, it seems that it simply put a stop to cell proliferation. As we discussed at the beginning of the chapter, it is not clear how often cells in normal human tissues run up against this limit; but if a self-renewing cell population does undergo replicative senescence, any rogue cell that undergoes a mutation that lets it carry on dividing will enjoy a huge competitive advantage—much more than if the same mutation had occurred in a cell in a nonsenescent population. Viewed in this light, replicative senescence might be expected to favor the development of cancer.

Mice have telomeres much longer than those of humans. Moreover, unlike humans, they keep telomerase active in their somatic cells, and mouse telomere therefore do not tend to shorten with increasing age of the organism. It is possible, however, to use gene knockout technology to make mice that lack functional telomerase. In these mice, the telomeres become shorter with every generation, but no untoward consequences are seen until, in the great-great-grandchildren of the initial mutants, the telomeres become so short that they disappear or cease to function. Beyond this point, the mice begin to show various abnormalities, including an increased incidence of cancer. This raises the possibility that natural telomere shortening helps to engender many human tumors.

In a population of telomere-deficient cells, loss of p53 opens an easy gateway to cancer

In contrast with most normal cells in humans, most human cancer cells express telomerase. This is thought to be the reason why, unlike normal cells, they tend to divide without limit in culture, and it is added evidence that telomere maintenance has a significant role in cancer.

Given that cancer cells contain telomerase, and obvious suggestion is that they arise from mutant precursors that have simply avoided shortening their telomeres, and so have never encountered the telomeric limit to cell division. There is, however, another possibility, highlighted by the observations in the telomerase-deficient mice. A cancer may derive from a cell that has experienced telomere shortening, but has suffered a mutation that lets it disregard the signals that normally arrest cell division when telomeres are too short. A mutation causing loss of p53 activity can have just this effect: occurring in one cell within a population arrested by telomere shortening, it can give that cell and its progeny an immediate competitive advantage over their nondividing neighbors. At the same time, the absence of p53 can bring on the gross chromosomal instability that is characteristic of human carcinomas, allowing the mutant cells to accumulate more mutations and evolve rapidly toward cancer. This is what appears to happen in the telomerase-deficient mice. If they lack not only telomerase but also one of their two copies of the *p53* gene, carcinomas become even more frequent. It is striking that these additional cancers are predominantly carcinomas—cancers of self-renewing epithelia—rather than the sarcomata and lymphomas usually seen in mice. Carcinomas constitute by far the commonest class of human tumors. The differences in telomere behavior could thereby account for the major difference observed between normal mice and humans in the predominant types of cancers that arise. Like human carcinoma cells, the mouse tumor cells are usually found to have inactivated their last remaining *p53* gene; they also show gross chromosomal abnormalities, with many breaks, fusions, and broken chromosome ends.

If this scenario applies to cancers in humans, the suggestion would be that telomerase becomes reactivated after, not before, the genetic catastrophe. The progressive genetic disruption following loss of p53 may be so severe that cells can rarely survive it for more than a few generations. A cell that reactivates telomerase expression will be able to halt the catastrophic cycle and regain enough chromosomal stability to survive. Its progeny will inherit a highly abnormal chromosome set, with many mutations and alterations; these damaged cells can then continue to accumulate further mutations at a more moderate rate, driving tumor progression. This model tallies with the findings that, in breast and colorectal cancers, gross chromosomal abnormalities appear to arise early during tumor development, before telomerase is reactivated. With telomerase turned on, these carcinomas still possess enough genetic instability—due to a loss of p53 or other mutations—to continue to evolve, thereby tending to become metastatic.

It is possible that many of the most common types of human carcinomas originate in the way we have just described. But there are certainly many other ways in which cancers can arise, and the true importance of replicative senescence and telomere behavior in human cancer remains to be determined. The uncertainties highlight how little we still know about the natural history of cancer, despite the dramatic advances in understanding of cancer molecular genetics. At the very least, however, the results suggest that telomere-deficient mice can provide a reasonable model for the study of common human carcinomas. Perhaps, this mouse model will also help us to devise cancer treatments that work as well in humans as they do in mice.

The mutations that lead to metastasis are still a mystery

Perhaps the most serious gap in our understanding of cancer concerns malignancy and metastasis. We have yet to clearly identify mutations that specifically permit cells to invade surrounding tissues, spread through the body, and form metastases. Indeed, it is not even clear exactly what properties a cancer cell must acquire to become metastatic. One extreme view would be that the ability of cancer cells in the body to metastasize requires no further genetic changes beyond those needed for loss of cell division control. An opposite view, more commonly espoused, is that metastasis is a difficult and complex task for a cell, requiring a mass of new mutations—so many, and so varied according to circumstance, that it is hard to discover what they are individually.

Which steps in metastasis are the most difficult is still a matter of debate. But where are experimental findings that throw some light on the issue. It is obvious, first of all, that metastasis presents different problems for different types of cells. For a leukemia cell, already roaming the body via the circulation, metastasis should be easier than for a carcinoma cell that has to escape from an epithelium. As we saw at the beginning of the chapter, it is helpful to distinguish two degrees of malignancy for carcinomas, representing two phases of tumor progression. In the first phase, the tumor cells escape the normal confines of their parent epithelium and begin to invade the tissue immediately beneath—becoming locally invasive. In the second phase, they travel to distant sites and settle to form colonies, a process known as metastasis.

Local invasiveness requires a breakdown of the mechanisms that normally hold epithelial cells together. In some carcinomas of the stomach and of the breast, the E-cadherin gene has been identified as a tumor suppressor gene. The primary function of the E-cadherin protein is in cell-cell adhesion, where this protein is embedded in two adjacent plasma membranes to bind epithelial cells together. When tumor cells lacking this adhesion molecule are placed in culture, and a functional E-cadherin gene is put back into them, they lose some of their invasive characteristics and begin to cohere more like normal cells. Loss of E-cadherin, therefore, may favor cancer by specifically contributing to local invasiveness.

The second stage of malignance, involving entry into the bloodstream or lymphatic vessels, travel via the circulation, and colonization of remote sites, is more enigmatic. The cells must, for example, cross barriers such as the basal lamina of the parent epithelium and of blood vessels. Do they need to acquire additional mutations to become able to do this? The answer is not clear. To discover what steps in metastasis present cells with the greatest difficulties—and thus might be aided by the acquisition of additional mutations—one can label cancer cells with a fluorescent dye, inject them into the circulation of a living animal, and monitor their fate. In these experiments, many cells are found to survive in the circulation, to become lodged in small vessels, and to get out of these into the surrounding tissue, regardless of whether they come from a metastatic or a nonmetastatic tumor. Most of the losses occur after this. Some cells die immediately; others survive entry into the foreign tissue but fail to grow; still others divide a few times and then stop. Here the metastasis-competent cells outperform their nonmetastatic relatives, suggesting that the ability to grow in the foreign tissue is a key property that cells must acquire to become metastatic.

To discover the changes that confer metastatic potential, one can use DNA microarrays to look for genes that are selectively switched on in cancer cells that have become highly malignant. These microarrays allow one to monitor the expression of thousands of genes at a time. One such study took human and mouse melanoma cells that had been selected for high metastatic potential and compared them with their poorly metastatic counterparts. Of the dozen or so genes that appeared to be selectively active in the malignant cells, one that showed increased expression repeatedly in the metastatic cells was *RhoC*—a member of a family of genes known to regulate cell motility. A greatly expanded use of such methods, made much more powerful with the availability of the human genome sequence, should soon give us a clearer picture of the molecular changes that allow tumor cells to metastasize.

Colorectal cancers evolve slowly via a succession of visible changes

At the beginning of this chapter, we saw that most cancers develop gradually from a single aberrant cell, progressing from benign to malignant tumors by the accumulation of a number of independent genetic accidents. We have discussed what some of these accidents are in molecular terms and seen how they contribute to cancerous behavior. We now examine one particular class of common human cancers more closely, using it to illustrate and enlarge upon some of the general principles and molecular mechanisms we have introduced, and to see how we can make sense of the natural history of the disease in terms of them. We take colorectal cancer as our example, where the steps of tumor progression have been followed *in vivo* and carefully studied at the molecular level.

Colorectal cancers arise from the epithelium lining the colon and rectum (the lower end of the gut). They are common, currently causing over 60000 deaths a year in the United States, or about 11% of total deaths from cancer. Like most cancers, they are not usually diagnosed until late in life (90% after the age of 55). However, routine examination of normal adults with a colonoscope (a fiber-optic device for viewing the interior of the colon and rectum) often reveals a small benign tumor, or adenoma, of the gut epithelium in the form of a protruding mass of tissue called a polyp. These adenomatous polyps are believed to be the precursors of a large proportion of colorectal cancers. Because progression of the disease is usually very slow, there is typically a period of 10-35 years in which the slowly growing tumor is detectable but has not yet turned malignant. Thus, when people are screened by colonoscopy in their fifties and the polyps are removed—a quick and easy surgical procedure—the subsequent incidence of colorectal cancer is very low—according to some studies, less than a quarter of what it would be otherwise.

Colon cancer provides a clear example of the phenomenon of tumor progression discussed previously. In polyps smaller than 1cm in diameter, the cells and the local details of their arrangement in the epithelium usually appear almost normal. The larger the polyp, the more likely it is to contain cells that look abnormally undifferentiated and form abnormally organized structures. Sometimes, two or more sectors can be distinguished within a single polyp, the cells in one sector appearing relatively normal, those in the other appearing frankly cancerous, as though they have arisen as a mutant subclone within the original clone of adenomatous cells. At later stages in the disease, the tumor cells become invasive, first breaking through the epithelial basal lamina, then spreading through the layer of muscle that surrounds the gut, and finally metastasizing to lymph nodes, liver, lung, and other tissues.

A few key genetic lesions are common to a majority of cases of colorectal cancer

What are the mutations that accumulate with time to produce this chain of events? Of those genes so far discovered to be involved in colorectal cancer, three—*K-Ras* (a member of the *Ras* gene family), *p53*, and a third gene, *APC*, to be discussed below—stand out as very frequently mutated. Others are involved in smaller number of colon cancers. Still other critical genes remain to be identified.

One approach to discovery of the mutations responsible for colorectal cancer is to screen the cells for abnormalities in genes already known or suspected to be involved in cancers elsewhere. This type of genetic screening has revealed that about 40% of colorectal cancers have a specific point mutation in *K-Ras*, activating it as an oncogene, and about 60% have inactivating mutations or deletions of *p53*.

Another approach to finding cancer-critical genes is to track down the genetic defects in those rare families that show a hereditary predisposition to colorectal cancer. The first of these hereditary colorectal cancer syndromes to be elucidated was a condition known as *familial adenomatous polyposis coli* (FAP), in which hundreds or thousands of polyps develop along the length of the colon. These make their appearance in early adult life, and if they are not removed, it is almost inevitable that one or more of them will progress to become malignant; on average, 12 years elapse from the first detection of polyps to the diagnosis of cancer. The disease can be traced to deletion or inactivation of *APC*, a gene on the long arm of chromosome 5. Individuals with FAP have inactivating mutations or deletions of the *APC* gene in all the cells of the body. Of the patients with colorectal cancer who do not have the hereditary condition—the vast majority of cases of the disease—more than 60% have similar mutations in the cells of the cancer but not in their other tissues; in other words, both copies of the *APC* gene have been lost during their lifetime. Thus, by a route similar to that which we have discussed for retinoblastoma, mutation of *APC* is identified as one of the central ingredients of colorectal cancer.

As explained earlier, tumor suppressor genes can also be tracked down, even where there is not known hereditary syndrome, by searching for genetic deletions in the tumor cells. A systematic scan of a large number of colorectal cancers reveals frequent losses of large sections of certain chromosomes, suggesting that those regions may harbor tumor suppressor genes. One of them is the region including *APC*. Another includes the *SMAD4* gene, which is mutated in perhaps 30% of colon cancers. At least one other important tumor suppressor gene is thought to lie in the same neighborhood as *SMAD4* on chromosome 18 but remains to be identified. Specific parts of other chromosomes also show frequent losses—or gains—in colorectal cancers and are now being searched for additional cancer-critical genes.

As our knowledge of genes and their functions has expanded, another fruitful approach has been to look for genes that interact with a known cancer-critical gene in the hope that these, too, may be targets for mutation. The APC protein is now known to be an inhibitory component of the Wnt signalling pathway. It acts by binding to β-catenin, another component of the pathway, and thereby preventing activation of TCF4, a gene regulatory protein that stimulates growth of the colonic epithelium when it has β-catenin bound to it. Loss of TCF4 causes a depletion of the gut stem-cell population, so that loss of the antagonist APC may cause overgrowth by the opposite effect. When the β-catenin gene was sequenced in a collection of colorectal tumors, it turned out that among the few that did not have *APC* mutations, a high proportion had activating mutations in β-catenin instead. Thus, it is the Wnt signalling pathway, rather than any single oncogene or tumor suppressor gene that it contains, that is critical for the cancer.

Defects in DNA mismatch repair provide an alternative route to colorectal cancer

In addition to the hereditary disease associated with *APC* mutations, there is a second, and actually commoner, kind of hereditary predisposition to colon carcinoma in which the course of events is quite different from the one we have described for FAP. In patients with this condition, called hereditary nonpolyposis colorectal cancer, or HNPCC, the probability of colon cancer increases without any increase in the number of colorectal polyps (adenomas). Moreover, the cancer cells in the tumors that develop are unusual, in as much as examination of their chromosomes in a microscope reveals a normal (or almost normal) karyotype and a normal (or almost normal) number of chromosomes. In contrast, the vast majority of colorectal tumors in non-HNPCC patients have gross chromosomal abnormalities, with multiple translocations, deletions, and other aberrations, and a total of 55 to 70 or more chromosomes instead of the normal 46.

The mutations that predispose an individual with HNPCC to colorectal cancer turn out to be in one of several genes that code for central components of the DNA mismatch repair system in humans, homologous in structure and function to the *mutL* and *mutS* genes in bacteria and yeast. Only one of the two copies of the involved gene is defective, so the inevitable DNA replication errors are efficiently removed in most of the patient's cells. However, as discussed previously for other tumor suppressor genes, these individuals are at risk, because the accidental loss or alteration of the remaining good gene copy will immediately elevate the spontaneous mutation rate by a hundred-fold or more. These genetically unstable cells presumably go speeding through the standard processes of mutation and natural selection that allow clones of cells to progress to a malignancy.

This type of genetic instability produces invisible changes in the chromosomes—most notably changes in individual nucleotides and short expansions and contractions of mono-and dinucleotide repeats such as AAAA…or CACACA…Once the phenomenon was recognized, mutations in mismatch repair genes were found in about 15% of the colorectal cancers occurring in normal people, with no inherited mutation. Again, the chromosomes in these cancers were unusual in having nearly normal karyotypes.

Thus genetic instability is found in practically all colorectal cancers, but it can be acquired in at least two very different ways. The majority of colorectal cancers have become unstable by rearranging their chromosomes—some perhaps as the combined result of a *p53* mutation and telomere shortening, as previously discussed. Others have been able to avoid this type of trauma; their genetic instability occurs on a much smaller scale, being caused by defect in DNA mismatch repair. The fact that many carcinomas show either chromosomal instability or defective mismatch repair—but rarely both—clearly demonstrates that genetic instability is not an accidental byproduct of malignant behavior, but a contributory cause; it is a property that most cancer cells need in order to become malignant, but one that they can acquire in several alternative ways.

The steps of tumor progression can be correlated with specific mutations

In what sequence do *K-Ras*, *p53*, *APC*, and the other identified colorectal cancer genes undergo their mutations, and what contribution does each of them make to the eventual unruly behavior of the cancer cell? There cannot be a simple answer to this question, because colorectal cancer can arise by more than one route. Thus, as we have just seen, in some cases the first mutation leading toward the cancer may be in a DNA mismatch repair gene; in other cases, it may be in a gene regulating growth. A general feature, such as genetic instability, can arise in a variety of ways, through mutations in different genes.

Nevertheless, certain patterns of events are particularly common. Thus, mutations inactivating the *APC* gene appear to be the first, or at least a very early, step in most cases. They can be detected already in small benign polyps at the same high frequency as in large malignant tumors. Loss of *APC* seems to increase the rate of cell proliferation in the colonic epithelium relative to the rate of cell loss, without affecting the way the cells differentiate or the details of the histological pattern they form.

Mutations activating the *K-Ras* oncogene—a member of the *Ras* family—appear to take place a little later than those in *APC*; they are rare in small polyps but common in larger ones that show disturbances of cell differentiation and histological pattern. When malignant colorectal carcinoma cells containing such *Ras* mutations are grown in culture, they show typical features of transformed cells, such as the ability to proliferate without anchorage to a substratum. Loss of cancer-critical genes on chromosome 18 and mutations in *p*53 may come later still. They are rare in polyps but common in carcinomas, suggesting that they may often occur late in the sequence. As discussed earlier, loss of *p*53 function is thought to allow the abnormal cells not only to avoid apoptosis and to divide, but also to accumulate additional mutations at a rapid rate by progressing through the cell cycle when they are not fit to do so, creating many abnormal chromosomes.

Each case of cancer Is characterized by its own array of genetic lesions

As we have just seen in colorectal cancer, the traditional classification of cancers is simplistic: a single one of the conventional categories of tumor will turn out—on close scrutiny—to be a heterogeneous collection of disorders with some features in common, but each characterized by its own array of genetic lesions. Many types of cancers that have been analyzed genetically show a large variety of genetic lesions and a great deal of variation from one case of the disease to another. In the form of lung cancer known as small-cell lung cancer, for example, one finds mutations not only in *Ras*, *p*53, and *APC*, but also in *Rb*, in members of the *Myc* gene family (in the form of amplification of the mumber of *Myc* gene copies), and in at least five other known proto-oncogenes and tumor suppressor genes. Different combinations of mutations are encountered in different patients and correspond to cancers that react differently to treatment.

In principle, molecular biology provides the tools to find out precisely which genes are amplified, which are deleted, and which are mutated in the tumor cells of any given patient. Such information may soon prove to be as important for the diagnosis and treatment of cancer as is the identification of microorganisms in patients with infectious diseases.

Summary

Studies of developing embryos and transgenic mice have helped to reveal the functions of many cancer-critical genes. Most of the genes found to be mutated in cancer, both oncogenes and tumor suppressor genes, code for components of the pathways that regulate the social and proliferative behavior of cells in the body—in particular, the mechanisms by which signals from a cell's neighbors can impel it to divide, differentiate, or die. Other cancer-critical genes are involved in maintaining the integrity of the genome and guarding against damage. The molecular changes that follow cancer to metastasize, however, escaping the parent tumor and growing in foreign tissues, are still largely unknown.

DNA viruses such as *papillomaviruses* can promote the development of cancer by sequestering the products of tumor suppressor genes—in particular, the Rb protein, which regulates cell division, and the p53 protein, which is thought to act as an emergency brake on cell division in cells that have suffered genetic damage and to call a halt to cell division in senescent cells with shortened telomeres.

The p53 protein has a dual role, regulating both progression through the cell cycle and the initiation of apoptosis. So loss or inactivation of *p*53, which occurs in about half of all human cancers, is doubly dangerous: it allows genetically damaged and senescent cells to continue to replicate their DNA, increasing the damage, and it allows them to escape apoptosis. The loss of p53 function may contribute to the genetic instability of many full-blown metastasizing cancers.

Generally speaking, the steps of tumor progression can be correlated with mutations that activate specific oncogenes and inactivate specific tumor suppressor genes. But different combinations of mutations are found in different forms of cancer and even in patients that nominally have the same form of the disease, reflecting the random way in which mutations occur. Nevertheless, many of the same types of genetic lesions are encountered repeatedly, suggesting that there is only a limited number of ways in which our defenses against cancer can be breached.

Glossary

adenoma [ˌædi'nəumə] n.（医）腺瘤
adenomatous [ˌædi'nɔmətəs] adj.（医）腺瘤（状）的
adhesion [əd'hiːʒən] n. 支持，附着力
anchorage ['æŋkəridʒ] n. 固定支座，固定，锚式固定
antagonist [æn'tægənist] n. 拮抗药（剂）
apoptosis [ˌæpəuptəsis] n.（细胞）凋亡
breast cancer n. 乳腺癌
cancerous ['kænsərəs] adj. 癌的，患癌症的
cancer-critical ['kænsə 'kritikəl] adj. 与肿瘤相关的
carcinogenesis [ˌkɑːsinəu'dʒenisis] n.（医）癌发生，致癌作用
carcinomas [ˌkɑːsi'nəuməs] n.（医）癌科
catastrophe [kə'tæstrəfi] n. 大灾难，大祸
β-catenin n. β-连环蛋白（一类细胞骨架蛋白）

cervix [ˈsəːviks] n.（解）颈部，子宫颈
colon [kɔˈlɔn] n. 结肠
colonization [ˌkɔlənaiˈzeiʃən] n. 移植，群集现象
colonoscope [kɔˈlɔnskəup] n. 结肠显微镜
colony [ˈkɔləni] n. 菌落，集落
colorectal [ˌkəuləˈrektəl] adj.（解）结肠直肠的
cyclin [ˈsaiklin] n. 细胞周期蛋白
degradation [ˌdegrəˈdeiʃən] n. 降解，退化，恶化
dinucleotide [daiˈnjuːkliəˌtaid] n.（生化）二核苷酸
embryonic [ˌembriˈɔnik] adj.（生）胚胎的，开始的
embryos [ˈembriəuz] n. 晶胚
epithelia [ˌepiˈθiːliə] n.（生）上皮细胞（epithclium 的复数形式）
epithelial [ˌepiˈθiːljəl] adj.（生）上皮的，（植）皮膜的
epithelium [ˌepiˈθiːljəm] n. 上皮，上皮细胞
extrachromosomal [ˈekstrəkrəuməˈsəuməl] adj.（生）（位于或发生在）染色体外的
fluorescent [fluəˈresənt] adj. 荧光的
genesis [ˈdʒenisis] n. 起源，生殖，发生
germ [dʒəːm] n. 微生物，细菌，胚原基
germinal [ˈdʒəːminl] adj. 幼芽的
glioblastoma [ˌglaiəublæsˈtəumə] n.（医）神经胶质胚细胞瘤
gut [gʌt] n.（复）内脏，（幽门到直肠间的）肠子
homeostasis [ˌhəumiəuˈsteisis] n. 动态平衡，稳态，内环境稳定
homolog [ˈhɔməlɔg] n. 相同器官，同系物
homozygous [ˌhɔməˈzaigəus] adj.（生）同型结合的，纯合子的
inhibitory [inˈhibitəri] adj. 禁止的，抑制的
karyotype [ˈkæriətaip] n.（生物）染色体组型
kinase [ˈkineis] n.（生化）激酶
knock out （基因）敲除
lamina [ˈlæminə] n. 薄层，叠层，层状体
leukemia [ljuːˈkiːmiə] n. 白血病
lymph node [limf nəud] n. 淋巴腺，淋巴结
lymphatic [limˈfætik] adj. 含淋巴的，淋巴腺的
lymphoma [limˈfəumə] n.（医）淋巴瘤
malignancy [məˈlignənsi] n. 恶意，恶性
melanoma [ˌmeləˈnəumə] n.（医）（恶性）黑素瘤，（良性）胎记瘤
metastases [məˈtæstəsiːz] n. metastasis 的复数
metastasis [məˈtæstəsis] n. 转移
metastasize [məˈtæstəˌsaiz] vi. 转移
metastatic [meteˈstætik] adj. 转移的，迁徙的
methylation [ˈmeθileiʃən] n. 甲基化，甲基化作用
microarray [ˌmaikrəuəˈrei] n. 微阵列
microorganism [maikrəuˈɔːgəniz(ə)m] n.（生）微生物
mutation [mju(ː)ˈteiʃən] n.（生物物种的）突变
noncancerous [ˌnɔnˈkænsərəs] adj. 非癌的，不是癌症的
oncogene [ˈɔnkəˌdʒiːn] n. 致癌基因，肿瘤基因
overexpression [ˈəuvərˈiksˈpreʃən] n. 过度表达
pancreas [ˈpænkriəs] n.（解）胰腺
papillomavirus [ˌpæpiˈləuməˈvaiərəs] n. 乳头状瘤病毒

phosphorylate [ˈfɔsfərileit] vt. （化）使磷酸化；n. 磷酸化
plasmid [ˈplæzmid] n. （生）质粒，质体
polyp [ˈpɔlip] n. 息肉
polyposis [ˌpɔliˈpəusis] n. （医）息肉病
proto-oncogene 原癌基因
receptor [riˈseptə] n. 受体
rectum [ˈrektəm] n. 直肠
replication [ˌrepliˈkeiʃən] n. 复制
retinoblastoma [ˌretinəublæsˈtəumə] n. （医）眼癌，视网膜母细胞瘤
sarcomas [sɑːˈkəuməs] n. （医）肉瘤，恶性毒瘤
senescence [siˈnesns] n. 衰老，老化
stem cell 干细胞
subclone [ˈsʌbkləun] n. 亚克隆
substratum [ˈsʌbˈstrɑːtəm,ˈsʌbˈstreitəm] n. 下层，底层，底基
suppressor [səˈpresə] n. 抑制物，抑制剂，抑制子，抑制基因
symptom [ˈsimptəm] n. （医）（植）症状，征兆
syndrome [ˈsindrəum] n. 症候群，综合征
telomerase [ˈteləməreiz] n. 端粒末端转移酶
telomere [ˈteləmiə] n. （生）端粒（在染色体端位上的着丝点）
telomeric [ˌteləˈmerik] adj. 调聚的
therapeutic [θerəˈpjuːtik] adj. 治疗的，治疗学的
transcription [trænsˈkripʃən] n. 转录
transducer [trænzˈdjuːsə] n. 转换器，传送器，传感器
transgenic [trænsˈdʒenik] adj. 转基因的
translocation [ˌtrænsləuˈkeiʃən] n. 迁移，移动
trauma [ˈtrɔːmə] n. （医）外伤，损伤
tumor [ˈtjuːmə] n. 瘤，肿瘤
tyrosine [ˈtirəsiːn] n. （生化）酪氨酸
ultraviolet [ˈʌltrəˈvaiəlit] adj. 紫外线的，紫外的；n. 紫外线
uterine [ˈjuːtərain] adj. （解）子宫的
wart [wɔːt] n. （医）疣

难句分析

1. Typically，in mice that are endowed with a *Myc* or *Ras* oncogene，some of the tissues that express the oncogene grow to an exaggerated size，and over time occasional cells undergo further changes to give rise to cancers.

   译文：普遍的情况是，在被转入癌基因 *My* 或 *Ras* 的鼠体内，一些表达癌基因的组织生长到一定大小，经过一段时间后有些细胞会发生转变而发展成为癌细胞。

2. These offspring develop cancers at a much higher rate than either parental strain，but again the cancers originate as scattered isolated tumors among noncancerous cells.

   译文：这些子代癌变的比率要比它们的双亲高得多，但是，这些癌细胞在非癌细胞中形成分散的肿瘤。

3. But genetic accidents can occur，such that the virus misuses its equipment for commandeering the DNA replication machinery，and instead of switching on rapid replication of its own genome，switches on persistent proliferation of the host cell.

   译文：但是，会发生一些遗传上的突发事件，从而使病毒为了利用宿主细胞的复制系统，

不是开启它自身基因组的快速复制，而是开启宿主细胞的永久性增殖。

4. As we discussed at the beginning of the chapter, it is not clear how often cells in normal human tissues run up against this limit; but if a self-renewing cell population does undergo replicative senescence, any rogue cell that undergoes a mutation that lets it carry on dividing will enjoy a huge competitive advantage—much more than if the same mutation had occurred in a cell in a nonsenescent population. Viewed in this light, replicative senescence might be expected to favor the development of cancer.

译文：正像我们在本章开始时所讨论的，还不清楚在人体正常细胞中的细胞多长时间能经历这种（因端粒缩短所受到的）限制；但是，如果一个能自我更新的细胞群体会产生复制的衰老细胞，那么，那些遭受使其继续分裂的突变的细胞将会拥有很大的竞争优势——要比那些发生同样突变的非（复制）衰老的细胞群有更多的优势。从这个观点来看，复制的衰老（的细胞）更可能发展成癌细胞。

（选自：Bruce Alberts, et al. Molecular Biology of the Cell. 4th eds. New York: Garland Science, 2002.）

The Basic Principles of Recombinant DNA Technology

Introduction to cloning

Let us suppose that we wish to construct a bacterium that produces human insulin. Naively it might be thought that all that is required is to introduce the human insulin gene into its new host. In fact, a foreign gene is not maintained in its new environment if it is simply inserted into the bacterium on a DNA fragment for such fragments are not replicated. The reason for this is that DNA polymerase which makes copies of the DNA does not initiate the process at random but at selected sites known as origins of replication. Invariably, fragments of DNA do not possess an origin of replication. Using recombinant DNA technology it is possible to ensure the replication of the insulin gene in its new host by inserting the gene into a cloning vector. A cloning vector is simply a DNA molecule possessing an origin of replication and which can replicate in the host cell of choice. Most cloning vectors used with microorganisms are extrachromosomal, autonomously replicating circles of DNA called plasmids. Viruses are used occasionally as vehicles for gene insertion into microorganisms but they are far more important for work with animal cells.

In order to insert foreign DNA into a plasmid, special enzymes known as restriction endonucleases are used. These enzymes cut large DNA molecules into shorter fragments by cleavage at specific nucleotide sequences called recognition sites, i. e. restriction endonucleases are highly specific deoxyribonuclease (DNase). Some of these enzymes cut the two helices a few base pairs apart generating two fragments with single-stranded protrusions called sticky ends because their bases are complementary. Fragments of the foreign DNA are inserted into plasmid vectors cut open with the same enzyme or one which produces a matching end. The resulting recombinants or chimaeras are introduced into the host cell by the process of transformation. In this process the DNA is mixed with a suspension of bacteria, which have been prepared under specialized conditions, and the DNA enters the cell by a mechanism (transformation) which still is poorly understood.

Plasmids as cloning vectors

Hundreds of different cloning vectors have been described and many of them have been constructed for special purposes. However all of them have three features in common.

① The DNA of the cloning vector ideally has only a single target site for any particular restriction endonuclease. If it has more than one target site, bits of the plasmid can be lost or rearranged during the cloning process. The more restriction endonucleases for which the plasmid has unique sites then the better it will be as a vector.

② All useful cloning vectors have one or more readily selectable genetic marker such as antibiotic resistance. The efficiency of the transformation process is very low and such markers are essential so that cells which have been transformed can be selected, e. g. on the basis of acquisition of ampicillin resistance.

③ As indicated above, one property of vectors is their possession of an origin of replication which ensures that they are propagated in the desired host cell. In some instances a plasmid is used as a cloning vector in two unrelated host cells, e. g. *Escherichia coli* and *Bacillus subtilis*. Such bifunctional vectors have more than one origin of replication. Some cloning vectors used in yeast carry a centromere to facilitate segregation at cell division.

Expression of cloned genes

Theoretically it is possible to clone in one organism any desired gene from another organism by the technique known as shotgunning. To do this the entire genome of the first organism is digested with a restriction endonuclease to produce a random mixture of fragments. These fragments are inserted into a plasmid vector and the recombinant plasmids transformed into the desired host cell. Since each recombinant plasmid will contain a different fragment of foreign DNA, it is a major task to select those transformed cells that carry the cloned gene of interest, in this case of the human insulin gene. Even when the right cells can be identified, it is highly probable that they would not synthesize human insulin. That is, the information contained in the cloned gene is not used to create a functional protein or, in common parlance, the gene is not expressed.

There are a number of steps in gene expression:

① transcription of DNA to mRNA;

② translation of the mRNA into a polypeptide sequence;

③ in some instances, post-translational modification of the protein.

In many eukaryotes there is an additional step before translation, the removal of noncoding sequences from mRNA by a process known as splicing.

Transcription: Transcription of DNA into mRNA is mediated by RNA polymerase. The process starts with RNA polymerase binding to recognition sites on the DNA which are called promoters. After binding, the RNA polymerase molecule travels along the DNA molecule until a termination signal is encountered. It follows that a gene which does not lie between a promoter and a termination signal will not be transcribed. Genes isolated in certain ways, such as by cDNA cloning or artificial synthesis (see below) do not have their own promoter and they must be inserted into a vector close to a promoter site. Even if a cloned gene carries its own promoter, this promoter may not function in the new host cell. In such circumstances the original promoter has to be replaced.

Translation: Translation of mRNA into protein is a complex process which involves interaction of the messenger with ribosomes. For translation to take place the mRNA must carry a ribosome binding site (RBS) in front of the gene to be translated. After binding, the ribosome moves along the mRNA and initiates protein synthesis at the first AUG codon it encounters and continues until it encounters a stop codon (UAA, UAG or UGA). If the cloned gene lacks a ribosome binding site, it is necessary to use a vector in which the gene can be inserted downstream from both a promoter and an RBS.

Splicing mRNA Genes from bacteria and viruses have a very simple structure in that all the genetic information in the mRNA between the initiation and stop codons is translated into protein. Many genes of eukaryotic organisms, including the human insulin gene, have a more complex structure. They are a mixture of coding regions or exons, which contribute to the final protein sequence, and noncoding regions or introns, which are not translated into protein. In eukaryotes, genes containing introns are transcribed into mRNA in the usual manner but then the corresponding intron sequences are spliced out. As bacteria cannot splice out introns, they cannot be used directly to express many genes from mammals or other eukaryotes.

One solution to the problem of introns is to clone the gene of interest in yeast (*Saccharomyces cerevisiae*), which can mediate splicing. Unfortunately cloning directly into yeast is much less efficient than cloning into a bacterium such as *E. coli* and this technique is not favored. A better approach is to start by isolating mRNA from the original organism. In the case of insulin the mRNA would be obtained from human pancreatic cells, as these cells are rich in insulin-specific mRNA from which introns have already been spliced out. Using the enzyme reverse transcriptase it is possible to convert the mRNA into a DNA copy. This copy DNA (cDNA), which carries the uninterrupted genetic information for insulin production, can then be cloned. It should be noted that such cDNA has blunt ends instead of sticky ends but it still can be inserted into plasmid vectors using blunt-end ligation.

An alternative way of avoiding the problem of introns is to synthesize an artificial gene in the test tube starting with deoxyribonucleotides. In practice various oligonucleotides are synthesized and ligated together before insertion into an appropriate vector. This approach has been used to clone genes encoding proteins up to 500 amino acids long but it demands that the entire amino acid sequence be known in advance.

Post-translational modifications: A number of proteins undergo post-translational modifications and insulin is one of these. Proteins that are destined to be transported out of the cell are synthesized with an extra 15-30 amino acids at the amino-terminus (N-terminus). These extra amino acids are referred to as a signal sequence and a common feature of these sequences is that they have a central core of hydrophobic amino acids flanked by polar or hydrophilic residues. During passage through the membrane the signal sequence is cleaved off. If our insulin gene was cloned by the cDNA method, the signal sequence would be present and, in *E. coli* at least, the insulin would be transported through the cytoplasmic membrane (exported). Using the synthetic gene approach a signal sequence would be present on the protein only if the nucleotide sequence corresponding to the protein's signal sequence has been incorporated at the time of gene construction. Sometimes it is desirable for the bacterium to export the protein, in which case signal sequence is incorporated; in other cases it may be desirable that proteins are retained within the cell.

In *E. coli* cells the presence of a signal sequence usually does not result in export of a protein into the growth medium. Instead it is directed to the periplasmic space, which is the space between the cytoplasmic membrane and the complex membrane-like cell wall. Unfortunately, many recombinant proteins are rapidly and extensively degraded in the periplasmic space due to the presence there of numerous proteolytic enzymes (proteases). In Gram positive species such as *Bacillus* and in eukaryotic microorganisms signal sequences direct the export of proteins into the growth medium.

The export process is even more complex in the protein synthesis and export occurs simultaneously on the inner surface of the cell membrane. If the rate of synthesis of an exported protein is very high, it is possible to envisage a situation in which the membrane becomes "jammed" with protein in the various stages of synthesis and export. This problem can be minimized by using as host cell for the recombinant plasmid one which has a very high surface area to volume ratio. Suitable organism are actinomycetes and filamentous fungi.

A small number of proteins, and again insulin is a good example, are synthesized as proproteins. This means that there is an additional amino acid sequence which dictates the final three-dimensional structure but which is deleted before the protein becomes fully functional. In the case of proinsulin, proteolytic attack cleaves out a stretch of 35 amino acids in the middle of the molecule to generate insulin. The peptide that is removed is known as the C fragment. The other, chains, A and B, remain cross-linked and thus locked in a stable tertiary structure by the disulphide bridges formed when the molecule originally folded as proinsulin. Bacteria have no mechanism for specifically cutting out the folding sequences from proproteins. If native insulin is to be made using recombinant bacteria, it is necessary to synthesize the A and B chains separately, purify them and join them together in vitro.

Another post-translational modification occurring naturally in the construction of many proteins is the addition of oligosaccharides to certain amino acid residues. This process of glycosylation occurs, for example, in the synthesis of the antiviral interferon-β and interferon-γ. Bacteria cannot glycosylate the products of cloned mammalian genes. Nonglycosylated proteins usually retain their pharmacological or biological activities but their stability in vitro and in vivo and their distribution in the animal body may be different. Yeast cells can glycosylate proteins but the pattern of glycosylation differs from that mediated by animal cells. Where the correct glycosylation is necessary the only option is to clone the corresponding gene in a mammalian cell.

Selection of recombinants

The task of isolating a desired recombinant from a population of transformed bacteria depends very much on the cloning strategy that has been adopted; for instance, if a synthetic gene has been cloned, not selection is necessary because every transformed cell will contain the correct sequence. When a cDNA derived from a purified or abundant mRNA is to be cloned, the task is relatively simple: only a small number of clones need to be screened. Isolating a particular single-copy gene sequence from a complete mammalian genetic library requires techniques in which hundreds of thousands of recombinants can be screened. A number of different methods have been devised to facilitate screening of recombinants. These include genetic methods, immunochemical methods and methods based on nucleic acid hybridization. The simplest example of a genetic method is the complementation of nutritional defects; for example, suppose a bacterial strain is available which has a mutation in gene encoding an enzyme involved in the biosynthesis of the amino acid histidine. Such a mutant strain will only grow in medium supplemented with histidine. By cloning DNA from a normal strain, i. e. one that can synthesize its own histidine, in the mutant strain and selecting those transformants which grow in the absence of histidine it is possible to isolate the gene of interest. Even yeast and mammalian genes can be selected in this way. In a procedure analogous to the described above, a clone carrying the mouse dihydrofolate reductase (DHFR) gene was selected from a population of recombinant plasmids containing cDNA derived from an unfractionated mouse cell mRNA preparation. In this instance the basis of selection was that mouse DHFR is more resistant to the inhibitor trimethoprim than *E. coli* DHFR. When trimethoprim was added to the medium on which transformed *E. coli* were grown, only those cells carrying the mouse DHFR gene survived.

Immunochemical methods are widely used to select particular recombinant clones but demand the availability of a specific antibody against the desired protein product. When such an antibody is available the method works well. Transformed cells are grown on agar in a conventional Petri dish and a duplicate set of colonies prepared in as second Petri dish. The cells in the duplicate set of colonies are lysed by exposure to chloroform vapor and the released proteins blotted onto an adsorbent matrix. In effects this produces a map of the original colonies in the form of their protein products. The matrix is then exposed to antibody which has been radioactively labelled in vitro. Positively reacting lysates are detected by washing surplus radiolabelled material off the matrix and making an autoradio-graphic image. Many different variations of the technique have been adopted. Detection methods based on nucleic acid hybridization are also widely used by these techniques have other applications as well.

Maximizing gene expression

If the objective of cloning a mammalian gene in a microorganism is to facilitate commercial production of the corresponding gene product, it follows that it is essential to maximize gene expression. Important factors are: ①the number of copies of the plasmid vector per unit cell (copy number); ②the strength of the promoter; ③the sequence of the RBS and flanking DNA; ④ codon choice in the cloned gene; ⑤ genetic stability of the recombinant; ⑥proteolysis.

The limiting factor in expression is the initiation of protein synthesis. Increasing the number of plasmids per cell increases the number of mRNA molecules transcribed from the cloned gene and this results in increased protein synthesis. Similarly, the stronger the promoter, the more mRNA molecules are synthesized. The nucleotide base sequence of the RBS and the length and sequence of the DNA between the RBS and the initiating AUG codon are so important that a single base change, addition or deletion can affect the level of translation up to 1000-fold.

Another important factor is related to the redundancy of the genetic code. There are several trinucleotide codons for most amino acids and different organisms favor different codons in their genes. If genes inserted into cells of another species utilize codons rare in the host cell, the host's biosynthetic machinery may be starved of charged tRNA. This could result in premature protein chain termination or a high error frequency in the amino acid sequence of the protein.

In any culture containing cells engaged in excess synthesis, be it of a protein or a low molecular weight molecule (e. g. an antibiotic), there is a strong selective pressure in favor of nonproducing cells. Cells which are not encumbered by extra production can invest their energy in cell division and hence rapidly outnumber producers. Non-producing cells frequently arise in transformed cultures as a result of spontaneous changes in gene sequence, e. g. deletions and gene rearrangements. To minimize selection against cells which can achieve high product levels it is wise to minimize recombinant gene expression until the final production vessel is reached. This can be achieved by using controllable promoters and plasmid vectors with controllable copy number. An example of a controllable promoter is that derived from the *E. coli* lactose operon: no transcription occurs unless lactose is added. Runaway plasmids are examples of vectors with controllable copy number. At low temperatures, e. g. below 30℃, the copy number may be as low as 10 plasmids per cell but when the temperature is raised to 37℃, the copy number increases to several hundred, or even several thousand.

The enzymatic breakdown of protein (proteolysis) does not affect transcription and translation but by degrading the desired product it influences the apparent rate of gene expression. Although proteolysis can be reduced, it is difficult to eliminate completely. One approach which is used widely is to "protect" the desired protein by fusing it to a normal cellular protein from which it must subsequently be released.

Glossary

actinomycete [ˌæktinəumaiˈsiːt] n. 放线菌类
ampicillin [ˌæmpiˈsilin] n. 氨苄西林
antibiotic resistance 抗生素耐受性
Bacillus subtilis 枯草芽孢杆菌，枯草杆菌
blunt end 平末端
centromere [ˈsentrəˌmiə] n. 着丝点，着丝粒
chloroform [ˈklɔ(ː)rəfɔːm] n. 氯仿；vt. 用氯仿麻醉
deoxyribonuclease [diːˈɔksiˌraibəuˈnjuːklieis] n. 脱氧核糖核酸酶（DNase）
Escherichia coli 大肠杆菌
extrachromosomal [ˈekstrəˌkrəuməˈsəuməl] adj. （位于或发生在）染色体外的
filamentous [filəˈmentəs] adj. 丝性的，丝状的，菌丝的
histidine [ˈhistidiːn, -din] n. 组氨酸，组织氨基酸
immunochemical [iˌmjuːnəuˈkemikəl] adj. 免疫化学的；n. 免疫化学药品
operon [ˈɔpəˌrɔn] n. 操纵子
pancreatic [ˌpæŋkriˈætik] adj. 胰腺的
periplasmic [ˈperiplæsmik] adj. 周质的，胞质的
pharmacological [ˌfɑːməkəˈlɔdʒikəl] adj. 药理学的
proinsulin [ˌprəuˈinsjulin] n. 胰岛素原
protrusion [prəˈtruːʒən] n. 伸出，突出
recognition site 识别位点
Saccharomyces cerevisiae 酿酒酵母
splicing [ˈsplaisiŋ] 粘接，缝接，拼接，剪接，结合
sticky end 黏性末端
termination signal 终止信号
transformant [trænsˈfɔːmənt,trænz-,trɑːn-] n. 转化株（指已经转化的细菌细胞）
transformation [ˌtrænsfəˈmeiʃən] n. 转化
trimethoprim [ˌtraiˈmeθəprim] n. 甲氧苄氨嘧啶，三甲氧苄二氨嘧啶
trinucleotide codon 三核苷酸密码
vessel [ˈvesl] n. 容器，器皿，脉管，导管

难句分析

1. Naively it might be thought that all that is required is to introduce the human insulin gene into its new host.
   Naively it might be thought that…中，it 为形式主语，that 引导的从句为主语从句；在 all that is required…中，that 为关系代词，它引导的定语从句修饰 all。全句可译为：可以朴实地认为所需的一切就是将人胰岛素的基因引进其新的宿主内。
2. A cloning vector is simply a DNA molecule possessing an origin of replication and which can replicate in the host cell of choice.

句中的 possessing an origin of replication 短语和 which can replicate in the host cell of choice 都是 molecule 的定语，前者为现在分词短语，后者为从句。全句可译为：克隆载体一般都具有复制起点并在选择的宿主细胞内复制 DNA 分子。

3. The more restriction endonucleases for which the plasmid has unique sites then the better it will be as a vector.

   句中 for which the plasmid has unique sites 为定语从句，it 代表 plasmid。全句可译为：如果在质粒上有特异作用位点的限制性内切酶越多，那么这种质粒作为载体就越好。

4. For translation to take place the mRNA must carry a ribosome binding site（RBS）in front of the gene to be translated.

   For translation to take place 为不定式的复合结构，在句中作状语。to be translated 为不定式被动语态，在句中作 gene 的定语。全句可译为：如果要使翻译进行，那么，在 mRNA 分子的被翻译部分的前面必须带有核糖体结合的位点。

5. This problem can be minimized by using as host cell for the recombinant plasmid one which has a very high surface area to volume ratio.

   using as a host cell for the recombinant plasmid one which has a very high surface area to volume ratio 为动名词短语作 by 的宾语。using 的宾语为 one，是代词，代替 cell。后面由 which 引导的从句是 one 的定语。全句可译为：将表面积和体积比很大的细胞用作重组体质粒的宿主细胞可以使此问题简化。

6. In any culture containing cells engaged in excess synthesis，be it of a protein or a low molecular weight molecule（e. g. an antibiotic)，there is a strong selective pressure in favor of nonproducing cells.

   containing cell 为 culture 的定语，engaged in 为 cell 的定语。be it of protein or a low molecular weight molecule 是让步状语从句。全句可译为：在进行含有大量合成的细胞培养中，不管是合成蛋白质或合成低分子量的分子（如抗生素），都存在一个对于产生细胞不利的选择压力。

（选自：李青山，安玉华，刘永新，陶小娟编．生物化学和分子生物学英语．长春：吉林大学出版社，1994.）

Laboratory Techniques Commonly Used in Immunology

Many laboratory techniques that are routine in research and clinical settings are based on the use of antibodies. In addition, many of the techniques of modern molecular biology have provided invaluable information about the immune system. We will describe the principles underlying some of the most commonly used laboratory methods in immunology. Details of how to carry out various assays may be found in laboratory manuals.

Laboratory methods using antibodies

The exquisite specificity of antibodies for particular antigens makes antibodies valuable reagents for detecting, purifying, and quantitating antigens. Because antibodies can be produced against virtually any type of macromolecule and small chemical antibody-based techniques may be used to study virtually any type of molecule in solution or in cells. The method for producing monoclonal antibodies has greatly increased our ability to generate antibodies of almost any desired specificity. Historically, many of the uses of antibody depended on the ability of antibody and specific antigen to form large immune complexes, either in solution or in gels, that could be detected by various optical methods. These methods were of great importance in early studies but have now almost entirely been replaced by simpler methods based on immobilized antibodies or antigens.

Quantitation of antigen by immunoassays

Immunologic methods of quantifying antigen concentration provide exquisite sensitivity and specificity and have become standard techniques for both research and clinical applications. All modern immunochemical methods of quantitation are based on having a pure antigen or antibody whose quantity can be measured by an indicator molecule. When the indicator molecule is labeled with a radioisotope, as first introduced by Rosalyn Yalow and colleagues, it may be quantified by instruments that detect radioactive decay events; the assay is called a radioimmunoassay (RIA). When the indicator molecule is covalently coupled to an enzyme, it may be quantified by determining with a spectrophotometer the rate at which the enzyme converts a clear substrate to a colored product; the assay is called an enzyme-linked immunosorbent assay (ELISA). Several variations of RIA and ELISA exist, but the most commonly used version is the sandwich assay. The sandwich assay used two different antibodies reactive with different epitopes on the antigen whose concentration needs to be determined (Fig. 1. 17). A fixed quantity of one antibody is attached to a series of replicate solid supports, such as plastic microtiter wells. Test solutions containing antigen at an unknown concentration or a series of standard solutions with know concentrations of antigen are added to the wells and allowed to bind. Unbound antigen is removed by washing, and the second antibody, which is enzyme linked or radiolabelled, is allowed to bind. The antigen serves as a bridge, so the more antigen in the test or standard solutions, the more enzyme-linked or radiolabelled second antibody will bind. The results from the standard solutions are used to construct a binding curve for second antibody as a function of antigen concentration, from which the quantities of antigen in the test solutions may be inferred. When this test is performed with two monoclonal antibodies, it is essential that these antibodies see nonoverlapping determinants on the antigen; otherwise, the second antibody cannot bind.

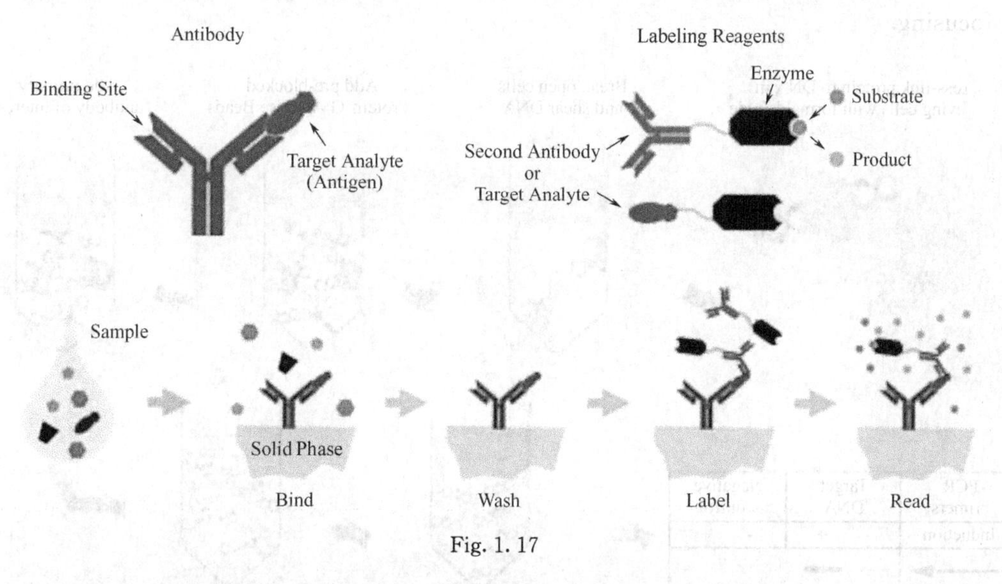

Fig. 1. 17

In an important clinical variant of immunbinding assays, samples from patients may be tested for the presence of antibodies that are specific for a microbial antigen [e. g. , antibodies reactive with proteins from human immunodeficiency virus (HIV) or hepatitis B virus] as indicators of infection. In this case, a saturating quantity of antigen is added to replicate wells containing plate-bound antibody, or the antigen is attached directly to the plate, and serial dilutions of the patient's serum are then allowed to bind. The amount of the patient's antibody bound to the immobilized antigen is determined by use of an enzyme-linked or radiolabelled second antihuman immunoglobulin (Ig) antibody.

Purification and identification of proteins

Antibodies can be used to purify proteins from solutions and to identify and characterize proteins. Two commonly used methods to purify proteins are immunoprecipitation and affinity chromatography. Western blotting is a widely used technique to determine the presence and size of a protein in a biologic sample.

Immunoprecipitation and affinity chromatography

Immunoprecipitation is a technique in which an antibody specific for one protein antigen in a mixture of proteins is used to isolate the specific antigen from the mixture. In most modern procedures, the antibody is attached to a solid-phase particle (e. g. , an agarose bead) either by direct chemical coupling or indirectly. (Fig. 1. 18) Indirect coupling may be achieved by means of an attached anti-antibody, such as rabbit antimouse Ig antibody, or by means of some other protein with specific affinity for the Fc portion of Ig molecules, such as protein A or protein G from staphylococcal bacteria. After the antibody-coated beads are incubated with the solution of antigen, unbound molecules are separated from the bead-antibody-antigen complex by washing. Specific antigen is then released (eluted) from the antibody by changing the pH or by other solvent conditions that reduce that affinity of binding. The purified antigen can then be analyzed by conventional chemical techniques. Alternatively, a small amount of radiolabelled protein can be purified and the characteristics of the macromolecule inferred from the behavior of the radioactive label in analytical separation techniques, such as sodium dodecyl sulfate-polyacrylamide gel electrophoresis (SDS-PAGE) or isoelectric focusing.

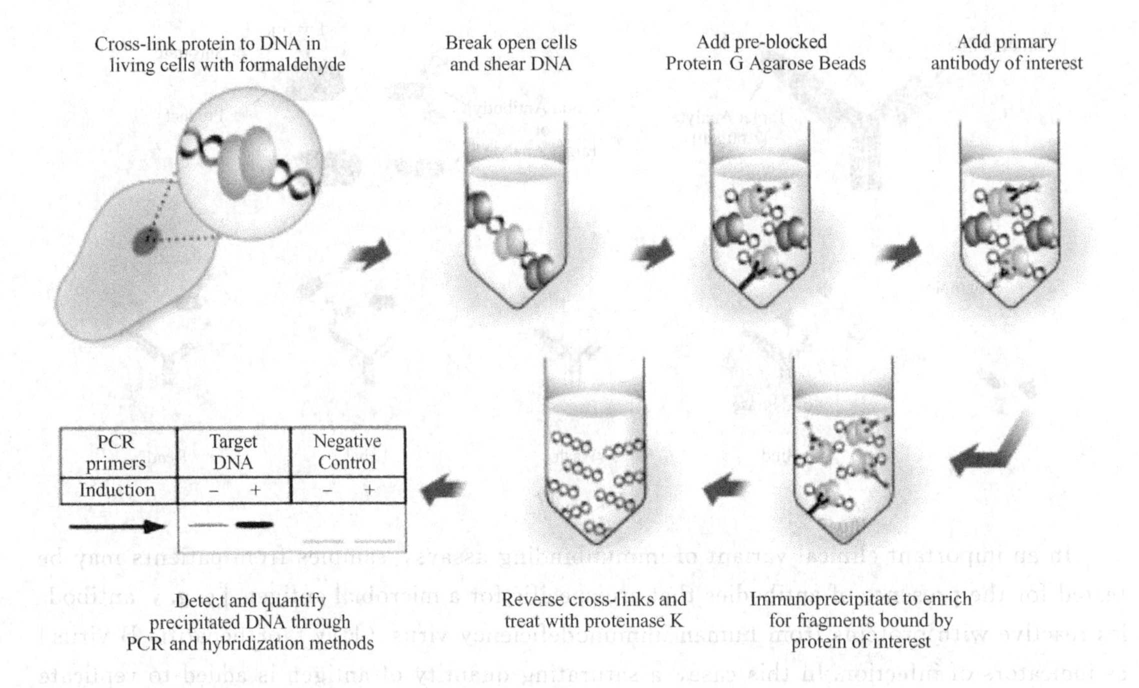

Fig. 1. 18 Sandwich enzyme-linked immunosorbent assay or radioimmunoassay

A fixed amount of one immobilized antibody is used to capture an antigen. The binding of a second, labeled antibody that recognizes a non-overlapping determinant on the antigen will increase as the concentration of antigen increases and thus allow quantification of the antigen

Affinity chromatography, like immunoprecipitation, uses antibodies attached to an insoluble support to remove and thereby purify antigens from a solution. Antibodies specific for the desired antigen are attached to a solid support, such as agarose beads packed into a column, either by direct coupling or indirectly, as described for immunoprecipitation. A complex mixture of antigens is passed through the beads to allow the antigen that is recognized by the antibody to bind. Unbound molecules are washed away, and the bound antigen is eluted by changing the pH or by exposure to a chemical that breaks that antigen-antibody bonds. The same method may be used to purify antibodies from culture supernatants or natural fluids, such as serum, by first attaching the antigen to beads and passing the supernatants or serum through.

Western blot

Western blot is used to determine the relative quantity and the molecular weight of a protein within a mixture of proteins or other molecules. The mixture is first subjected to analytical separation, typically by SDS-PAGE, so that the final positions of different proteins in the gel are a function of their molecular size. The array of separated proteins is then transferred from the separating polyacrylamide gel to a support membrane by capillary action (blot) or by electrophoresis such that the membrane acquires a replica of the array of separated macromolecules present in the gel. SDS is displaced from the protein during the transfer process, and native antigenic determinants are often regained as the protein refolds. The position of the protein antigen on the membrane can then be detected by binding of labelled antibody specific for that protein, thus providing information about antigen size and quantity. If radiolabelled antibody probes are used, the proteins on the blot are visualized by autoradiographic exposure of film. More recently, antibody probes are labelled with enzymes that generate chemiluminescent signals and leave images on photographic film. The sensitivity and specificity of this technique can be increased by starting with immunoprecipitated proteins instead of crude protein mixtures. This sequential technique is especially useful for detecting protein-protein interactions. For example, the physical association of two different proteins in the membrane of a lymphocyte can be established by immunoprecipitating a membrane extract by use of an antibody specific for one of the proteins and probing a Western blot of the immunoprecipitate by use of a labelled antibody specific for the second protein that may have been co-immunoprecipitated along with the first protein. A variation of the Western blot technique is routinely used to detect the presence of anti-HIV antibodies in patients' sera. In this case, a defined mixture of HIV proteins is separated by SDS-PAGE and blotted onto a membrane, and the membrane is incubated with dilutions of the test serum. The blot is then probed with a second labelled antihuman Ig to detect the presence of HIV-specific antibodies that were in the serum and bound to the HIV proteins.

The technique of transferring proteins from a gel to a membrane is called Western blot as a biochemist's joke. Southern is the last name of the scientist who first blotted DNA from a separating gel to a membrane, a technique since called Southern blot. By analogy, Northern blot was applied to the technique of transferring RNA from a gel to a membrane, and Western blot was applied to protein transfer. A more detailed description of Southern blot and Northern blot is presented later.

Labelling and detection of antigens in cells and tissues

Antibodies specific for antigens expressed on or in particular cells types are commonly used to identify these cells in tissues or cells suspensions and to separate these cells from mixed populations. In these methods, the antibody can be radiolabelled, enzyme linked. Or, most commonly, fluorescently labelled, and a detection system is used that can identify the bound antibody.

Flow cytometry and fluorescence-activated cell sorting

The tissue lineage, maturation stage, or activation status of a cell can often be determined by analyzing the cell surface or intracellular expression of different molecules. This technique is commonly done by staining the cell with fluorescently labelled probes that are specific for those molecules and measuring the quantity of fluorescence emitted by the cell. The flow cytometer is a specialized instrument that can detect fluorescence on individual cells in a suspension and thereby determine the number of cells expressing the molecule to which a fluorescent probe binds. Suspensions of cells are incubated with fluorescently labelled probes, and the amount of probe bound by each cell in the population is measured by passing the cells one at a time through a fluorimeter with a laser-generated incident beam. The relative amounts of a particular molecule on different cell populations can be compared by staining each population with the same probe and determining the amount of fluorescence emitted. In preparation for flow cytometric analysis, cell suspension are stained with the fluorescent probes of choice. Most often, these probes are fluorochrome-labelled antibodies specific for a cell surface molecule. Alternatively, cytoplasmic molecules can be stained by temporarily permeabilizing cells and permitting the labelled antibodies to enter through the plasma membrane. In addition to antibodies, various fluorescent indicators of cytoplasmic ion concentrations and reduction-oxidation potential can also be detected by flow cytometry. Cell cycle studies can be performed by flow cytometric analysis of cells stained with fluorescent DNA-binding probes such as propidium iodide. Modern flow cytometers can routinely detect three or more different-colored fluorescent signals, each attached to a different antibody or other probe. This technique permits simultaneous analysis of the expression of many different combinations of molecules by a cell. In addition to detecting fluorescent signals, flow cytometers also measure the forward and side light-scattering properties of cells, which reflect cell size and internal complexity, respectively. This information is often used to distinguish different cell types. For example, compared with lymphocytes, neutrophils cause greater side scatter because of their cytoplasmic granules, and monocytes cause greater forward scatter because of their size.

A fluorescence-activated cells sorter (FACS) is an adaptation of the flow cytometer that allows one to separate cell populations according to which and how much fluorescent probe they bind. This technique is accomplished by differentially deflecting the cells with electromagnetic fields whose strength and direction are varied according to the measured intensity of the fluorescence signal. A more rapid but less rigorous separation can be accomplished without a FACS by allowing cells to attach to antibodies bound to plates (panning) or to magnetic beads that can be pulled out of solution by a strong magnet.

Immunofluorescence and immunohistochemistry

Antibodies can be used to identify the anatomic distribution of an antigen within a tissue or within compartments of a cell. To do so, the tissue or cell is incubated with an antibody that is labelled with a fluorochrome or enzyme, and the position of the label, determined with a suitable microscope, is used to infer the position of the antigen. In the earliest version of this method, called immunofluorescence, the antibody was labelled with a fluorescent dye and allowed to bind to a monolayer of cells or to a frozen section of a tissue. The stained cells or tissues were examined with a fluorescence microscope to locate the antibody. Although sensitive, the fluorescence microscope is not an ideal tool for identifying the detailed structures of the cell or tissue because of a low signal-to-noise ratio. This problem, has been overcome by new technologies including confocal microscopy, which uses optical sectioning technology to filter out unfocused fluorescent light, and two-photon microscopy, which prevents out-of-focus light from forming. Alternatively, antibodies may be coupled to enzymes that convert colorless substrates to colored insoluble substances that precipitate at the position of the enzyme. A conventional light microscope may then be used to localize the antibody in a stained cells or tissue. The most common variant of this method uses the enzyme horseradish peroxidase, and the method is commonly referred to as the immunoperoxidase technique. Another commonly used enzyme is alkaline phosphatase. Different antibodies coupled to different enzymes may be used in conjunction to produce simultaneous two-color localizations of different antigens. In other variations, antibody can be coupled to an electron-dense probe such as colloidal gold, and the location of antibody can be determined subcellularly by means of an electron microscope, a technique called immunoelectron microscopy. Different-sized gold particles have been use for simultaneous localization of different antigens at the ultrastructural level.

In all immunomicroscopic methods, signals may be enhanced by use of sandwich techniques. For example, instead of attaching horseradish peroxidase to a specific mouse antibody directed against the antigen of interest, it can be attached to a second anti-antibody (e.g., rabbit antimouse Ig antibody) that is used to bind to the first, unlabelled antibody. When the label is attached directly to the specific, primary antibody, the method is referred to as direct; when the label is attached to a secondary or even tertiary antibody, the method is indirect. In some cases, molecules other than antibody can be used in indirect methods. For example, staphylococcal protein A, which binds to IgG, or avidin, which binds to primary antibodies labelled with biotin, can be coupled to fluorochromes or enzymes.

Measurement of antigen-antibody interactions

In many situations, it is important to know the affinity of an antibody for an antigen. For example, the usefulness of a monoclonal antibody as an experimental or therapeutic reagent depends on its affinity. Antibody affinities for antigen can be measured directly for small antigens (e.g., haptens) by a method called equilibrium dialysis. In this method, a solution of antibody is confined within a "semipermeable" membrane of porous cellulose and immersed in a solution containing the antigen (Semipermeable in this context means that small molecules, such as antigen, can pass freely through the membrane pores but that macromolecules, such as antibody, cannot). If no antibody is present within the membrane-bound compartment, the antigen in the bathing solution enters until the concentration of antigen within the membrane-bound compartment becomes exactly the same as that outside. Another way to view the system is that at dynamic equilibrium, antigen enters and leaves the membrane-bound compartment at exactly the same rate. However, when antibody is present inside the membrane, the net amount of antigen inside the membrane at equilibrium increases by the quantity that is bound to antibody. This phenomenon occurs because only unbound antigen can diffuse across the membrane, and at equilibrium, it is the unbound concentration of antigen that must be identical inside and outside the membrane. The extent of the increase in antigen inside the membrane depends on the antigen concentration, on the antibody concentration, and on the dissociation constant of the binding interaction. By measuring the antigen and antibody concentrations, by spectroscopy or by other means, K_d can be calculated.

An alternative way to determine K_d is by measuring the rates of antigen-antibody complex formation and dissociation. These rates depend, in part, on the concentrations of antibody and antigen and on the affinity of the interaction. All parameters except the concentrations can be summarized as rate constants, and both the on-rate constant (K_{on}) and the off-rate constant (K_{off}) can be calculated experimentally by determining the concentrations and the actual rates of association or dissociation, respectively. The ratio of K_{off}/K_{on} allows one to cancel out all the parameters not related to affinity and is exactly equal to the dissociation constant K_d. Thus, one can measure K_d at equilibrium by equilibrium dialysis or calculate K_d from rate constants measured under nonequilibrium conditions.

Analysis of gene structure and expression

Many of the advances in immunology during the last two decades have been a direct result of the advent of powerful techniques for characterizing the structure and expression of genes, synthesizing genes at will, and manipulating genes in cells and animals. In the following section, we outline some of the molecular genetic techniques that are widely used to study the immune system.

Southern blot hybridization

Southern blot hybridization, introduced by E. M. Southern, is used to characterize the organization of DNA surrounding a specific nucleic acid sequence, such as a particular gene. In a typical experiment, genomic DNA is chemically extracted from the nuclei of isolated cells or from whole tissues. At this point, the DNA will be present in extremely long segments and must be broken down into small fragments for analysis. This digestion is accomplished by enzymatic cleavage with restriction endonucleases, which cleave double-stranded DNA only at positions of particular symmetric nucleotide sequences, usually about 6 bp in length. Such sequences are called restriction sites and are present at the same positions in every cell, barring somatic mutation or DNA rearrangement. Complete digestion with a particular enzyme leads to cleavage of the genomic DNA, which gives rise to an array of DNA fragments ranging from about 0.5 to 10 kb in size. Different restriction endonucleases produce different arrays of restriction fragments. But each enzyme generally produces the same fragments from the DNA of every cell from an individual. Such is not the case, of course, when one examines genes that rearrange in different ways in different clones of cells, such as the Ig and T cell receptor genes.

To analyze the fragments, the digested DNA is separated according to size by electrophoresis in an agarose gel. The separated fragments are transferred by capillary action (blot) from the gel to a membrane of nitrocellulose or nylon. Each fragment is then attached in place to the membrane by heating or ultraviolet irradiation. The net result is that the array of fragments is arranged by size on the membrane. Any particular nucleic acid sequence, such as a gene of interest, will be present on one or a few unique-sized fragments, depending only on the choice of restriction endonuclease and the distances between the relevant restriction sites that flank the gene or are located within the gene. The analysis is completed by ascertaining the size of the restriction fragment that contains the gene of interest, which is accomplished by nucleic acid hybridization. Double-stranded DNA can be "melted" into single-stranded DNA by changing the temperature and solvent conditions. When the temperature is lowered, single-stranded DNA will reanneal to form double-stranded DNA. The rate at which reannealing of a sequence occurs is determined by the concentration of DNA containing complementary nucleic acid sequences present in the system. In the Southern blot hybridization technique, the DNA on the membrane is melted and then allowed to reanneal in the presence of a solution containing a large excess of single-stranded DNA (probe) with a sequence complementary to the gene of interest. The probe is typically labelled with radioactive phosphorus (e. g. P). It preferentially anneals only to the melted DNA on the membrane that contains the complementary sequence. After excess probe is removed by washing, the location of the bound probe is determined by autoradiography and compared with the positions of DNA fragments of known size.

Restriction sites are generally located at the same positions in the genomes of all individuals of a species. On occasion, this generalization is not true, and a particular restriction site is present only in or near some allelic forms of a gene. This variability in the presence of a particular restriction site leads to variability in the length of the restriction fragment that I detected by Southern blot hybridization if a probe is used that hybridizes near the variable restriction site. The length of the fragment is inherited as a mendelian allele, and its variation among individuals is described as a restriction fragment length polymorphism (RFLP). Southern blot for RFLPs is now commonly used to study the inheritance of nearby (linked) genes.

Northern blot hybridization and RNase protection assay

Nucleic acid electrophoresis, blot, and hybridization with DNA probes have also been used to analyze messenger RNA (mRNA) molecules. In this case, mRNA is isolated from the cell of interest and subjected to electrophoresis without digestion; mRNA is already single-stranded, and the gel electrophoresis runs under denaturing conditions to prevent internal hybridization. The position of probe binding can indicate the size of the mRNA (rather than the size of a restriction fragment of DNA), and the extent of probe binding correlates with the abundance of mRNA present in the cell. This technique for RNA analysis is now universally referred to as Northern blot to contrast it with Southern blot for analysis of DNA.

A more sensitive technique for the detection of mRNA transcribed from defined genes is the RNase protection assay. In this technique, cellular RNA is hybridized with an excess of a radioactively labelled RNA probe of defined length that is complementary to part of the sequence of a gene of interest. RNase is then added to the RNA mixture to digest only single-stranded RNA. The binding of the RNA probe to mRNA from the cell forms double-stranded RNA duplexes that are protected from digestion by the RNase. After digestion of single-stranded cellular and unhybridized probe RNA, the radioactively labelled RNA duplexes are separated by PAGE and visualized by autoradiography. The predicted position of the radioactive band on the gel, corresponding to a particular mRNA, will be determined by the length of the RNA probe, and the intensity of the band will correspond to the amount of the specific mRNA in the original sample. Multiple RNA probes of different lengths, each complementary to a different gene, are commonly used to detect the presence and quantity of several different mRNA in a sample simultaneously. For example, the relative amounts of multiple cytokine gene mRNA, made by a helper T cell population can be determined in one RNase protection assay.

Polymerase chain reaction

The polymerase chain reaction (PCR) is a rapid and simple method for copying and amplifying specific DNA sequences up to about 1 kb in length. PCR is widely used as a preparative and analytical technique in all branches of molecular biology. To use this method, it is necessary to know the sequence of a short region of DNA on each end of the larger sequence that is to be copied; these short sequences are used to specify oligonucleotides primers. The method consists of repetitive cycles of DNA melting, DNA annealing, and DNA synthesis. Double-stranded DNA containing the sequence to be copied and amplified is mixed with a large molar excess of two single-stranded DNA oligonucleotides (the primers). The first primer is identical to the 5′ end of the sense strand of the DNA to be copied, and the second primer is identical to the antisense strand at the 3′ end of the sequence (Because double-stranded DNA is antiparallel, the second primer is also the inverted complement of the sense strand). The PCR reaction is initiated by melting the double-stranded DNA at high temperature and cooling the mixture to allow DNA annealing. During annealing, the first primer (present in large molar excess) will hybridize to the 3′ end of the antisense strand, and the second primer (also present in large molar excess) will hybridize to the 3′ end of the antisense strand, and the second primer (also present in large molar excess) will hybridize to the 3′end of the sense strand. The annealed mixture is incubated with DNA polymerase Ⅰ and all four deoxynucleotide triphosphates (A, T, G, and C) to allow new DNA to be synthesized. DNA polymerase Ⅰ will extend the 3′end of each bound primer to synthesize the complement of the single-stranded DNA templates. Specifically, the original sense strand is used as a template to make a new antisense strand, and the original antisense strand is used as a template to make a new sense strand. This ends cycle 1. Cycle 2 is initiated when the reaction mixture is remelted and then allowed to reanneal with the primers. In the DNA synthetic step of the second cycle, each strand synthesized in the first cycle serves as an additional template after having hybridized with the appropriate primers (It follows that the number of templates in the reaction doubles with each cycle, hence the name "chain reaction") . The second cycle is completed when DNA polymerase Ⅰ extends the primers to synthesize the complement of the templates. PCR reactions can be conducted for 20 or more cycles, with the sequence flanked by the primers doubled at each step. These reactions are routinely automated by temperature-controlled cyclers to regulate melting, annealing, and DNA synthesis and with a DNA polymerase Ⅰ isolated from thermostable bacteria that can with stand the temperatures used for melting of DNA.

PCR has many uses. For example, by choosing suitable primers, exon 2 of an unknown HLA-DR B allele (i. e. , the exon that encodes the polymorphic B1 protein domain) can be amplified, the amplified DNA lighted into a suitable vector, and the DNA sequence determined without ever isolating the original gene from genomic DNA. This objective was the original purpose for which PCR was invented. PCR is also widely used for the cloning of known genes or genes related to known genes (e. g. , by choosing primers from highly conserved regions of a gene family). PCR can be used to detect or, in some cases, even to quantify the presence of particular DNA sequences in a sample (e. g. , the presence of viral sequences in a clinical specimen). Because PCR can amplify DNA only when the two primers are near each other (i. e. , within about 1 kb), the technique is useful to detect a specific gene recombination even by choosing primers complementary to sequences that are near one another only in the rearranged DNA. This approach is used to detect antigen receptor gene recombination events and chromosomal translocations. In the reverse transcriptase-PCR (RT-PCR) method, mRNA transcripts of a particular gene are detected by first using reverse transcriptase to make a complementary DNA copy of the mRNA before PCR amplification. Relative quantification of RNA between samples can be achieved by an adaptation of RT-PCR called real-time RT-PCR, in which fluorescent probes are used to follow the accumulation of amplified product during each PCR cycle (i. e. , in real time) as opposed to the end point detection by conventional quantitative PCR methods.

Transgenic mice and targeted gene knockouts

Two important methods for studying the functional effects of specific gene products in vivo are the creation of transgenic mice that overexpress a particular gene in a defined tissue and the creation of gene knockout mice, in which a targeted disruption is used to ablate the function of a particular gene. Both techniques have been widely used to analyze many biological phenomena, including the maturation, activation, and tolerance of lymphocytes.

To create transgenic mice, foreign DNA sequences, called transgenes, are introduced into the pronuclei of fertilized mouse eggs, and the eggs are implanted into the oviducts of pseudopregnant females. Usually, if a few hundred copies of a gene are injected into pronuclei, about 25% of the mice that are born are transgenic. One to 50 copies of the transgene insert in tandem into a random site of breakage in a chromosome and are subsequently inherited as a simple mendelian trait. Because integration usually occurs before DNA replication, most (about 75%) of the transgenic pups carry the transgene in all their cells, including germ cells. In most cases, integration of the foreign DNA does not disrupt endogenous gene function. Also, each founder mouse carrying the transgene is a heterozygote, from which homozygous lines can be bred.

The great value of transgenic technology is that it can be used to express genes in particular tissues by attaching coding sequences of the gene to regulatory sequences that normally drive the expression of genes selectively in that tissue. For instance, lymphoid promoters and enhancers can be used to overexpress genes, such as rearranged antigen receptor genes, in lymphocytes, and the insulin promoter can be used to express genes in the B cells of pancreatic islets. Transgenes can also be expressed under the control of promoter elements that respond to drugs or hormones, such as tetracycline or estrogens. In these cases, transcription of the transgene can be controlled at will by administration of the inducing agent.

A powerful method for developing animal models of single-gene disorders, and the most definitive way of establishing the obligatory function of a gene in vivo, is the creation of knockout mice by targeted mutation or disruption of the gene. This technique relies on the phenomenon of homologous recombination. If an exogenous gene is inserted into a cell, for instance, by electroporation, it can integrate randomly into the cell's genome. However, if the gene contains sequences that are homologous to an endogenous gene, it will preferentially recombine with and replace endogenous sequences. To select for cells that have undergone homologous recombination, a drug-based selection strategy is used. The fragment of homologous DNA to be inserted into a cell is placed in a vector typically containing a neomycin resistance gene and a viral thymidine kinase (TK) gene. This targeting vector is constructed in such a way that the neomycin resistance gene is always inserted into the chromosomal DNA, but the TK gene is lost whenever homologous recombination (as opposed to random insertion) occurs. The vector is introduced into cells, and the cells are grown in neomycin and ganciclovir, a drug that is metabolized by thymidine kinase to generate a lethal product. Cells in which the gene is integrated randomly will be resistant to neomycin but will be killed by ganciclovir, whereas cells in which homologous recombination has occurred will be resistant to both drugs because the TK gene will not be incorporated. This positive-negative selection ensures that the inserted gene in surviving cells has undergone homologous recombination with endogenous sequences. The presence of the inserted DNA in the middle of an endogenous gene usually disrupts the coding sequences and ablates expression or function of that gene. In addition, targeting vectors can be designed such that homologous recombination will lead to the deletion of one or more exons of the endogenous gene.

To generate a mouse carrying a targeted gene disruption or mutation, a targeting vector is used to first disrupt the gene in a marine embryonic stem (ES) cell line. ES cells are pluripotent cells derived from mouse embryos that can be propagated and induced to differentiate in culture or that can be incorporated into a mouse blastocyst, which may be implanted in a pseudopregnant mother and carried to term. Importantly, the progeny of the ES cells develop normally into mature tissues that will express the exogenous genes that have been transfected into the ES cells. Thus, the targeting vector designed to disrupt a particular gene is inserted into ES cells, and colonies in which homologous recombination has occurred (on one chromosome) are selected with drugs, as described before. The presence of the desired recombination is verified by analysis of DNA with techniques such as Southern blot hybridization or PCR. The selected ES cells are injected into blastocysts, which are implanted into pseudopregnant females. Mice that develop will be chimeric for a heterozygous disruption or mutation, that is, some of the tissues will be derived from the ES cells and others from the remainder of the normal blastocyst. The germ cells are also usually chimeric, but because these cells are haploid, only some will contain the chromosome copy with the disrupted (mutated) gene. If chimeric mice are mated with normal (wild-type) animals and either sperm or eggs containing the chromosome with the mutation fuse with the wild-type partner, all cells in the offspring derived from such a zygotes will be heterozygous for the mutation (so-called germline transmission). Such heterozygous mice can be mated to yield animals that will be homozygous for the mutation with a frequency that is predictable by simple mendelian segregation. Such knockout mice are deficient in expression of the targeted gene.

Homologous recombination can also be used to replace a normal gene sequence with another gene, thereby creating a knock-in mouse strain. In a sense, knock-in mice are mice carrying a transgene at a defined site in the genome rather than in a random site as in conventional transgenic mice. This strategy is used when it is desirable to have the expression of the transgene regulated by certain endogenous DNA sequences, such as a particular enhancer or promoter region. In this case, the targeting vector contains an exogenous gene encoding desired product as well as sequences homologous to an endogenous gene that are needed to target the site of recombination.

Although the conventional gene-targeting strategy has proved to be of great usefulness in immunology research, the approach has some potentially significant drawbacks. Firstly, the mutation of one gene during development may be compensated for by altered expression of other gene products, and therefore the function of the targeted gene may be obscured. Secondly, in a conventional gene knockout mouse, the importance of a gene in only one tissue or at only one time during development cannot be easily assessed. Thirdly, a functional selection marker gene, such as the neomycin resistance gene, is permanently introduced into the animal genome, and this alteration may have unpredictable results on the phenotype of the animal. An important refinement of gene knockout technology that can overcome many of these drawbacks takes advantage of the bacteriophage-derived *Cre/loxP* recombination system. The Cre enzyme is a DNA recombinase that recognizes a 34-bp sequence motif called *loxP*, and the enzyme mediates the deletion of gene segments flanked by two *loxP* sites in the same orientation. To generate mice with *loxP*-tagged genes, targeting vectors are constructed with one *loxP* site flanking the neomycin resistance gene at one end and a second *loxP* site flanking the sequences homologous to the target at the other end. These vectors are transfected into ES cells, and mice carrying the *loxP*-flanked but still functional target gene are generated as described for conventional knockout mice. A second strain of mice carrying a *Cre* transgene is then bred with the strain carrying the *loxP*-flanked targeted gene, can be restricted to certain tissues or specified times by the use of *Cre* transgene constructs with different promoters. For example, selective deletion of a gene only in helper T cells can be accomplished by using a *Cre* transgenic mouse in which Cre is driven by a CD4 promoter. Alternatively, a steroid-inducible promoter can be used so that Cre expression and subsequent gene deletion only occur after mice are given a dose of dexamethasone. Many other variations on this technology have been devised to created conditional mutants. *Cre/loxP* technology can also be used to create knock-in mice. In this case, *loxP* sites are placed in the targeting vector to flank the neomycin resistance gene and the homologous sequences, but they do not flank the replacement (knock-in) gene sequences. Therefore, after *Cre*-mediated deletion, the exogenous gene remains in the genome at the targeted site.

Glossary

affinity chromatography 亲和色谱（法）
agarose [ˈɑːgərəus] n. 琼脂糖
alkaline phosphatase 碱性磷酸（酯）酶
allelic [əˈliːlik] adj.（遗传学）等位基因的
anatomic [ˌænəˈtɔmik] adj. 解剖的，解剖学上的
anneal [əˈniːl] n. 复性，对合，退火
antiparallel [ˌæntiˈpærəlel] adj. 反平行的
autoradiography [ˈɔːtəuˌreidiˈɔgrəfi] n. 放射自显影法，放射自显影术
avidin [ˈævidin] n. 卵白素，抗生物素蛋白
biotin [ˈbaiətin] n. 维生素 H，生物素
blastocyst [ˈblæstəusist] n. 胚泡
capillary action 毛细（管）吸引（作用），毛细作用
chemiluminescent adj. 化合光的，化学发光的
chimeric [kaiˈmerik] adj. 嵌合体的
colloidal gold 胶态金，胶体金
compartment [kəmˈpɑːtmənt] n. 室，间隔，层
confocal microscopy 共聚焦显微镜
denature [diːˈneitʃə] v. 使变性
deoxynucleotide triphosphate 碱磷酸去氧核苷酸
dexamethasone [ˌdeksəˈmeθəzəun] n. 地塞米松，氟美松（抗炎药）
dissociation constant 解离常数
DNA polymerase Ⅰ DNA 聚合酶Ⅰ
duplex [ˈdjuːpleks] n. 复式，双显性组合，二倍体
electroporation n. 电穿孔术，电穿孔（法）
embryo [ˈembriəu] n. 胚胎，胎儿，胚芽；adj. 胚胎的，初期的
endogenous [enˈdɔdʒənəs] adj. 内生的，内源的，内生性，内源性
endonuclease [ˌendəˈnjuːkliˌeis] n. 核酸内切酶
enzyme-linked immunosorbent assay (ELISA) 酶联免疫吸附测定法
equilibrium dialysis 平衡透析，平衡透析法

estrogen [ˈestrədʒən] n. 雌激素
exogenous [ekˈsɔdʒinəs] adj. 外生性，外源性，外源的，外生的
exon n. 编码序列，外显子，输出子
flow cytometry 流式细胞计数
fluorescence-activated cells sorter (FACS) 荧光激活细胞分类器
fluorochrome [ˈfluːərəˌkrəum] n. 荧光染料，荧色物
ganciclovir n. 羟甲基无环鸟苷，更昔洛韦
germ cell 生殖细胞，胚细胞
germline n. 种系
granule [ˈgrænjuːl] n. 小粒，颗粒，细粒
hapten [ˈhæpten] n. 半抗原
heterozygote [hetərəuˈzaigəut] n. 杂合子，杂合体，异型合子
hormone [ˈhɔːməun] n. 荷尔蒙，激素
horseradish peroxidase 辣根过氧化物酶
immunofluorescence [ˌimjunəufluəˈresns] n. 免疫荧光
immunoglobulin (Ig) [iˈmjuːnəuˈglɔbjulin] n. 免疫球蛋白
immunoprecipitation [ˌimjunəupriˌsipiˈteiʃən] 免疫沉淀作用
insulin [ˈinsjulin] n. 胰岛素
isoelectric focusing 等电聚焦
knockout [ˈnɔˈkaut] n. 敲除
magnet [ˈmægnit] n. 磁体，磁铁
Mendelian segregation 孟德尔分离定律
neomycin [ni(:) əuˈmaisin] n. 新霉素
nylon [ˈnailən] n. 尼龙
offspring [(us)ˈɔːf-, ˈɔfspriŋ] n. (单复数同形) 子孙，后代，产物
oligonucleotide [ˌɔligəuˈnjuːkliəutaid, -tid] n. 低(聚)核苷酸，寡核苷酸
optical [ˈɔptikəl] adj. 眼的，视力的，光学的
oviduct [ˈəuvidʌkt] n. 输卵管
oxidation [ɔksiˈdeiʃən] n. 氧化
pancreatic islet n. 胰岛
parameter [pəˈræmitə] n. 参数，参量
phosphorus [ˈfɔsfərəs] n. 磷
pluripotent [pluəˈripətənt] adj. 多向性(的)，多能性(的)
polymerase chain reaction (PCR) 聚合酶链反应
primer [ˈpraimə] n. 引物
pronuclei [prəˈnjuːklei] n. (pronucleus 的复数) (动) 原核，前核
pseudopregnant adj. 假妊娠的，假孕的
quantitation n. 定量，测数量
radioimmunoassay (RIA) [ˈreidiəuˌimjunəuˈæsei] n. 放射性免疫测定
radioisotope [ˈreidiəuˈaisətəup] n. 放射性同位素
radiolabeled 放射标记的，放射示踪的
real-time RT-PCR 实时逆转录 PCR
rearrangement [ˌriːəˈreindʒmənt] n. (基因) 重排，重(新)排(列)，重新布置
reduction [riˈdʌkʃən] n. 还原，减少，缩减，复位术
restriction site 限制(酶切)位点，限制性(酶切)部位
reverse transcriptase-PCR (RT-PCR) 逆转录 PCR
RNase [ˌɑːˈrenˌeis, -ˌeiz] n. 核糖核酸酶
sensitivity [ˈsensiˈtiviti] n. 敏感，灵敏(度)，灵敏性

signal-to-noise n. 信噪比
sodium dodecyl sulfate-polyacrylamide gel electrophoresis (SDS-PAGE) 十二烷基硫酸钠-聚丙烯酰胺凝胶电泳
Southern blot 索瑟恩印迹，DNA 印迹
specimen ['spesimin, -mən] n. 标本，样品，样本，待试验物
sperm [spəːm] n. 精液，精子
staphylococcal bacteria [ˌstæfiləu'kɔkəl bæk'tiəriə] 葡萄球菌
symmetric [si'metrik] adj. 相称性的，均衡的
tandem ['tændəm] adj. 一前一后排列的，串联的，串
tetracycline [tetrə'saiklin,-lain] n. 四环素
thermostable [ˌθəːməu'steibəl] adj. 耐热的，热稳定的
transgenic mice 转基因小鼠
two-photon microscopy 双光子显微镜
ultraviolet irradiation 紫外线照射，紫外线照射法
Western blot 蛋白质印迹法，免疫印迹分析
zygote ['zaigəut] n. (生物) 受精卵，接合子，接合体

难句分析

1. The exquisite specificity of antibodies for particular antigens makes antibodies valuable reagents for detecting, purifying, and quantitating antigens.
   译文：抗体对特定抗原的高度特异性使得抗体成为抗原检测、纯化以及定量的有价值的试剂。
2. Immunoprecipitation is a technique in which an antibody specific for one protein antigen in a mixture of proteins is used to isolate the specific antigen from the mixture. In most modern procedures, the antibody is attached to a solid-phase particle (e.g., an agarose bead) either by direct chemical coupling or indirectly.
   译文：免疫共沉淀技术是一种抗体特异识别在一群蛋白质混合物中的一种蛋白抗原的一项技术，这种技术常用于从蛋白质中分离某种特殊抗原。在大多数的现代方法中，抗体通常通过直接或间接的化学连接被固定在一个特定的固相介质（例如琼脂糖树脂）上。
3. The tissue lineage, maturation stage, or activation status of a cell can often be determined by analyzing the cell surface or intracellular expression of different molecules. This technique is commonly done by staining the cell with fluorescently labelled probes that are specific for those molecules and measuring the quantity of fluorescence emitted by the cell.
   译文：细胞的组织系统，成熟时期或者活动时期可以通过分析细胞表面或细胞内分泌的不同分子来检测到。这种技术常用在固定细胞中加入特异分子的荧光标签探针，并且通过测量那些细胞所发散出来的荧光来达到检测目的。
4. Many of the advances in immunology during the last two decades have been a direct result of the advent of powerful techniques for characterizing the structure and expression of genes, synthesizing genes at will, and manipulating genes in cells and animals.
   译文：在过去 20 年中许多在免疫技术中的进步已经对于基因表达和结构的特性，按着自己的意愿合成基因，以及使基因在细胞和动物中有很强的多样性有着强有力的技术支持。
5. For example, by choosing suitable primers, exon 2 of an unknown HLA-DR B allele (i.e., the exon that encodes the polymorphic B1 protein domain) can be amplified, the amplified DNA lighted into a suitable vector, and the DNA sequence determined without ever isolating the original gene from genomic DNA.

译文：例如，通过选择合适的引物，一个不知道的 HLA-DR B 等位基因外显子 2，(即外显子编码多样 B1 蛋白的范围) 可以被扩增，这个扩增的基因被转化进合适的载体，同时这个基因将不用通过原始的基因组 DNA 的原始基因而被决定。

6. The great value of transgenic technology is that it can be used to express genes in particular tissues by attaching coding sequences of the gene to regulatory sequences that normally drive the expression of genes selectively in that tissue.

译文：转录技术的巨大价值在于它可以用来通过编码基因的序列来调控某些基因在特定组织中的表达，而这一表达可以通过特定的调控序列而加以控制。

(选自：Abul K Abbas，Andrew H Lichtman. Cellular and Molecular Immunology 5th eds. Philadelphia：Saunders，2003.)

Viable Offspring Derived from Fetal and Adult Mammalian Cells

Fertilization of mammalian eggs is followed by successive cell divisions and progressive differentiation, first into the early embryo and subsequently into all of the cell types that make up the adult animal. Transfer of a single nucleus at a specific stage of development, to an enucleated unfertilized egg, provided an opportunity to investigate whether cellular differentiation to that stage involved irreversible genetic modification. The first offspring to develop from a differentiated cell were born after nuclear transfer from an embryo-derived cell line that had been induced to become quiescent. Using the same procedure, we now report the birth of live lambs from three new cell populations established from adult mammary gland, fetus and embryo. The fact that a lamb was derived from an adult cell confirmed that differentiation of that cell did not involve the irreversible modification of genetic material required for development to term. The birth of lambs from differentiated fetal and adult cells also reinforces previous speculation that by inducing donor cells to become quiescent it will be possible to obtain normal development from a wide variety of differentiated cells.

It has long been known that in amphibians, nuclei transferred from adult keratinocytes established in culture support development to the juvenile, tadpole stage. Although this involves differentiation into complex tissues and organs, no development to the adult stage was reported, leaving open the question of whether a differentiated adult nucleus can be fully reprogrammed. Previously we reported adult nucleus can be fully reprogrammed. Previously we reported the birth of live lambs after nuclear transfer from cultured embryonic cells that had been induced into quiescence. We suggested that inducing the donor cell to exit the growth phase causes changes in chromatin structure that facilitate reprogramming of gene expression and that development would be normal if nuclei are used from a variety of differentiated donor cells in similar regimes. Here we investigate whether normal development to term is possible when donor cells derived from fetal or adult tissue are induced to exit the growth cycle and enter the G0 phase of the cell cycle before nuclear transfer.

Three new populations of cells were derived from①a day-9 embryo, ②a day-26 fetus and ③mammary gland of a 6-year-old ewe in the last trimester of pregnancy. Morphology of the embryo-derived cells is unlike both mouse embryonic stem (ES) cells and the embryo-derived cells used in our previous study. Nuclear transfer was carried out according to one of our established protocols and reconstructed embryos transferred into recipient ewes. Ultrasound scanning detected 21 single fetuses on day 50-60 after oestrus. On subsequent scanning at about 14-day intervals, fewer fetuses were observed, suggesting either mis-diagnosis or fetal loss. In total, 62% of fetuses were lost, a significantly greater proportion than the estimate of 6% after natural mating. Increased prenatal loss has been reported after embryo manipulation or culture of unreconstructed embryos. At about day-110 of pregnancy, four fetuses were dead, all from embryo-derived cells, and postmortem analysis was possible after killing the ewes. Two fetuses had abnormal liver development, but no other abnormalities were detected and there was no evidence of infection.

Eight ewes gave birth to live lambs. All three cell populations were represented. One weak lamb, derived from the fetal fibroblasts, weighed 3.1 kg and died within a few minutes of birth, although postmortem analysis failed to find any abnormality or infection. At 12.5%, perinatal loss was not dissimilar to that occurring in a large study of commercial sheep, when 8% of lambs died within 24 h of birth. In all cases the lambs displayed the morphological characteristics of the breed used to derive the nucleus donors and not that of the oocyte donor. This alone indicates that the lambs could not have been born after inadvertent mating of either the oocyte donor or recipient ewes. In addition, DNA microsatellite analysis of the cell populations and the lambs at four polymorphic loci confirmed that each lamb was derived from the cell population used as nuclear donor. Duration of gestation is determined by fetal genotype, and in all cases gestation was longer than the breed mean. By contrast, birth weight is influenced by both maternal and fetal genotype. The birth weight of all lambs was within the range for single lambs born to Blackface ewes on our farm (up to 6.6 kg) and in most cases was within the range for the breed of the nuclear donor. There are no strict control observations for birth weight after embryo transfer between breeds, but the range in weight of lambs born to their own breed on our farms is 1.2-5.0 kg, 2-4.9 kg and 3-9 kg for the Finn Dorset, Welsh Mountain and Poll Dorset genotypes, respectively. The attainment of sexual maturity in the lambs is being monitored.

Development of embryos produced by nuclear transfer depends upon the maintenance of normal ploidy and creating the conditions for developmental regulation of gene expression. These responses are both influenced by the cell-cycle stage of donor and recipient cells and the interaction between them. A comparison of development of mouse and cattle embryos produced by nuclear transfer to oocytes or enucleated zygotes suggests that a greater proportion develop if the recipient is an oocyte. This may be because factors that bring about reprogramming of gene expression in a transferred nucleus are required for early development and are taken up by the pronuclei during development of the zygote.

If the recipient cytoplasm is prepared by enucleation of an oocyte at metaphase Ⅱ, it is only possible to avoid chromosomal damage and maintain normal ploidy by transfer of diploid nuclei, but further experiments are required to define the optimum cell-cycle stage. Our studies with cultured cells suggest that there is an advantage if cells are quiescent. In earlier studies, donor cells were embryonic blastomeres that had not been induced into quiescence. Comparisons of the phases of the growth cycle showed that development was greater if donor cells were in mitosis or in the G1 phase of the cycle, rather than in S or G2 phases. Increased development using donor cells in G0, G1 or G2 phases. Increased development using donor cells in G0, G1 or mitosis may reflect greater access for reprogramming factors present in the oocyte cytoplasm, but a direct comparison of these phases in the same cell population is required for a clearer understanding of the underlying mechanisms.

Together this result indicate that nuclei from a wide range of cell types should prove to be totipotent after enhancing opportunities for reprogramming by using appropriate combinations of these cell-cycle stages. In turn, the dissemination of the genetic improvement obtained within elite selection herds will be enhanced by nuclear transfer from cells derived from adult animals. In addition, gene targeting in livestock should now be feasible by nuclear transfer from modified cell populations and will offer new opportunities in biotechnology. The techniques described also offer an opportunity to study the possible persistence and impact of epigenetic changes, such as imprinting and telomere shortening, which are known to occur in somatic cells during development and senescence, respectively.

The lamb born after nuclear transfer from a mammary gland cell is, to our knowledge, the first mammal to develop from a cell derived from an adult tissue. The phenotype of the donor cell is unknown. The primary culture contains mainly mammary epithelial (over 90%) as well as other differentiated cell types, including myoepithelial cells and fibroblasts. We cannot exclude the possibility that there is a small proportion of relatively undifferentiated stem cells able to support regeneration of the mammary gland during pregnancy. Birth of the lamb shows that during the development of that mammary cell there was no irreversible modification of genetic information required for development to term. This is consistent with the generally accepted view that mammalian differentiation is almost all achieved by systematic, sequential changes in gene expression brought about by interactions between the nucleus and the changing cytoplasmic environment. (Fig. 1. 19)

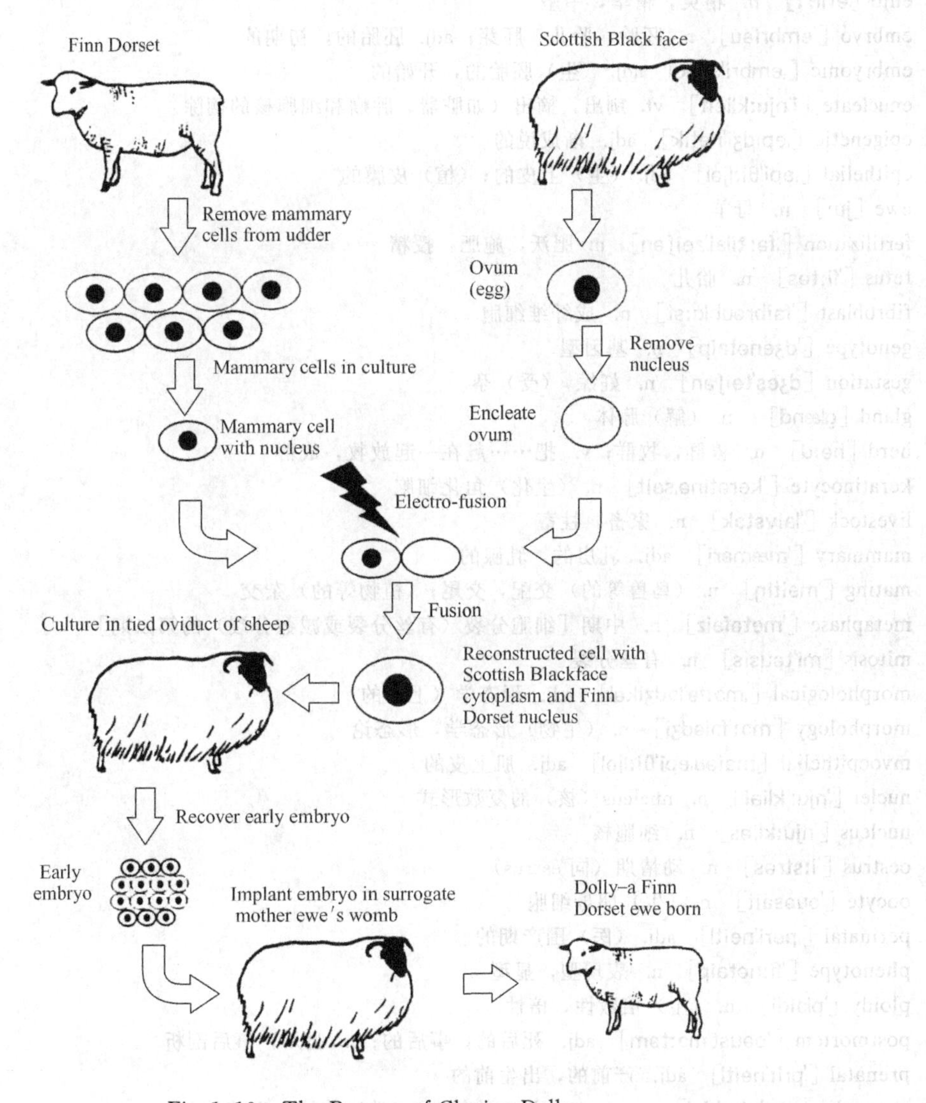

Fig. 1. 19 The Process of Cloning Dolly

Dolly, the first cloned animal, was born on July 5th, 1996 and was introduced to the public on February 23rd of the next year. It was said to be an exact genetic copy of its mother, which was the critical point in global cloning research

Glossary

amphibian [æm'fibiən] adj. 两栖类的，水陆两用的；n. 两栖动物
blastomere ['blæstəumiə] n.（分）裂球（受精卵分裂形成的细胞之一）
cellular ['seljulə] adj. 细胞的
chromatin ['krəumətin] n. 染色质
cytoplasm ['saitəuplæzm] n.（生）细胞质
differentiation [ˌdifəˌrenʃi'eiʃən] n. 鉴别，分化，歧化
diploid ['diplɔid] adj. 二倍体的；n. 二倍体
dissemination [diˌsemi'neiʃən] n. 播种，传播，散步
division [di'viʒən] n. 分裂，切开，部分
donor ['dəunə] n. 供体
elite [ei'li:t] n. 精英，精华，中坚
embryo ['embriəu] n. 胚胎，胎儿，胚芽；adj. 胚胎的，初期的
embryonic [ˌembri'ɔnik] adj.（生）胚胎的，开始的
enucleate [i'nju:klieit] vt. 剜出，摘出（如脏器、肿瘤和细胞核的摘除）
epigenetic [ˌepidʒi'netik] adj. 渐成说的
epithelial [ˌepi'θi:ljəl] adj.（生）上皮的；（植）皮膜的
ewe [ju:] n. 母羊
fertilization [ˌfə:tilai'zeiʃən] n. 肥沃，施肥，授精
fetus ['fi:təs] n. 胎儿
fibroblast ['faibrəublɑ:st] n. 成纤维细胞
genotype ['dʒenətaip] n. 基因型
gestation [dʒes'teiʃən] n. 妊娠，（受）孕
gland [glænd] n.（解）腺体
herd [hə:d] n. 兽群，牧群；v. 把……赶在一起放牧，成群
keratinocyte ['kerətinəˌsait] n.（生化）角化细胞
livestock ['laivstɔk] n. 家畜，牲畜
mammary ['mæməri] adj. 乳房的，乳腺的
mating ['meitiŋ] n.（鸟兽等的）交配，交尾；（植物等的）杂交
metaphase ['metəfeiz] n. 中期［细胞分裂（有丝分裂或减数分裂）的第二期］
mitosis [mi'təusis] n. 有丝分裂
morphological [ˌmɔ:fə'lɔdʒikəl] adj. 形态学（上）的
morphology [mɔ:'fɔlədʒi] n.（生物）形态学，形态论
myoepithelial [maiəuˌepi'θi:ljəl] adj. 肌上皮的
nuclei ['nju:kliai] n. nucleus（核）的复数形式
nucleus ['nju:kliəs] n. 细胞核
oestrus ['i:strəs] n. 动情期（同 estrus）
oocyte ['əuəsait] n.（生）卵母细胞
perinatal [ˌperi'neitl] adj.（医）围产期的
phenotype ['fi:nətaip] n. 表现型，显型
ploidy ['plɔidi] n.（生）倍数性，倍性
postmortem ['pəust'mɔ:təm] adj. 死后的，事后的；n. 验尸，事后剖析
prenatal ['pri:'neitl] adj. 产前的，出生前的
pronuclei [prə'nju:klei] n.（pronucleus 的复数）（动）前核，生殖核
protocol ['prəutəkɔl] n. 程序，方案
quiescent [kwai'esənt] adj. 不动的，静止的

recipient [ri'sipiənt] n. 受（血）者，受体
regime [rei'ʒiːm] n. 政权，政治制度，食物疗法，养生法
reinforce [ˌriːin'fɔːs] vt. 加强，增援，补充，增加……的数量，修补，加固；vi. 求援，得到增援；n. 加固物
replication [ˌrepli'keiʃən] n. 复制
senescence [si'nesns] n. 衰老，变老
somatic [səu'mætik] adj. 躯体的，体壁的
speculation [ˌspekju'leiʃən] n. 思索，构思，推测
stem cell 干细胞
tadpole ['tædpəul] n.（动）蝌蚪
telomere ['teləmiə] n.（生）端粒
totipotent [təu'tipətənt] adj.（动）能由（分）裂球变成胚胎的，全能性的，两性的
trimester [trai'mestə] n. 三个月，三月期（通常用于妊娠）
transfer [træns'fəː] n. 转移；v. 转移
zygote ['zaigəut] n.（生物）受精卵，接合子，接合本

难句分析

1. Transfer of a single nucleus at a specific stage of development, to an enucleated unfertilized egg, provided an opportunity to investigate whether cellular differentiation to that stage involved irreversible genetic modification.

   译文：将一个处于特殊发育阶段的单个细胞核转移到一个去核的未受精的卵中，可以研究细胞分化到该阶段是否与不可逆的遗传修饰有关。

2. The fact that a lamb was derived from an adult cell confirmed that differentiation of that cell did not involve the irreversible modification of genetic material required for development to term.

   译文：一只山羊由一个成年细胞发育而来的事实证明了那个细胞的分化与发育到某个时期的遗传物质的不可逆修饰无关。

3. It has long been known that in amphibians, nuclei transferred from adult keratinocytes established in culture support development to the juvenile, tadpole stage. Although this involves differentiation into complex tissues and organs, no development to the adult stage was reported, leaving open the question of whether a differentiated adult nucleus can be fully reprogrammed.

   译文：人们早已知晓，在两栖类动物中，成年的角化细胞经过培养可以发育到幼年的蝌蚪阶段。虽然这个（细胞）能分化成复杂的组织和器官，但还没有它发育到成年阶段的报道，这就提出了一个问题——一个已分化的成年细胞核是否能完全改变它的发育进程。

4. This is consistent with the generally accepted view that mammalian differentiation is almost all achieved by systematic, sequential changes in gene expression brought about by interactions between the nucleus and the changing cytoplasmic environment.

   译文：这（一结论）与通常可接受的观点是一致的，即哺乳动物的分化几乎都是由系统的、连续的基因表达的改变所完成的，这种基因表达的改变是由细胞核与变化的细胞质环境之间相互作用引起的。

（选自：Wilmut I, Schnieke A E, Mc Whir J, Kind A J, Campbell K H. Nature, 1997, 385 (27): 810-812.）

CRISPR/Cas9 for Genome Editing: Progress, Implications and Challenges

Benefiting from the rapid development of high-throughput sequencing technology and bioinformatics, researchers make great progress on gene mapping in a short time. Currently, a major challenge faced by researchers is how to reveal the molecular mechanism of genes influencing individual phenotypes. A good way to elucidate the function of a gene is to shut it down or overexpress it in living organisms, which is previously complicated and time consuming. A new approach named "genome editing" emerged and widely used in the studies of functional genomics, transgenic organisms and gene therapy during the past decades. Genome editing is built on engineered, programmable and highly specific nucleases, which can induce site specific changes in the genomes of cellular organisms through a sequence-specific DNA-binding domain and a nonspecific DNA cleavage domain. Subsequent cellular DNA repair process generates desired insertions, deletions or substitutions at the loci of interest. Multiple artificial nuclease systems have been developed for genome editing. Zinc-finger nucleases (ZFNs) are one of widely applied engineered nucleases. ZFNs contain a common Cys_2-His_2 DNA-binding domain and a DNA cleavage domain of the *FokI* restriction endonuclease. Another popular genome editing platform is transcription activator-like effector nucleases (TALENs), which are derived from a natural protein of plant pathogenic bacteria *Xanthomonas*. The DNA-binding domain of TALENs is composed of 33-35 conserved amino acid repeated motifs, each of which recognizes a specific nucleotide. Through shuffling repeated amino acid recognition motifs, TALENs can be programmed to target-specific DNA sequence. Recently, clustered regularly interspaced short palindromic repeats (CRISPR) / CRISPR-associated (Cas) protein 9 system provides an alternative to ZFNs and TALENs for genome editing. Distinct from the protein-guided DNA cleavage of ZFNs and TALENs, CRISPR/Cas9 depends on small RNA for sequence-specific cleavage. Because only programmable RNA is required to generate sequence specificity, CRISPR/Cas9 is easily applicable and develops very fast over the past year. Here, we review the molecular mechanism, applications and challenges of CRISPR/Cas9-mediated genome editing and clinical therapeutic potential of CRISPR/Cas9 in future.

CRISPR/Cas9-mediated genome modification

In bacteria and archaea, CRISPR/Cas was discovered as an acquired immune system against viruses and phages through CRISPR RNA (crRNA) -based DNA recognition and Cas nucleases-mediated DNA cleavage. CRISPR/Cas is observed in nearly 40% genomes of sequenced bacteria and nearly 90% genomes of sequenced archaea. CRISPR locus consists of a series of conserved repeated sequences interspaced by distinct nonrepetitive sequences named spacers [Fig. 1. 20 (A)] . In CRISPR/Cas system, invading foreign DNA is processed by Cas nuclease into small DNA fragments, which are then incorporated into CRISPR locus of host genomes as the spacers. In response to viruses and phage infections, the spacers are

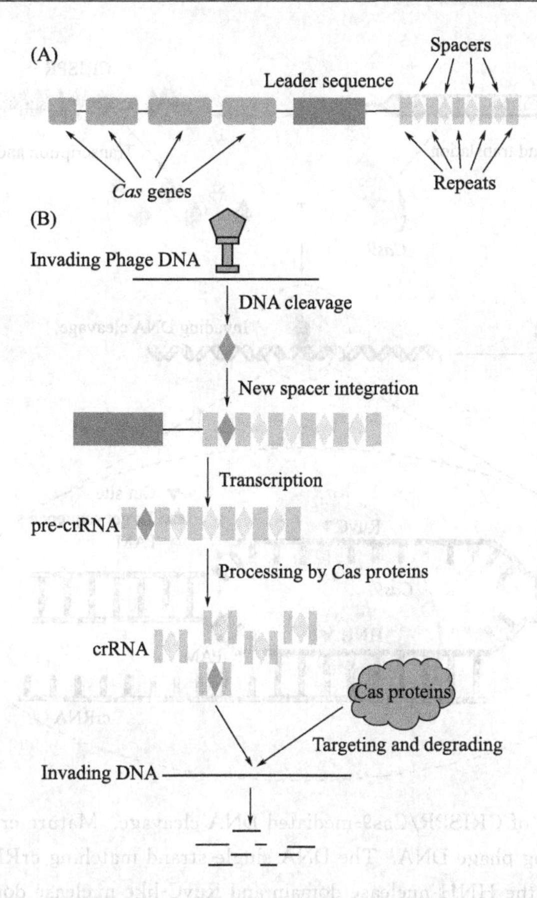

Fig. 1. 20 Overview of CRISPR/Cas bacterial immune system. (A) A typical structure of CRISPR locus; (B) illustration of new spacer acquisition and invading DNA cleavage.

used as transcriptional templates for producing crRNA, which guides Cas to cleave target DNA sequences of invading viruses and phages [Fig. 1. 20 (B)] . More than 40 different Cas protein families have been reported, playing important roles in crRNA biogenesis, spacers incorporation and invading DNA cleavage. Based on the sequences and structures of Cas protein, CRISPR/Cas system is primarily classified into three types, Ⅰ, Ⅱ and Ⅲ. The type Ⅱ CRISPR/Cas system only needs a single Cas protein Cas9, which contains a HNH nuclease domain and a RuvC-like nuclease domain. CRISPR/Cas9 has been demonstrated to be a simple and efficient tool for genome editing. CRISPR/Cas9-mediated genome editing depends on the generation of double-strand break (DSB) and subsequent cellular DNA repair process. In endogenous CRISPR/Cas9 system, maturecrRNA is combined with transactivating crRNA (tracr-RNA) to form a tracrRNA : crRNA complex that guides Cas9 to a target site. TracrRNA is partially complementary to crRNA and contributes to crRNA maturation. At the target site, CRISPR/Cas9-mediated sequence-specific cleavage requires a DNA sequence protospacer matching crRNA and a short protospacer adjacent motif (PAM). After binding to the target site, the DNA single-strand matching crRNA and opposite strand are cleaved, respectively, by the HNH nuclease domain and RuvC-like nuclease domain of Cas9, generating a DSB at the target site (Fig. 1. 21) . For easy application in genome editing, researchers designed a delicate guide RNA (gRNA), which was a chimeric RNA con-

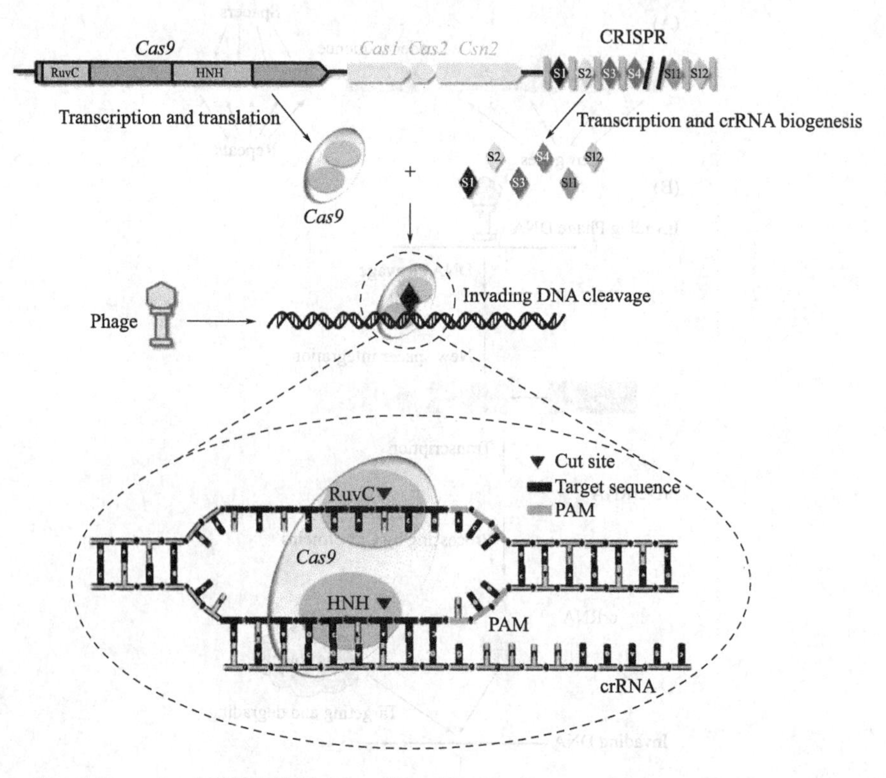

Fig. 1. 21 Schematic of CRISPR/Cas9-mediated DNA cleavage. Mature crRNA guides Cas9 to the target site of invading phage DNA. The DNA single-strand matching crRNA and opposite strand are cut, respectively, by the HNH nuclease domain and RuvC-like nuclease domain of Cas9, generating a DSB at the target site. The specificity of CRISPR/Cas9-mediated DNA cleavage requires target sequence matching crRNA and a 3 nt PAM locating at downstream of the target sequence.

taining all essential crRNA and tracrRNA components. Multiple CRISPR/Cas9 variants have been developed, recognizing 20 or 24 nt sequences matching engineered gRNA and 2-4 nt PAM sequences at target sites. Therefore, CRISPR/Cas9 can theoretically target a specific DNA sequence with 22-29 nt, which is unique in most genomes. However, recent studies observed that CRISPR/Cas9 had high tolerance to base pair mismatches between gRNA and its complementary target sequence, which was sensitive to the numbers, positions and distribution of mismatches. For instance, the CRISPR/Cas9 of Streptococcus pyogenes appeared to tolerate up to six base pair mismatches at target sites. The DSB generated by CRISPR/Cas9 will trigger cellular DNA repair processes, including nonhomologous end-joining (NHEJ) -mediated error-prone DNA repair and homology directed repair (HDR) -mediated error-free DNA repair. NHEJ-mediated DNA repair can rapidly ligate the DSB but generate small insertion and deletion mutations at target sites. These mutations can help us to disrupt or abolish the function of target genes or genomic elements. For instance, Gratz et al. generated frame-shifting indels at the *yellow* locus of Drosophila genome through CRISPR/Cas9-induced DNA cleavage following by NHEJ-mediated DNA repair. DSB can also initiate HDR-mediated DNA repair, which is more complicated than NHEJ-mediated DNA repair. HDR-mediated error-free DNA repair requires a homology-containing donor DNA sequence as repair template. Through co-injection of Cas9, two gRNA targeting, respectively, the 5′ and 3′ sequences of the *yellow* locus, and a single-strand oligodeoxynucleotide template, Gratz et al. successfully replaced the *yellow* locus with a 50 nt attP recombination site in Drosophila genome. Comparing with ZFNs and TALENs, there are several advantages for CRISPR/Cas9. ZFNs and TALENs are built on protein guided DNA cleavage, which needs complex and time-consuming protein engineering, selection and validation. In contrast, CRISPR/Cas9 only needs a short programmable gRNA for DNA targeting, which is relative cheap and easy to design and produce. Through using Cas9 and several gRNA with different target sites, CRISPR/Cas9 is able to simultaneously induce genomic modifications at multiple independent sites. This technology can accelerate the generation of transgenic animals with multiple gene mutations, and disrupt multiple genes or a whole gene family to investigate gene function and epistatic relationships.

Applications

Genome editing

CRISPR/Cas9 provides a robust and multiplexable genome editing tool, enabling researchers to precisely manipulate specific genomic elements, and facilitating the function elucidation of target genes in biology and diseases. Through co-delivery of plasmids expressing Cas9 and crRNA, CRISPR/Cas9 has been used to induce specific genomic modifications in human cells. Through integrating multiple distinct gRNA with Cas9 in a CRISPR array, CRISPR/Cas9 can simultaneously induce multiple mutations in mammalian genomes. In addition to mammalian genomes, CRISPR/Cas9 also demonstrates its potentiality in the genome editing of zebrafish, mice, drosophila, caenorhabditis elegans, Bombyx mori and bacteria. For instance, Bassett et al. provided an improved RNA injection-based CRISPR/Cas9 system, which was highly efficient for creating desired mutagenesis in Drosophila genome. Through directly injecting Cas9 mRNA and gRNA into embryo, they successfully induced mutagenesis at target sites in up to 88% of injected flies. The generated mutations were stably transmitted to 33% of total offspring through the germline. CRISPR/Cas9 is also used to induce desired genomic alterations in plants for generating specific traits, such as valuable phenotypes or disease resistance. To validate the application of CRISPR/Cas9 in plants, Jiang et al. transferred green fluorescence protein gene into Arabidopsis and tobacco genomes, and bacterial blight susceptibility genes into rice genome. Miao et al. illustrated the robustness and efficiency of CRISPR/Cas9 in the genome editing of rice. Through modification of crop genomes, CRISPR/Cas9 can be used to improve crop quality as a new breeding technique in future.

Transcription regulation

Gene transcription regulation in living organisms is very useful for gene function and transcriptional network studies. Through disrupting transcription-related functional sites, CRISPR/Cas9 can regulate the transcription of specific genes. However, this process is irreversible due to permanent DNA modifications. Recently, a modified CRISPR/Cas9 system named CRISPR inference (CRISPRi) is develped for RNA-guided transcription regulation. Qi etal. generated a catalytically defective Cas9 (dCas9) mutant without nucleases activity. dCas9 was co-expressed with gRNA to form a recognition complex, which could interfere with transcriptional elongation, RNA polymerase and transcription factor binding. With two gRNA targeting, respectively, a red fluorescent protein (RFP) gene and a green fluorescent protein (GFP) gene, Qi et al. observed that CRISPRi could simultaneously repress the expression of RFP and GFP without crosstalk in *Escherichia coli*. However, the degree of gene expression repression achieved by CRISPRi was modest in mammalian cells. Gilbert et al. fused repressive or activating effector domains to dCas9, which together with gRNA could implement precise and stable transcriptional control of target genes, including transcription repression and activation. Chen et al. illustrated the performance of CRISPRi for individually or simultaneously regulating the transcription of multiple genes. CRISPRi provides a novel highly specific tool for switching gene expression without genetically altering target DNA sequence.

Gene therapy

Precisely genome editing has the potential to permanently cure diseases through disrupting endogenous disease-causing genes, correcting disease-causing mutations or inserting new protective genes. Using ZFNs-induced HDR, Urnov et al. corrected disease-causing gene mutation in human cell for the first time. Subsequently, ZFNs were used to correct the gene mutations causing sickle-cell disease and hemophilia B. Through disabling virulence genes or inserting protective genes, ZFNs have been used to induce resistance to virus infection in human cells and enhance the efficiency of immunotherapies. As the newest engineered nucleases, CRISPR/Cas9 provides a novel highly efficient genome editing tool for gene therapy studies. For instance, Ebina et al. disrupted the long-terminal repeat promoter of HIV-1 genome using CRISPR/Cas9, which significantly decreased HIV-1 expression in infected human cells. The integrated proviral viral genes in host cell genomes can also be removed by CRISPR/Cas9. With the rapid development of induced pluripotent stem (iPS) cells technology, engineered nucleases are applied to genome manipulation of iPS cells. The unlimited self-renewing and multipotential differentiation capacity of iPS cells make them very useful in disease modeling and gene therapy. Using CRISPR/Cas9, Horri et al. created an iPS cell model for immunodeficiency, centromeric region instability, facial anomalies syndrome (ICF) causing by DNMT3B gene mutation. In this study, iPS cells were transfected with plasmids expressing Cas9 and gRNA, which disrupted the function of DNMT3B in transfected iPS cells. Using the same hPSC lines and delivery method, Ding et al. compared the efficiencies of CRISPR/Cas9 and TALENs for genome editing of iPS cells. They observed that CRISPR/Cas9 was more efficient than TALENs. However, it is still a long road to clinically applying CRISPR/Cas9 for gene therapy. We must ensure the high specificity of CRISPR/Cas9 for target sites and eliminate possible off target mutations with negative effects. Careful selection of target sites, delicate gRNA design and genome-wide search of potential off-target sites are mostly required.

Challenges

Despite the great potential of CRISPR/Cas9 in genome editing, there are some important issues that need to be addressed, such as off-target mutations, PAM dependence, gRNA production and delivery methods of CRISPR/Cas9.

Off-target mutations

Off-target mutations are one major concern about CRISPR/Cas9-mediated genome editing. Compared with ZFNs and TALENs, CRISPR/Cas9 presents relative high risk of off-target mutations in human cells. Large genomes often contain multiple DNA sequences that are identical or highly homologous to target DNA sequences. Besides target DNA sequences, CRISPR/Cas9 also cleaves these identical or highly homologous DNA sequences, which leads to mutations at undesired sites, called off-target mutations. Off-target mutations can result in cell death or transformation. To reduce the cellular toxicity of CRISPR/Cas9, more and more efforts are paid to eliminate the off-target mutations of CRISPR/Cas9. To ensure the specificity of CRISPR/Cas9, it is better to select the target sites with the fewest off-target sites and mismatches between gRNA and its complementary sequence. Xiao et al. recently developed a flexible searching tool CasOT, which could identify potential off-target sites across whole genomes. The dosage of CRISPR/Cas9 is another factor affecting off-target mutations and should be carefully controlled. Methylation of target DNA sequences appeared not to affect the specificity of CRISPR/Cas9. Additionally, converting Cas9 into nickase can help to reduce off-target mutations, while maintaining the efficiency of on-target cleavage implemented by CRISPR/Cas9.

PAM dependence

Theoretically, CRISPR/Cas9 can be applied to any DNA sequence through engineered programmable gRNA. However, the specificity of CRISPR/Cas9 requires a 2-5 nt PAM sequence locating at immediately downstream of the target sequence, besides gRNA/target sequence complementarity. The identified PAM sequences vary among different Cas9 orthologs, such as NGG PAM from Streptococcus pyogenes, NGGNG and NNAGAAW PAM from Streptococcus thermophiles and NNNNGATT PAM from Neisseria meningitidis. Recently, Hsu et al. reported a NAG PAM, which had only ~20% efficiency of NGG PAM for guiding DNA cleavage. On the one hand, the PAM dependent manner of CRISPR/Cas9-mediated DNA cleavage constrains the frequencies of targetable sites in genomes. For instance, it is possible to find a target site per 8 nt for NGG PAM and NAG PAM, while per 32 and 256 nt for NGGNG PAM and NNAGAAW PAM. On the other hand, PAM dependence also increases the specificity of CRISPR/Cas9. The off-target mutations of CRISPR/Cas9 requiring long PAM should be less than that of CRISPR/Cas9 requiring short PAM.

gRNA production

gRNA production is another important issue for CRISPR/Cas9-mediated genome editing. Due to extensive posttranscriptional processing and modification of mRNA transcribed by RNA polymerase Ⅱ, it is currently difficult to apply RNA polymerase Ⅱ for gRNA production. RNA polymerase Ⅲ, U3 and U6 snRNA promoters are currently used to produce gRNA in vivo. However, U3 and U6 snRNA genes are ubiquitously expressed housekeeping genes, which cannot be used to generate tissue and cell-specific gRNA. The lack of commercially available RNA polymerase Ⅲ also limits the application of U3- and U6-based gRNA production. Gao et al. designed an artificial gene RGR, the transcribed mRNA of which contained desired gRNA and ribozyme sequences at both ends of gRNA. After self-catalyzed cleavage, mature gRNA were produced and successfully induced sequence-specific cleavage in vitro and in yeast.

Delivery methods

Questions also remain regarding the delivery methods of CRISPR/Cas9 into organisms. DNA and RNA injection-based techniques are used for CRISPR/Cas9 delivery, such as injection of plasmids expressing Cas9 and gRNA and injection of CRISPR components as RNA. The efficiencies of delivery methods depend on the types of target cells and tissues. More attentions should be paid to develop novel robust delivery methods for CRISPR/Cas9.

Conclusion

Genome editing is initially applied to Drosophila melanogaster, and rapidly extends to a broad range of organisms. An ideal genome editing tool should have simple, efficient and low-cost assembly of nucleases that can target any site without off-target mutations in genomes. CRISPR/Cas9 has the potential to become a reliable and facile genome editing tool, after addressing some issues. Benefiting from the simplicity and adaptability of CRISPR/Cas9, it opens the door for revealing gene function in biology and correcting gene defects in diseases. Further studies are necessary to explore the characteristic and improve the performance of CRISPR/Cas9, especially the specificity, off-target effects and delivery methods of CRISPR/Cas9. For instance, recent genome-wide deeply sequencing results will be helpful for selecting suitable target sites and designing highly specific gRNA.

Glossary

abolish [ə'bɑːliʃ] vt. 消灭，撤销，废除

adaptability [əˌdæptə'biləti] n. 适应性，合用性

Arabidopsis [əræ'bidɔpsɪs] n. 拟南芥

archaea [ɑ'kiə˞] n. 古生菌

biogenesis [ˌbɑiou'dʒenisis] n. 生源论，生物合成

Bombyx mori n. [医] 家蚕：广泛用于实验遗传学的商品蚕

breeding ['bridɪŋ] n. 生育，(动物的) 饲养，教养 (尤指行为或礼貌方面)；v. 繁殖，生育 (breed 的现在分词)，孕育，导致

Caenorhabditis [siːnɔːhæb'daitis] n. [医] 新杆状线虫

centromeric adj. 着丝粒的

chimeric [kai'merɪk] adj. 嵌合的

cleavage ['klividʒ] n. 分裂，乳沟，[胚] 卵裂

clustered ['klʌstərd] adj. 丛生的，群集的，成群的

commercially [kə'məːʃəli] adv. 商业上，通商上

disrupt [dis'rʌpt] vt. 破坏，使瓦解，使中断；adj. 混乱的，分裂的，中断的

Drosophila [drə'sɑːfələ] n. [医] 果蝇属

elucidate [i'lusiˌdet] vt. 阐明，解释

endogenous [en'dɑːdʒənəs] adj. 内源性的，内生的

endonuclease [ˌendou'njuːkliːˌeis] n. 核酸内切酶

epistatic［epə'stætik］ n. 上位的，强性的
fluorescence［flʊ'rɛsəns，flɔ-，flo-］ n. 荧光，荧光性
germline 种系
homologous［hoʊ'mɑːləgəs］ adj. 同源的，类似的，一致的
investigate［ɪn'vɛstɪˌget］ vt. 调查，研究，审查；vi. 作调查
loci［'loʊsai］ n. 场所，所在地，轨迹（locus 的复数）
manipulate［mə'nipjəˌlet］ vt. 操纵，处理，调整
methylation［meθi'leiʃn］ n. 甲基化作用
multiplexable adj. 多路共用的
mutagenesis［ˌmjuːtə'dʒenəsis］ n. 诱变，突变形成，变异发生
Neisseria meningitidis 脑膜炎奈瑟菌
oligodeoxynucleotide［ɒlɪgoudiːɔksin'juːkliːəʊtaid］ n.［医］寡聚脱氧核苷酸
organism［'ɔrgənizəm］ n. 生物，有机体
orthologs 直系同源，同源序列，同源基因
palindromic［ˌpælin'drɔmik］ adj. 回文，［医］复发的，再发的
pathogenic［'pæθə'dʒenik］ adj. 致病的
phenotype［'finətaip］ n. 外表型，显型
pluripotent［plʊr'ripətənt］ adj. 多能（性）的
programmable［'proʊgræməbl］ adj. 可设计的，可编程的
proviral［ˌproʊ'vairəl］ adj. 前病毒的
simultaneously［saiməl'teniəsli］ adv. 同时地
Streptococcus pyogenes 酿脓链球菌
Streptococcus thermophiles 嗜热链球菌
switching［switʃiŋ］ n. 开关，交换，转换；v. 转换，转变，改变（switch 的现在分词）
tolerate［'tɑːləreit］ vt. 忍受，容忍（不同意或不喜欢的事物）
traits［t'reits］ n. 个性，特征（trait 的名词复数）
Xanthomonas n. 黄［单胞］杆菌属
zebrafish［ziːb'rɑːfiːʃ］ n. 斑马鱼

难句分析

1. Genome editing is built on engineered，programmable and highly specific nucleases，which can induce site specific changes in the genomes of cellular organisms through a sequence-specific DNA-binding domain and a nonspecific DNA cleavage domain.
   译文：基因组编辑是建立在工程、可编程和高特异性的核酸酶，其可以通过序列特异性DNA 结合结构域和一个非特异性的 DNA 裂解域诱导细胞生物的基因组位点特异性变化。
2. The DSB generated by CRISPR/Cas9 will trigger cellular DNA repair processes，including nonhomologous end-joining（NHEJ）-mediated error-prone DNA repair and homology directed repair（HDR）-mediated error-free DNA repair.
   译文：利用 CRISPR/Cas9 产生 DSB 将触发细胞 DNA 修复过程，包括非同源末端连接（NHEJ）介导的易错修复和 DNA 同源性修复（HDR）介导的无差错修复。

（选自：Feng Zhang，Yan Wen，Xiong Guo. Human Molecular Genetics，2014，23（R1）：R40-R46.）

Epigenetics in Humans: An Overview

Introduction

Various definitions of epigenetics have emerged since Waddington first coined the term in the 1940s. Generally, epigenetic refers to chemical alterations to DNA or associated histone proteins that change the structure of chromatin and modulate the readability of genomic regions, but do not involve alterations in the DNA sequence. Importantly, these modifications are heritable and can be stably transmitted through many cell divisions, but can also be reset. The classic epigenetic modifications include DNA methylation, post-translational modifications of histone proteins, silencing of the extra copy of the X chromosome in women, and genomic imprinting. In addition, proteins (e. g., DNA methyltransferases, histone deacetylases and methyltransferases, heterochromatin-associated proteins) and protein complexes [e. g., polycomb group (PcG) proteins] with epigenetic modifying capabilities have been included under the umbrella definition of epigenetics. More recently, with the identification of the RNA interference machinery in the late 1990s and several classes of functional noncoding RNAs [e. g., microRNA (miRNA), small-interfering RNA, long noncoding RNA], a new layer of gene regulation has been added to the definition. In this opinion, we touch upon the basic mechanisms of several epigenetic modifications as well as highlight some of the current literature.

DNA methylation

DNA methylation (DNAm) is a highly stable heritable covalent modification that alters DNA without changing its sequence. It involves the addition of a methyl (CH_3) group to the fifth carbon of a cytosine nucleotide predominantly in the context of a CpG dinucleotide. Generally in normal cells, regions of repetitive DNA are methylated which is proposed to be important for genomic stability. DNAm is responsible for silencing parasitic DNA sequences and the inactive X chromosome, genomic imprinting and tissue-specific, and developmental specific silencing/activation of gene transcription. CpG-rich regions known as CpG islands (CGIs) found at the 5-prime regulatory regions of more than 50% of human genes are generally unmethylated in normal cells, with some germ-line and tissue-specific genes being notable exceptions. Methylation in those regions is generally repressive by directly preventing binding of transcription factors or facilitating the binding of methyl-CpG-binding proteins, which also prevent transcription factor binding. DNAm patterns are cell lineage-specific, established during embryonic development and then maintained throughout adulthood.

A new epigenetic discovery is the hydroxymethylcytosine mark, the genomic distribution and function of which remains to be determined. Hydroxymethylcytosine cannot be distinguished from methylcytosine using enzyme-based techniques to study DNAm or procedures that use bisulfite treatment of DNA. However, affinity-based methods such as MeDIP can distinguish between the two marks depending on the antibody used.

DNA methyltransferases

DNAm is mediated by DNA methyltransferases (DNMT1, DNMT3A and DNMT3B, DNMT3L) in eukaryotes. DNMT1 is responsible for maintaining DNAm patterns during replication. DNMT3A and DNMT3B invoke de-novo methylation, particularly during embryogenesis. DNMTs are overexpressed in many tumor types and may be at least partly responsible for hypermethylation observed in tumor suppressor genes. However, it is becoming increasingly recognized that upregulation of DNMTs is only observed in subsets of patients. For example, in a recent study of 765 colorectal carcinomas, DNMT3B protein was increased in only 15% of cases and, therefore, other mechanisms modulating DNMT activity must exist, such as by splice variants, trans-acting factors that target DNMT mRNA or miRNAs.

Methodological advances for investigating the DNA methylomes

A rapidly evolving field is genome-wide profiling to study the human methylome, described in detail in recent reviews. Microarray-based methylation profiling has been widely used and the most popular methodologies involve either restriction enzyme-based enrichment of methylated or unmethylated DNA, affinity-based enrichment of methylated DNA or bisulfite conversion-based microarray. More recently, next generation sequence-based profiling (NGS) provides single base resolution of methylation. The three major techniques to generate bisulfite sequencing libraries are BS-seq, methyl C-seq, and reduced representation bisulfite sequencing (RRBS), which are described in detailed reviews by Lister and Ecker and Ansorge. MethylC-seq and RRBS have advantages over BS-seq in that they reduce the complexity of the genome prior to sequencing, which facilitates less sequencing and lower costs. However, those methods involve enzymatic cleavage and so are biased toward CpG-rich regions and may, therefore, miss less CpG dense biologically relevant regulatory regions. Targeted approaches for reduced representation such as "padlock" probes do not yet facilitate genome-wide analysis. The most widely used platforms for NGS are the 454 Genome Sequencer FLX instrument (Roche Applied Science, Indianapolis, Indiana, USA), the Illumina (Solexa, Walnut, California, USA) Genome Analyzer, the Applied Biosystems ABI SOLiD system, and the HeliScope Single Molecule Sequencer (Helicos BioSciences, Cambridge, Massachu- setts, USA) . Emerging technologies not yet commercially available include real-time sequencing using SMRT technology (Pacific Biosciences, San Francisco, Califor- nia, USA) and nanospore-based methods that bypass the requirement for bisulfite treatment.

Rapid evolution of NGS has facilitated the first single base-resolution human methylome, published in a landmark paper in 2009 using MethyC-seq. Widespread differences in the distribution of cytosine methylation were observed in two human cell lines: H1 human embryonic stem cells and IMR90 fetal lung fibroblasts. Abundant non-CG DNAm was observed in the embryonic stem cell line, but not in differentiated cells and this article emphasizes the highly dynamic nature of DNAm, previously thought to be much more static. It was suggested that non-CG methylation in stem cells may play a role in the maintenance of pluripotency. Higher throughput and lower cost sequencing technologies and improved bioinformatics tools will open up the field for further comparative analyses of normal and disease methylomes.

DNA methylation and disease

Disruption of DNAm patterns has been observed in a growing number of disease processes, cancer being the most rigorously investigated. The dogma was that gene-specific hypermethylation leads to transcriptional repression, which is generally the case for hypermethylation occurring in promoters. Recently, however, it is recognized that hypermethylation occurring in the body of genes can lead to transcriptional activation.

Cancer cells are characterized by global hypomethylation accompanied by de-novo hypermethylation in CGI associated with genes, which can increase during progression from preneoplastic lesions to metastatic tumors, often leading to silencing of tumor suppressor genes or miRNA genes. The list of tumor suppressor genes silenced by DNAm in neoplasia is ever-expanding and these are too numerous to mention in this review, but an important unanswered question is why particular subsets of CGI become hypermethylated in cancer. One simple explanation could be that there are particular sequences in the genome that are more "susceptible" to becoming methylated. Long-range epigenetic silencing (LRES) mechanisms (discussed in more detail below) also provide a feasible explanation for concordant methylation of groups of loci.

A role for DNAm in many other diseases such as autoimmunity, developmental and neurological disorders, and diseases related to imprinting or X-chromosome inactivation has been less intensively studied, but nevertheless is becoming increasingly accepted. A recent article reported hypomethylation at multiple maternally methylated imprinted regions of imprinted genes in Beckwith-Wiedemann syndrome. DNAm also controls gene dosage reductions during X-chromosome inactivation in women and when disrupted can lead to developmental disorders such as fragile X syndrome.

Environmental effects

During embryogenesis widespread, almost complete CpG demethylation occurs and must then be re-established during early development, which necessitates the availability of nutritionally derived methyl donors like methionine and co-factors like folic acid. Diseases such as coronary artery disease, schizophrenia, and other congenital abnormalities have been associated with inadequate establishment of DNAm due to nutritional deficiency prenatally. Numerous other environmental factors, including stress or exposure to chemicals such as fungicides and pesticides can alter epigenetic components of the genome.

One of the largest groups of environmental factors that humans are exposed to daily is endocrine disrupters that alter hormone production and/or signaling, promoting conditions such as reproductive failure, infertility, or cancer. The distribution of DNAm in the developing embryo is tightly controlled, and disruption of normal methylation patterns by exposure to environmental factors such as endocrine disruptors during that time can result in developmental or transgenerational abnormalities, or adult-onset diseases. Interplay between genetics, the environment, and epigenetics may play a critical role in the pathobiology of diabetes. Recent studies suggest a role for DNAm in the regulation of insulin production in mice and humans, as the insulin promoter is methylated in embryonic murine cells, but demethylated in both mouse and human insulin-producing cells.

Long-range epigenetic silencing

The drive to identify genes that are methylated in cancer has mainly focused on discrete CGI-associated genes, but it is becoming increasingly recognized that epigenetic mechanisms can act over large megabase regions containing multiple genes that are coordinately suppressed. Concordant DNAm of adjacent CGIs encompassing a large genomic region was first reported in colorectal cancer and since in other cancers and is speculated to be a common phenomenon in malignancies.

Earlier this year, Coolen et al. reported that LRES is common in prostate cancer cells and that regions of LRES and loss of heterozygosity (LOH) overlap significantly, although the mechanistic link between the two phenomena is unclear. Global deacetylation was accompanied by combinations of repressive marks, including DNAm. LRES is much more abundant in cancer than in normal cells and leads to a major reduction in the accessible genome potentially available for normal transcriptional regulation. Another recent report identified an 800 kb region spanning more than 50 transcripts, encompassing three clusters of protocadherin genes, on chromosome 5q31. 3 that is hypermethylated in Wilms' tumors and was associated with transcriptional silencing.

One popular hypothesis for LRES is the "silencing and seeding" theory, based on the hypothesis that active transcription protects CGI-associated genes from de-novo DNAm, loss of which allows methylation to spread from adjacent loci. This introduces the concept of CGI "shores": low CpG density regions located close (within 2 kb) to CGI, which exhibit high tissue-specific DNAm. These regions have been called tissue or cancer differentially methylated regions (T-DMRs and C-DMRs) in normal cells and cancer cells, respectively. Hypermethylation involving C-DMRs in colon cancer extended into a nearby CGI in 24% of cases, which may partly explain CGI hypermethylation in cancer. This group used the same approach to discover CGI "shores" that are differentially methylated between human-induced pluripotent cells, embryonic stem cells, and fibroblasts.

Epigenetic control at the local level and across large genomic regions has mostly been studied at a one-dimensional level. However, there is a paradigm shift toward three-dimensional genome regulation. Recently, Hsu et al. reported epigenetic repression of large chromosomal regions through DNA looping in response to estrogen stimulation. Fourteen distinct loci were coordinately repressed by recruitment of H3K27me3 and DNAm.

Integration of linear and three-dimensional genomic and epigenomic studies using methods such as chromosome conformation capture (3C) and recently developed Hi-C has great potential to provide a much more comprehensive understanding of epigenetic control in normal tissues and reprogramming in carcinogenesis and other diseases.

Histone modifications

The functional unit of chromatin in eukaryotes, the nucleosome, is composed of 147 bp of DNA wrapped around an octamer of core histone proteins [two units each of histone 2A (H2A), H2B, H3, and H4]. Histones are small proteins (11-17 kD) with an overall positive charge that have affinity for the negatively charged DNA. A linker histone (H1 or H5) associates with the entry and exit of the DNA from the nucleosome. Covalent posttranslational modifications of the histone tails (e. g. , acetylation, methylation, ubiquitylation, and phosphorylation) change the structure and function of chromatin by modifying the interactions between these proteins and DNA. Whereas histone acetylation and phosphorylation are always associated with a transcriptionally active state, histone methylation and ubiquitylation may convey a repressive or activating signal depending on the residue modified. Recent informatic studies suggest that histone modifications act in a combinatorial and cooperative fashion to dictate gene activity. An example of this cross-talk among modifications has been observed by Zippo et al. They show that H3S10p at the FOSL1 enhancer induces H4K16ac within the enhancer and this initiates several protein interactions that ultimately results in the release of the promoter-stalled RNA polymerase Ⅱ (Pol Ⅱ) as well as increases its processivity. Further, a relationship exists between histone modifications and DNAm and it has become apparent that they may be dependent on each other. For example, recognition of an unmethylated lysine on H3 (H3K4) by DNMT3L is necessary prior to Dnmt3a recruitment and de-novo methylation. Additionally, cross-talk among epigenetic modifications is necessary for proper chromatin structure and nuclear organization.

Polycomb group proteins

First described in *Drosophila melanogaster* 50 years ago, PcG proteins are essential for normal development in multicellular organisms and target hundreds of developmentally important genes, impacting the epigenetic landscape. These most recent reviews describe polycomb repressive complexes (PRC1 and PRC2) and their roles in chromatin remodeling and transcriptional regulation.

Disruption of the epigenome caused by deregulated expression and/or binding of PcG proteins is involved in cellular transformation in cancer. PcG proteins were also shown to regulate expression of Ink4a/Arf, which may play a role in regeneration of pancreatic islet β-cells, which has implications for diabetes pathogenesis, although those studies were performed in mice.

Deciphering the mechanisms for recruitment of PRCs to their target genes has been somewhat elusive in humans, even though polycomb response elements (PREs) had been identified in *Drosophila*. Earlier this year, a putative PRE was identified between HOXD11 and HOXD12 in embryonic stem cells to which PRC1 and PRC2 were recruited. Alternative mechanisms for recruiting PRCs may involve noncoding RNAs and PcG-associated proteins. Recently, a new class of approximately 3300 RNAs called large intergenic noncoding (linc) RNAs was described, 20% of which are associated with PRC2. Furthermore, PRC2 target genes were reactivated by knockdown of their associated lincRNAs. There are a number of proteins that have been reported to associate with PcGs, but four very recent articles describe the association of PRC2 with the Jumonji C-containing protein JARID2. Evidence from those studies implicates a role for JARID2 in PcG recruitment, but that other factors such as ncRNAs and/or additional proteins are also required.

miRNA

Organization, biogenesis, and function

miRNA is a class of noncoding RNAs that are transcribed primarily by Pol Ⅱ. The long transcript is 3-prime polyadenylated and 5-prime capped. This primary transcript forms one or several stable hairpins due to base complementarity within the sequence. Upon transcription, the primary transcript is processed in the nucleus by the RNAse Ⅲ enzyme Drosha, which associates with the hairpin together with DGCR8 in a complex referred to as the microprocessor complex. DGCR8 binds to double-stranded RNA and directs Drosha to cleave the hairpin 11 bp away from the double-stranded/single-stranded junction. Cleavage of the primary miRNA generates the precursor (pre) miRNA, which is approximately 70 nucleotides in length with a 3' OH overhang. The 3' overhang is recognized by a complex formed by exportin5 and RanGTP, which shields the double-stranded RNA from degradation and catalyzes the export of the premiRNA from the nucleus. Once in the cytoplasm, another RNAse Ⅲ enzyme, Dicer, associates with the premiRNA and cleaves it to generate the mature approximately 22 nt miRNA duplex. The miRNA then associates with the protein Argonaute 2 to form the RNA-induced silencing complex (RISC). One strand of the duplex is retained and used by the RISC complex to target one or several mRNAs, most often at their 3' UTRs, leading to downregulation of target mRNAs by translational repression (imperfect complementarity of miRNA and target mRNA) or degradation (perfect complementarity); the other strand is targeted for degradation. The current miRNA count is nearing a thousand in humans (miRbase; http://www.mirbase.org/cgi-bin/browse.pl). This estimate comes mostly from computational predictions and only a few have known functions. Approximately, 50% of miRNAs in humans are found in clusters containing two to eight members and encompassed within no more than 50 kb (typically deriving from a single transcript). A functional relationship of clustered miRNAs has been recently documented. In that study, the Human Protein Reference Dataset was used to retrieve information on interacting proteins and by using a protein-centered perspective observed that proteins within a network tended to be regulated by miRNAs in the same cluster. Furthermore, miRNA also seems to regulate cellular pathways by co-targeting several members of a network. To that effect, Tsang et al. assessed the incidence of pathway- specific targeting by specific miRNAs with the use of computational predictions. They found many instances in which several members of a signaling pathway all shared complementarity to a specific miRNA: several of these predicted interactions were corroborated with published experimental results.

Epigenetic regulation by miRNAs

Controversy exists as to whether or not miRNAs should be considered part of the epigenetic program given that the best characterized function of these small RNAs is posttranscriptional gene regulation (e.g., translational repression). However, recent data place these ncRNAs as pivotal regulators of DNAm and histone modifications by directly or indirectly targeting epigenetic modifiers such as histone deacetylases and methyltransferases as well as DNA methyltransferases. The knowledge gained thus far in this area comes from studies using mainly nonhuman models and cellular systems, although some work with human cell lines and primary malignancies has supported findings in other species. It has long been recognized that the RNAi machinery is involved in regulating several epigenetic modifications in lower organisms. In mammals, Dicer deficiency results in decreased DNAm. This was shown to be the result of reduced expression of the DNA methyltransferases. Upon further investigation, this group found increased levels of a negative regulator of the DNMTs, namely, retinoblastoma-like 2 protein (RBL2) . Increased levels of RBL2 were due to a deficiency of miR290. Interestingly, RBL2 is also targeted by nonendogenous miRNAs as is the case with Kaposi's sarcoma-associated herpes virus (KSHV) . During latency, a miRNA that targets RBL2 is transcribed from the KSHV's genome creating an environment with high DNMT expression and hypermethylation of a transcriptional activator involved in viral reactivation.

miRNA and malignancies

Global DNA hypomethylation together with localized hypermethylation of CGIs (particularly that of tumor suppressor genes) is a hallmark of tumorigenesis. Recently, Garzon et al. linked global hypomethylation to a direct action of miR29a on the transcripts of DNMT3A and 3B, whereas DNMT1 levels were down-regulated indirectly by way of downregulation of the transactivator SP1. Beyond being hypermethylated in their promoter regions, the aberrantly silenced tumor-suppressor genes are often characterized by the presence of high levels of silencing histone modifications. The histone methyltransferase and PcG protein EZH2 is overexpressed in several types of cancer. EZH2 catalyzes the trimethylation of H3K27 and its mRNA is targeted by miR101. Two recent studies showed an inverse correlation between EZH2 and miR101 in human prostate tumors. Yet another characteristic of some cancers is histone deacetylase (HDAC) overexpression that promotes cell proliferation and survival. Noonan et al. found that miR449a is down-regulated in prostate cancer tissues. Using a 3'-UTR luciferase assay, they established that HDAC-1 is targeted and regulated by miR449a and they propose this miRNA deficiency in cancer cells to be a mechanism by which HDAC becomes overexpressed in human cancers.

Conclusion

The field of epigenetics is rapidly changing and ever-expanding. It has become evident that chromatin is not a static entity, but rather very dynamic. We now know that the different epigenetic modifications, the proteins and ncRNAs involved in orchestrating the acquisition and/or removal of these signals are tissue and cell-specific, genomic region-specific, spatial and temporally regulated, dependent on each other, labile to the environment, affected by health status, and able to be hijacked by exogenous agents making it quite difficult to draw an accurate picture of the epigenome. Due to this, studies in which single epigenetic modifications are determined are becoming less informative. This has led the way for the advent of high throughput technologies, data depositories, and bioinformatic approaches, which are providing the tools necessary to make testable genome-wide predictions.

Glossary

acetylation [əseti'leiʃən] n. 乙酰化作用
autoimmunity [ɔːtə'ʊimjuːniti] n. 自身免疫
bisulfite ['baisəlfit] n. 重亚硫酸盐，亚硫酸氢盐
carcinogenesis [kɑːsinəʊ'dʒenisis] n. 致癌作用
chromatin ['krəʊmətin] n. 核染色质
chromosome ['krəʊməsəʊm] n. 染色体
colorectal carcinoma 结肠癌
coronary artery disease 冠状动脉疾病
co-factor ['kəʊ'fæktə] n. 辅助因子
CpG island CpG 岛（基因组中富含 CpG 的单拷贝非甲基化基因座）
cytoplasm ['saitəʊplæzəm] n. 细胞质
cytosine ['saitəʊsiːn] n. 胞嘧啶
deacetylase [diːsiti'leis] n. 脱乙酰基酶
demethylate [di'meθəleit] vi. 脱甲基，去甲基
de-novo [diː'nəuvəu] n. 从头，再次
diabetes [ˌdaiə'biːtiːz] n. 糖尿病
dinucleotide [dai'njuːkliətaid] n. 二核苷酸
Drosophila [drə'sɔfilə] n. 果蝇
embryogenesis [ˌembriəʊ'dʒenəsis] n. 胚胎发育，胚胎形成，胚胎发生
endocrine ['endəukrin] n. 内分泌，内分泌腺；adj. 内分泌的
epigenetics [ˌepidʒ'netiks] n. 表观遗传学
epigenome ['epaidʒənəm] n. 表观基因组
estrogen ['iːstrədʒən] n. 雌激素
eukaryote [ju'kæriəʊt] n. 真核细胞
fibroblast ['faibrəblæst] n. 纤维组织母细胞，成纤维细胞
germ-line 种系
heterochromatin [ˌhetərə'krəʊmətin] n. 异染色质
heterozygosity [hetərəuzai'gɔsəti] n. 杂合性，异型结合性
histone ['histəʊn] n. 组蛋白
hydroxymethylcytosine [haidrɔksimeθil'sitəʊzin] n. 羟甲基胞嘧啶
hypermethylation [haipəmeθil'leiʃən] n. 超甲基化
luciferase [luː'sifəreis] n. 荧光素酶
mammal ['mæml] n. 哺乳动物
metastasis [mə'tæstəsis] n. 转移，转移瘤（形容词 metastatic）
methionine [me'θaieniːn] n. 甲硫氨酸，蛋氨酸
methylation [meθi'leiʃn] n. 甲基化，甲基化作用
methylcytosine [meθil'sitəʊzin] n. 甲基胞嘧啶
methylome 甲基化组
methyltransferase [meθil'trænsfəreis] n. 甲基转移酶
methyl ['meθil] n. 甲基
microarray [ˌmaikrəʊə'rei] n. 微矩阵
neoplasia [ˌniːəʊ'pleiʒə] n. 瘤形成
nucleosome ['njuːkliəsəʊm] n. 核小体
nucleotide ['njuːkliətaɪd] n. 核苷酸
octamer ['ɔktɛɪmə] n. 八聚物

pancreatic islet 胰岛，朗格罕氏岛
pathobiology [pæθəbai'ɔlədʒi] n. 病理学
phosphorylation [fɔsfɔri'leiʃən] n. 磷酸化，磷酸化作用
pluripotency [p'luəripətənsi] n. 多能性
polycomb [pɒ'likəʊm] n. 多梳蛋白
preneoplasia [p'riːniːəʊplziə] n. 肿瘤形成前（形容词 preneoplastic）
protocadherin [p'rəʊtəʊkədhərin] n. 原钙黏蛋白
retinoblastoma [retinəʊblæs'təʊmə] n. 眼癌，视网膜神经胶质瘤
schizophrenia [ˌskitsə'friːniə] n. 精神分裂症
transactivator [træn'zæktiveitə] n. 反式激活因子
ubiquitylation 泛素化
UTR（Untranslated region）非编码区

难句分析

1. DNAm is responsible for silencing parasitic DNA sequences and the inactive X chromosome，genomic imprinting and tissue-specific，and developmental specific silencing/activation of gene transcription.

   译文：DNA 甲基化负责寄生 DNA 序列和失活 X 染色体沉默化、基因组印迹以及组织或发育特异性基因转录的沉默或激活。

2. Microarray-based methylation profiling has been widely used and the most popular methodologies involve either restriction enzyme-based enrichment of methylated or unmethylated DNA，affinity-based enrichment of methylated DNA or bisulfite conversion-based microarray.

   译文：基于微矩阵的甲基化谱分析已经得到广泛应用，其中最流行的方法包括基于限制性内切酶的甲基化或非甲基化 DNA 富集、基于亲和度的甲基化 DNA 富集或基于重亚硫酸盐转化的微矩阵等。

3. Cancer cells are characterized by global hypomethylation accompanied by de-novo hypermethylation in CGI associated with genes，which can increase during progression from preneoplastic lesions to metastatic tumors，often leading to silencing of tumor suppressor genes or miRNA genes.

   译文：癌症细胞具有全面超甲基化的特征并伴随有 CpG 岛相关基因的从头超甲基化，而且这种作用在癌症从癌前病变到转移性肿瘤发展的过程中会进一步增强。

4. The distribution of DNAm in the developing embryo is tightly controlled，and disruption of normal methylation patterns by exposure to environmental factors such as endocrine disruptors during that time can result in developmental or transgenerational abnormalities，or adult-onset diseases.

   译文：DNA 甲基化的分布状态在发育的胚胎中是受到严格控制的，如果在特定时刻暴露于某些环境因素如内分泌干扰物下会扰乱正常的甲基化模式，这将导致发育或隔代异常以及成年期疾病等。

5. One popular hypothesis for LRES is the “silencing and seeding” theory，based on the hypothesis that active transcription protects CGI-associated genes from de-novo DNAm，loss of which allows methylation to spread from adjacent loci.

   译文：一个流行的关于大范围表观遗传沉默（LRES）的假设是“沉默和播种”理论，这也是基于这样一个假设，即有效的基因转录能够保护 CpG 岛相关基因不被从头 DNA 甲

基化，而转录沉默则会使甲基化从邻近的位点蔓延过来。

6. PcG proteins were also shown to regulate expression of Ink4a/Arf，which may play a role in regeneration of pancreatic islet β-cells，which has implications for diabetes pathogenesis，although those studies were performed in mice.

译文：多梳家族（PcG）蛋白也被证实能够调节 lnk4a/Arf 基因的表达，这可能在胰岛 β 细胞的再生中发挥了一定的作用，尽管这些研究是在小鼠中进行的，但仍为糖尿病发病机理的研究提供了启示。

7. The 3’ overhang is recognized by a complex formed by exportin5 and RanGTP，which shields the double-stranded RNA from degradation and catalyzes the export of the premiRNA from the nucleus.

译文：3’突出部分能够被输出蛋白 5（exportin5）和 RanGTP 组成的复合物识别，这个复合物能够保护双链 RNA 不被降解，并且催化前体 miRNA 从细胞核转移出来。

8. However，recent data place these ncRNAs as pivotal regulators of DNAm and histone modifications by directly or indirectly targeting epigenetic modifiers such as histone deacetylases and methyltransferases as well as DNA methyltransferases.

译文：但是，近期一些数据认为这些非编码 RNA 通过直接或间接地作用于表观遗传修饰物，如组蛋白去乙酰化酶和甲基转移酶以及 DNA 甲基转移酶等，成为 DNA 甲基化和组蛋白修饰重要的调节因子。

9. We now know that the different epigenetic modifications，the proteins and ncRNAs involved in orchestrating the acquisition and/or removal of these signals are tissue and cell-specific，genomic region-specific，spatial and temporally regulated，dependent on each other，labile to the environment，affected by health status，and able to be hijacked by exogenous agents making it quite difficult to draw an accurate picture of the epigenome.

译文：我们现在知道，不同的表观遗传修饰，参与协调获得和（或）移除这些信号的蛋白质和非编码 RNA 具有组织和细胞特异性及基因组区域特异性的，并且在时间和空间维度上被调节，它们相互依赖，受环境的不确定性和健康状况的影响，还会被外界的各种试剂所操纵，这些都使得描绘一个精确的表观遗传组变得十分困难。

（选自：Rivera RM，Bennett LB. Epigenetics in humans：an overview. Curr Opin Endocrinol Diabetes Obes. 2010；17（6）：493-9.）

Making New Bodies Mechanisms of Developmental Organization

Between fertilization and birth, the developing organism is known as an embryo. The concept of an embryo is a staggering one. As an embryo, you had to build yourself from a single cell. You had to respire before you had lungs, digest before you had a gut, build bones when you were pulpy, and form orderly arrays of neurons before you knew how to think. One of the critical differences between you and a machine is that a machine is never required to function until after it is built. Every multicellular organism has to function even as it builds itself. Most human embryos die before being born. You survived.

Multicellular organisms do not spring forth fully formed. Rather, they arise by a relatively slow process of progressive change that we call development. In nearly all cases, the development of a multicellular organism begins with a single cell—the fertilized egg, or zygote, which divides mitotically to produce all the cells of the body. The study of animal development has traditionally been called embryology, after that phase of an organism that exists between fertilization and birth. But development does not stop at birth, or even at adulthood. Most organisms never stop developing. Each day we replace more than a gram of skin cells (the older cells being sloughed off as we move), and our bone marrow sustains the development of millions of new red blood cells every minute of our lives. Some animals can regenerate severed parts, and many species undergo metamorphosis (such as the transformation of a tadpole into a frog, or a caterpillar into a butterfly). Therefore, in recent years it has become customary to speak of developmental biology as the discipline that studies embryonic and other developmental processes.

The Cycle of Life

For animals, fungi, and plants, the sole way of getting from egg to adult is by developing an embryo. The embryo is where genotype is translated into phenotype, where inherited genes are expressed to form the adult. The developmental biologist usually finds the transient stages leading up to the adult to be the most interesting. Developmental biology studies the building of organisms. It is a science of becoming, a science of process.

One of the major triumphs of descriptive embryology was the idea of a generalizable animal life cycle. Modern developmental biology investigates the temporal changes of gene expression and anatomical organization along this life cycle. Each animal, whether earthworm or eagle, termite or beagle, passes through similar stages of development: fertilization, cleavage, gastrulation, organogenesis, birth, metamorphosis, and gametogenesis. The stages of development between fertilization and hatching (or birth) are collectively called embryogenesis.

1. Fertilization involves the fusion of the mature sex cells, the sperm and egg, which are collectively called the gametes. The fusion of the gamete cells stimulates the egg to begin development and initiates a new individual. The subsequent fusion of the gamete nuclei (the male and female pronuclei, each of which has only half the normal number of chromosomes

characteristic for the species) gives the embryo its genome, the collection of genes that helps instruct the embryo to develop in a manner very similar to that of its parents.

2. Cleavage is a series of extremely rapid mitotic divisions that immediately follow fertilization. During cleavage, the enormous volume of zygote cytoplasm is divided into numerous smaller cells called blastomeres. By the end of cleavage, the blastomeres have usually formed a sphere, known as a blastula.

3. After the rate of mitotic division slows down, the blastomeres undergo dramatic movements and change their positions relative to one another. This series of extensive cell rearrangements is called gastrulation, and the embryo is said to be in the gastrula stage. As a result of gastrulation, the embryo contains three germ layers (endoderm, ectoderm, and mesoderm) that will interact to generate the organs of the body.

4. Once the germ layers are established, the cells interact with one another and rearrange themselves to produce tissues and organs. This process is called organogenesis. Chemical signals are exchanged between the cells of the germ layers, resulting in the formation of specific organs at specific sites. Certain cells will undergo long migrations from their place of origin to their final location. These migrating cells include the precursors of blood cells, lymph cells, pigment cells, and gametes (eggs and sperm).

5. In many species, the organism that hatches from the egg or is born into the world is not sexually mature. Rather, the organism needs to undergo metamorphosis to become a sexually mature adult. In most animals, the young organism is a called a larva, and it may look significantly different from the adult. In many species, the larval stage is the one that lasts the longest, and is used for feeding or dispersal. In such species, the adult is a brief stage whose sole purpose is to reproduce. In silkworm moths, for instance, the adults do not have mouthparts and cannot feed; the larva must eat enough so that the adult has the stored energy to survive and mate. Indeed, most female moths mate as soon as they enclose from the pupa, and they fly only once—to lay their eggs. Then they die.

6. In many species, a group of cells is set aside to produce the next generation (rather than forming the current embryo). These cells are the precursors of the gametes. The gametes and their precursor cells are collectively called germ cells, and they are set aside for reproductive function. All other cells of the body are called somatic cells. This separation of somatic cells (which give rise to the individual body) and germ cells (which contribute to the formation of a new generation) is often one of the first differentiations to occur during animal development. The germ cells eventually migrate to the gonads, where they differentiate into gametes. The development of gametes, called gametogenesis, is usually not completed until the organism has become physically mature. At maturity, the gametes may be released and participate in fertilization to begin a new embryo. The adult organism eventually undergoes senescence and dies, its nutrients often supporting the early embryogenesis of its offspring and its absence allowing less competition. Thus, the cycle of life is renewed.

Patterns of cleavage

E. B. Wilson, one of the pioneers in applying cell biology to embryology, noted in 1923, "To our limited intelligence, it would seem a simple task to divide a nucleus into equal parts. The cell, manifestly, entertains a very different opinion." Indeed, different organisms undergo cleavage in distinctly different ways, and the mechanisms for these differences remain at the frontier of cell and developmental biology. Cells in the cleavage-stage cells are called blastomeres. In most species (mammals being the chief exception), both the initial rate of cell division and the placement of the blastomeres with respect to one another are under the control of proteins and mRNAs stored in the oocyte. Only later do the rates of cell division and the placement of cells come under the control of the newly formed organism's own genome. During the initial phase of development, when cleavage rhythms are controlled by maternal factors, cytoplasmic volume does not increase. Rather, the zygote cytoplasm is divided into increasingly smaller cells—first in half, then quarters, then eighths, and so forth. Cleavage occurs very rapidly in most invertebrates, probably as an adaptation to generate a large number of cells quickly and to restore the somatic ratio of nuclear volume to cytoplasmic volume. The embryo often accomplishes this by abolishing the gap periods of the cell cycle (the G1 and G2 phases), when growth can occur. A frog egg, for example, can divide into 37000 cells in just 43 hours. Mitosis in cleavage-stage Drosophila embryos occurs every 10 minutes for more than 2 hours, forming some 50000 cells in just 12 hours.

The pattern of embryonic cleavage peculiar to a species is determined by two major parameters: (1) the amount and distribution of yolk protein within the cytoplasm, which determine where cleavage can occur and the relative sizes of the blastomeres; and (2) factors in the egg cytoplasm that influence the angle of the mitotic spindle and the timing of its formation.

In general, yolk inhibits cleavage. When one pole of the egg is relatively yolk-free, cellular divisions occur there at a faster rate than at the opposite pole. The yolk-rich pole is referred to as the vegetal pole; the yolk concentration in the animal pole is relatively low. The zygote nucleus is frequently displaced toward the animal pole. At one extreme are the eggs of sea urchins, mammals, and snails. These eggs have sparse, equally distributed yolk and are thus isolecithal (Greek, "equal yolk") . In these species, cleavage is holoblastic (Greek holos, "complete"), meaning that the cleavage furrow extends through the entire egg. With little yolk, these embryos must have some other way of obtaining food. Most will generate a voracious larval form, while mammals will obtain their nutrition from the maternal placenta.

At the other extreme are the eggs of insects, fish, reptiles, and birds. Most of their cell volumes are made up of yolk. The yolk must be sufficient to nourish these animals throughout embryonic development. Zygotes containing large accumulations of yolk undergo meroblastic cleavage (Greek meros, "part"), wherein only a portion of the cytoplasm is cleaved. The cleavage furrow does not penetrate the yolky portion of the cytoplasm because the yolk platelets impede membrane formation there. Insect eggs have yolk in the center (i. e. , they are centrolecithal), and the divisions of the cytoplasm occur only in the rim of cytoplasm, around the periphery of the cell (i. e. , superficial cleavage) . The eggs of birds and fish have only one small area of the egg that is free of yolk (telolecithal eggs), and therefore the cell divisions occur only in this small disc of cytoplasm, giving rise to discoidal cleavage. These are general rules, however, and even closely related species have evolved different patterns of cleavage in different environments.

Yolk is just one factor influencing a species' pattern of cleavage. There are also, as Conklin had intuited, inherited patterns of cell division superimposed on the constraints of the yolk. The importance of this inheritance can readily be seen in isolecithal eggs. In the absence of a large concentration of yolk, holoblastic cleavage takes place. Four major patterns of this cleavage type can be described: radial, spiral, bilateral, and rotational holoblastic cleavage.

The primary germ layers and early organs

The end of preformationism did not come until the 1820s, when a combination of new staining techniques, improved microscopes, and institutional reforms in German universities created a revolution in descriptive embryology. The new techniques enabled microscopists to document the epigenesis of anatomical structures, and the institutional reforms provided audiences for these reports and students to carry on the work of their teachers. The work of Christian Pander, Heinrich Rathke, and Karl Ernst von Baer transformed embryology into a specialized branch of science.

Studying the chick embryo, Pander discovered that the embryo was organized into germ layers—three distinct regions of the embryo that give rise through epigenesis to the differentiated cells types and specific organ systems (Fig. 1. 22). These three layers are found in the embryos of most animal phyla:

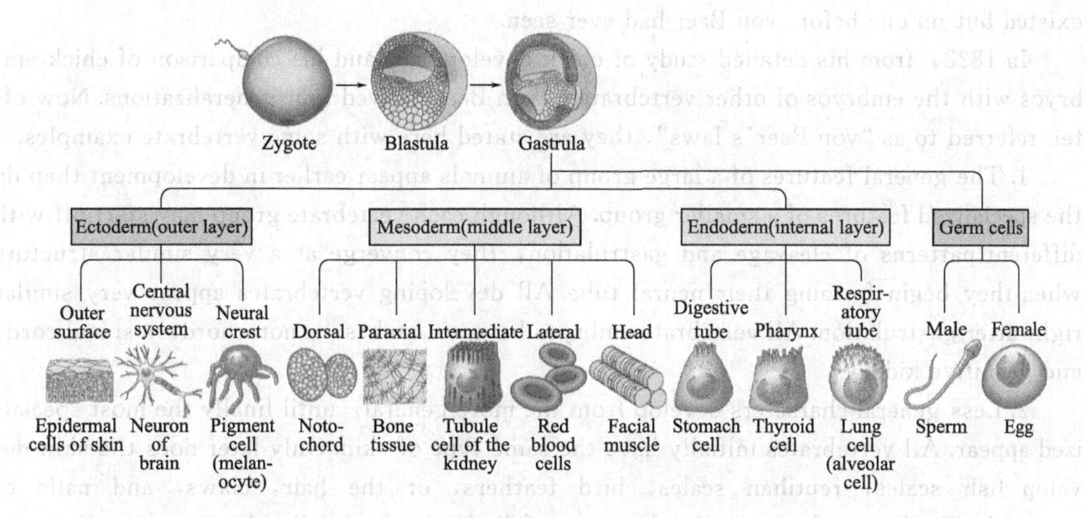

Fig. 1. 22 The dividing cells of the fertilized egg form three distinct embryonic germ layers. Each of the germ layers gives rise to myriad differentiated cell types (only a few representatives are shown here) and distinct organ systems. The germ cells (precursors of the sperm and egg) are set aside early in development and do not arise from any particular germ layer.

1. The ectoderm generates the outer layer of the embryo. It produces the surface layer (epidermis) of the skin and forms the brain and nervous system.

2. The endoderm becomes the innermost layer of the embryo and produces the epithelium of the digestive tube and its associated organs (including the lungs).

3. The mesoderm becomes sandwiched between the ectoderm and endoderm. It generates the blood, heart, kidney, gonads, bones, muscles, and connective tissues.

Pander also demonstrated that the germ layers did not form their respective organs autonomously. Rather, each germ layer "is not yet independent enough to indicate what it truly is; it still needs the help of its sister travelers, and therefore, although already designated for different ends, all three influence each other collectively until each has reached an appropriate level". Pander had discovered the tissue interactions that we now call induction. No vertebrate tissue is able to construct organs by itself, it must interact with other tissues.

Meanwhile, Rathke followed the intricate development of the vertebrate skull, excretory systems, and respiratory systems, showing that these became increasingly complex. He also showed that their complexity took on different trajectories in different classes of vertebrates. For instance, Rathke was the first to identify the pharyngeal arches. He showed that these same embryonic structures became gill supports in fish and the jaws and ears (among other things) in mammals.

The four principles of Karl Ernst von Baer

Karl Ernst von Baer extended Pander's studies of the chick embryo. He recognized that there is a common pattern to all vertebrate development—that each of the three germ layers generally gives rise to the same organs, whether the organism is a fish, a frog, or a chick. He discovered the notochord, the rod of mesoderm that separates the embryo into right and left halves and instructs the ectoderm above it to become the nervous system. He also discovered the mammalian egg, that minuscule, long-sought cell that everyone believed existed but no one before von Baer had ever seen.

In 1828, from his detailed study of chick development and his comparison of chick embryos with the embryos of other vertebrates, von Baer derived four generalizations. Now often referred to as "von Baer's laws", they are stated here with some vertebrate examples.

1. The general features of a large group of animals appear earlier in development than do the specialized features of a smaller group. Although each vertebrate group may start off with different patterns of cleavage and gastrulation, they converge at a very similar structure when they begin forming their neural tube. All developing vertebrates appear very similar right after gastrulation. All vertebrate embryos have gill arches, a notochord, a spinal cord, and primitive kidneys.

2. Less general characters develop from the more general, until finally the most specialized appear. All vertebrates initially have the same type of skin. Only later does the skin develop fish scales, reptilian scales, bird feathers, or the hair, claws, and nails of mammals. Similarly, the early development of limbs is essentially the same in all vertebrates. Only later do the differences between legs, wings, and arms become apparent.

3. The embryo of a given species, instead of passing through the adult stages of lower animals, departs more and more from them. The pharyngeal arches start off the same in all vertebrates. But the arch that becomes the jaw support in fish becomes part of the skull of reptiles and becomes part of the middle ear bones of mammals. Mammals never go through a fishlike stage.

4. Therefore, the early embryo of a higher animal is never like a lower animal, but only like its early embryo. Human embryos never pass through a stage equivalent to an adult fish or bird. Rather, human embryos initially share characteristics in common with fish and avian embryos. Later in development, the mammalian and other embryos diverge, none of them passing through the stages of the others.

Recent research has confirmed von Baer's view that there is a "phylotypic stage" at which the embryos of the different phyla of vertebrates all have a similar physical structure. At this same stage there appears to be the least amount of difference among the *genes* expressed by the different groups within the same vertebrate phylum.

Glossary

anatomical [ænə'tɔmik(ə)l] adj. 解剖的，解剖学的，结构上的
beagle ['bi:gl] n. 猎兔犬
blastomere ['blæstə(ʊ)miə] n. 卵裂球，胚叶细胞
blastula ['blæstjʊlə] n. 囊胚
bone marrow n. 骨髓
caterpillar ['kætəpilə] n. 毛虫
centrolecithal [ˌsentrə'lesiθəl] adj. 中（央卵）黄的
cytoplasm ['saɪtə(ʊ)plæz(ə)m] n. 细胞质
earthworm ['ə:θwɜ:m] n. ［无脊椎］蚯蚓
embryogenesis [ˌembriə(ʊ)'dʒenisis] n. 胚胎形成，胚胎发生
embryo ['ɛmbrio] n. 胚芽，（生）胚，胚胎，初期
epigenesis [ˌepi'dʒenisis] n. 新生论，［胚］渐成说，外力变质
fertilization [ˌfə:tlə'zeʃən] n. 受精，受精过程，受精行为［现象］，受孕，受胎
fungi ['fʌŋgai] n. 真菌，菌类，蘑菇（fungus 的复数）
gamete ['gæmi:t] n. 配子
gametogenesis [gəˌmi:tə(ʊ)'dʒenisis] n. 配子形成，配子发育
gastrulation [ˌgæstrʊ'leiʃən] n. 原肠形成
germ layer n. 胚芽层
gill [gil] n. 鳃；菌褶
gonad ['gəʊnæd] n. 性腺
holoblastic [ˌhɔləʊ'blæstik] adj. 全裂的（指某些有少卵黄的卵细胞）
isolecithal adj. 等卵黄的
kidney ['kidni] n. 肾
larva ['lɑ:və] n. 幼虫，幼体
lymph [limf] n. 淋巴（液）
mammal ['mæm(ə)l] n. 哺乳动物
marrow ['mærəʊ] n. 髓，骨髓，精华，活力
metamorphosis [ˌmetə'mɔ:fəsis] n. 变态
microscopist [mai'krɔskəpist] n. 显微镜工作者，显微镜学家
mitotic [mai'tɔtik] adj. 有丝分裂的，间接核分裂的
multicellular [mʌlti'seljʊlə] adj. 多细胞的
mouthparts n. 口器
neural tube n. 神经管
notochord ['nəʊtə(ʊ)kɔ:d] n. 脊索
organism ['ɔrgənizəm] n. 有机体，生物体，微生物，有机体系，有机组织
phenotype ['fi:nə(ʊ)taip] n. 表型，表现型，显型
pronucleus [prəʊ'nju:kli:əs] n. 前核，生殖核
silkworm ['silkwɜ:m] n. 蚕
slough [slʌf; slaʊ] n. 蜕下的皮（或壳）；vi. 脱落，蜕皮
somatic cell n. 体细胞
sperm [spə:m] n. 精子

tadpole [ˈtædpəul] n. 蝌蚪
telolecithal [ˌteləuˈlesiθəl] adj. 端卵黄的
termite [ˈtəːmait] n. [昆] 白蚁
vertebrate [ˈvəːtibrət] adj. 脊椎的，脊椎动物的；n. 脊椎动物
zygote [ˈzaigəut] n. 接合子，受精卵

难句分析

1. Each day we replace more than a gram of skin cells (the older cells being sloughed off as we move), and our bone marrow sustains the development of millions of new red blood cells every minute of our lives.
   译文：每天大于1g的皮肤细胞将会被新的细胞替代（老去的细胞会在我们移动的时候脱落），而我们的骨髓每一分钟都能维持数百万个新的红细胞的发育。
2. One of the major triumphs of descriptive embryology was the idea of a generalizable animal life cycle. Modern developmental biology investigates the temporal changes of gene expression and anatomical organization along this life cycle. Each animal, whether earthworm or eagle, termite or beagle, passes through similar stages of development: fertilization, cleavage, gastrulation, organogenesis, birth, metamorphosis, and gametogenesis.
   译文：描述性胚胎学的主要成就之一是可概括出一条普遍适用于动物生命周期的概念。现代发育生物学研究了基因表达和组织结构是如何沿着这条生命周期而变化的。所有动物，无论是蚯蚓还是鹰，白蚁或是猎犬，都会经历类似的发育阶段：受精，卵裂，原肠胚形成，器官形成，出生，变态和配子形成。
3. The subsequent fusion of the gamete nuclei (the male and female pronuclei, each of which has only half the normal number of chromosomes characteristic for the species) gives the embryo its genome, the collection of genes that helps instruct the embryo to develop in a manner very similar to that of its parents.
   译文：配子核（雄性和雌性生殖核，每个生殖核都只有物种正常染色体的一半数量）后续的融合为胚胎提供了基因组，这些基因帮助并指导胚胎以非常类似于父母方式进行发育。
4. Cleavage is a series of extremely rapid mitotic divisions that immediately follow fertilization. During cleavage, the enormous volume of zygote cytoplasm is divided into numerous smaller cells called blastomeres. By the end of cleavage, the blastomeres have usually formed a sphere, known as a blastula.
   译文：卵裂是一系列非常快速的有丝分裂，随即进行受精作用。在卵裂过程中，大量受精卵的细胞质被分裂成许多很小的细胞，称为卵裂球。卵裂结束时，卵裂球通常形成一个球体，称为囊胚。
5. This series of extensive cell rearrangements is called gastrulation, and the embryo is said to be in the gastrula stage. As a result of gastrulation, the embryo contains three germ layers (endoderm, ectoderm, and mesoderm) that will interact to generate the organs of the body.
   译文：这一系列广泛的细胞重排被称为原肠胚形成，这时的胚胎被认为处于原肠胚阶段。原肠胚形成之后，胚胎分为三个胚层（内胚层、外胚层和中胚层），三者相互作用后产生身体的器官。

6. In many species, a group of cells is set aside to produce the next generation (rather than forming the current embryo). These cells are the precursors of the gametes. The gametes and their precursor cells are collectively called germ cells, and they are set aside for reproductive function. All other cells of the body are called somatic cells.
   译文：在许多物种中，一部分细胞被留出来以产生下一代（而不是形成当前的胚胎）。这些细胞是配子的前体。配子和它们的前体细胞统称为生殖细胞，它们具有生殖功能。体内其他细胞称为体细胞。
7. He recognized that there is a common pattern to all vertebrate development that each of the three germ layers generally gives rise to the same organs, whether the organism is a fish, a frog, or a chick.
   译文：他认识到所有脊椎动物的发育共用一种模式，即三个胚层中的每个胚层通常发育成相同的器官，无论生物体是鱼、青蛙还是鸡。
8. Chemical signals are exchanged between the cells of the germ layers, resulting in the formation of specific organs at specific sites. Certain cells will undergo long migrations from their place of origin to their final location. These migrating cells include the precursors of blood cells, lymph cells, pigment cells, and gametes (eggs and sperm)
   译文：化学信号在胚层的细胞之间交换后会在特定部位形成特定的器官。某些细胞从其原始位置到其最终位置会经历很长迁移。这些能够迁移的细胞包括血细胞的前体、淋巴细胞、色素细胞和配子（卵子和精子）。

（选自：Scott F. Gilbert and Michael J. F. Barresi. Developmental Biology, 11th Edition, 2016, Sinauer Associates, Inc.）

Neuropathological Alterations in Alzheimer Disease

Abstract

The neuropathological hallmarks of Alzheimer disease (AD) include "positive" lesions such as amyloid plaques and cerebral amyloid angiopathy, neurofibrillary tangles, and glial responses, and "negative" lesions such as neuronal and synaptic loss. Despite their inherently cross-sectional nature, postmortem studies have enabled the staging of the progression of both amyloid and tangle pathologies, and, consequently, the development of diagnostic criteria that are now used worldwide. In addition, clinicopathological correlation studies have been crucial to generate hypotheses about the pathophysiology of the disease, by establishing that there is a continuum between "normal" aging and AD dementia, and that the amyloid plaque build-up occurs primarily before the onset of cognitive deficits, while neurofibrillary tangles, neuron loss, and particularly synaptic loss, parallel the progression of cognitive decline. Importantly, these cross-sectional neuropathological data have been largely validated by longitudinal in vivo studies using modern imaging biomarkers such as amyloid PET and volumetric MRI.

Introduction

The neuropathological changes of Alzheimer disease (AD) brain include both positive and negative features. Classical positive lesions consist of abundant amyloid plaques and neurofibrillary tangles, neuropil threads, and dystrophic neurites containing hyperphosphorylated tau, that are accompanied by astrogliosis, and microglial cell activation. Congophilic amyloid angiopathy is a frequent concurrent feature. Unique lesions, found primarily in the hippocampal formation, include Hirano bodies and granulovacuolar degeneration. In addition to these positive lesions, characteristic losses of neurons, neuropil, and synaptic elements are core negative features of AD. Each of these lesions has a characteristic distribution, with plaques found throughout the cortical mantle, and tangles primarily in limbic and association cortices. The hierarchical pattern of neurofibrillary degeneration among brain regions is so consistent that a staging scheme based on early lesions in the entorhinal/perirhinal cortex, then hippocampal Ammon subfields, then association cortex, and finally primary neocortex is well accepted as part of the 1997 NIA-Reagan diagnostic criteria. Neuronal loss and synapse loss largely parallel tangle formation, although whether tangles are causative of neuronal loss or synaptic loss remains uncertain.

Although all these neuropathological characteristics are useful diagnostic markers, the cognitive impairment in patients with AD is closely associated with the progressive degeneration of the limbic system, neocortical regions, and the basal forebrain. This neurodegenerative process is characterized by early damage to the synapses with retrograde degeneration of the axons and eventual atrophy of the dendritic tree and perikaryon. Indeed, the loss of synapses in the neocortex and limbic system is the best correlate of the cognitive impairment in patients with AD.

In addition to the lesions detected by classical histopathological stains, including silver stains for tangles and plaques or immunostaining and quantitative analysis (or quantitative EM) for synaptic alterations, several lines of investigation now support the view that increased levels of soluble amyloid-β_{1-42} (Aβ) oligomers, might lead to synaptic damage and neurodegeneration. In experimental models, it has been shown that transsynaptic delivery of Aβ, for example from the entorhinal cortex to the molecular layer of the dentate gyrus, promotes neurodegeneration characterized by synapse loss and alterations to calbindin-positive neurons. This is accompanied by circuitry dysfunction and aberrant innervation of the hippocampus by NPY-positive fibers among others. The Aβ oligomers secreted by cultured neurons inhibit long term potentiation (LTP), damage spines and interfere with activity-regulated cytoskeleton associated protein (Arc) distribution. Together, these studies indicate that Aβ oligomers ranging in size from 2 to 12 subunits might be responsible for the synaptic damage and memory deficits in AD. Similar neurotoxic Aβ oligomers found in vitro and in APP transgenic models have been also identified in the CSF and in the brains of patients with AD. These studies have shown that Aβ oligomers progressively accumulate in the brains of AD patients, although their relationship to the severity of the cognitive impairment remains uncertain.

In summary, in recent years, the concept of neurodegeneration in AD has been expanded from the idea of general neuronal loss and astrogliosis to include earlier alterations such as synaptic and dendritic injury and disturbances in the process of adult neurogenesis, circuitry dysfunction, and aberrant innervation. All of these factors are important targets to consider when developing neuroprotective treatments for AD.

Macroscopic features

Although the gross visual examination of the AD brain is not diagnostic, a typical symmetric pattern of cortical atrophy predominantly affecting the medial temporal lobes and relatively sparing the primary motor, sensory, and visual cortices, is considered strongly suggestive of AD being the condition underlying the patient's dementia. As a result of this pattern of cortical thinning, the lateral ventricules, particularly their temporal horns, can appear prominently dilated. This pattern is stereotypic and can be recognized early in the clinical course of the disease by MRI scan. Cerebrovascular disease, usually in the form of small vessel occlusive disease caused by chronic hypertension and other vascular risk factors, is a condition that frequently accompanies aging in general and also AD in particular. Thus, it is relatively common to find some cortical microinfarcts, lacunar infarcts in the basal ganglia, and demyelination of the periventricular white matter. The presence of cortical petechial microbleeds or even evident lobar hemorrhages, particularly in the posterior parietal and occipital lobes, should lead to the suspicion of a concurrent severe cerebral amyloid angiopathy. Unless there is a concomitant Parkinson's disease or dementia with Lewy bodies, the substantia nigra shows a normal coloration; in contrast, the locus coeruleus is affected in the early stages of AD.

Microscopic features

Neurofibrillary tangles

Composition

The neurofibrillary tangles (NFTs) were first described by Alois Alzheimer in his original autopsy case report as intraneuronal filamentous inclusions within the perikaryal region of pyramidal neurons. Ultrastructural studies on AD brain specimens revealed that NFTs are primarily made of paired helical filaments (PHFs), that is, fibrils of 10 nm in diameter that form pairs with a helical tridimensional conformation at a regular periodicity of 65 nm. A small proportion of fibrils within the NFTs do not form pairs, but give the appearance of straight filaments without the periodicity of PHFs. Occasional hybrid filaments, with a sharp transition between a paired helical segment and a straight segment, have also been described within NFTs. Recently, modern high-resolution molecular microscopy techniques have revealed the presence of twisted ribbon-like assemblies of tau fibrils in vitro, thus challenging the PHF concept. Regardless of the morphology of their structural units, the major constituent of NFTs was found to be the microtubule-associated protein tau, which is aberrantly misfolded and abnormally hyper-phosphorylated. Invariably accompanying NFTs are the neuropil threads, which are thought to result from the breakdown of dendrites and axons of the tangle-bearing neurons.

Morphological characteristics

The NFTs are argyrophilic and can be shown by silver impregnation methods such as the Gallyas technique. An alternative method to examine NFTs is their staining with fluorescent dyes such as Thioflavin-S, which recognize the β-sheet pleated structure of the paired helical filaments, or by immunostaining with anti-tau antibodies (Fig. 1. 23) . Three morphological stages have been distinguished: (1) Pre-NFTs or diffuse NFTs are defined by a diffuse, sometimes punctate, tau staining within the cytoplasm of otherwise normal-looking neurons, with well-preserved dendrites and a centered nucleus; (2) Mature or fibrillar intraneuronal NFTs (iNFTs) consist of cytoplasmic filamentous aggregates of tau that displace the nucleus toward the periphery of the soma and often extend to distorted-appearing dendrites and to the proximal segment of the axon; (3) extra-neuronal "ghost" NFTs (eNFTs) result from the death of the tangle-bearing neurons and are identifiable by the absence of nucleus and stainable cytoplasm. Both silver and Thioflavin-S stains, as well as some phosphotau antibodies such as AT8 and PHF1, preferentially identify the iNFTs and the eNFTs. By contrast, other phosphoepitopes (e. g. , pThr153, pSer262, pThr231) and a certain conformational epitope recognized by the antibodies MC1 and Alz50 also recognize pre-NFTs, suggesting that the misfolding of the tau molecule and its phosphorylation in certain sites represent an early step prior to tau aggregation. Interestingly, the immuno-reactivity for a caspase-cleaved form of tau with a faster rate of fibrillization than the full length molecule in vitro co-localize with Alz50 immunoreactivity in pre-NFTs, suggesting that the caspase-mediated cleavage of the carboxy-terminal region of the tau molecule is also a necessary step prior to further aggregation.

Topographical distribution

The spatiotemporal pattern of progression of NFTs (and neuropil threads in parallel) is

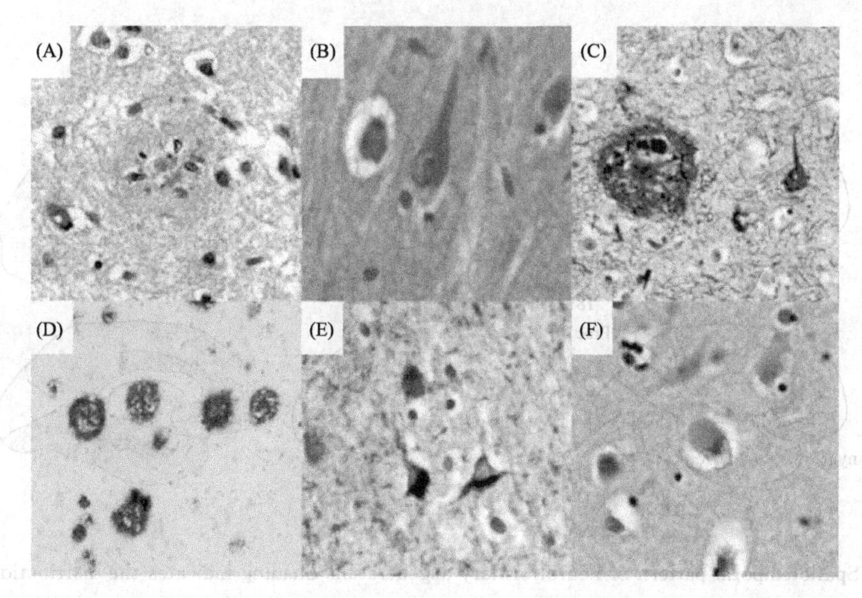

Fig. 1. 23 Photomicrographs of the core pathological lesions observed in Alzheimer and Lewy body diseases. (A) Plaque evident on routine H&E stained section of frontal cortex; (B) tangle in a hippocampal pyramidal neuron on routine H&E stained section; (C) silver stain highlights both a plaque and a tangle; (D) immunohistochemistry against Ab highlights plaques; (E) immunohistochemistry against tau highlights tangles; (F) a cortical Lewy body can be seen in a layer V neuron on a routine H&E stained section of frontal cortex.

rather stereotypical and predictable. Briefly, the neurofibrillary degeneration starts in the allocortex of the medial temporal lobe (entorhinal cortex and hippocampus) and spreads to the associative isocortex, relatively sparing the primary sensory, motor, and visual areas. In their clinicopathological study, Braak H and Braak E distinguished six stages that can be summarized in three: entorhinal, limbic, and isocortical (Fig. 1. 24). The first NFTs consistently appear in the transentorhinal (perirhinal) region (stage Ⅰ) along with the entorhinal cortex proper, followed by the CA1 region of the hippocampus (stage Ⅱ). Next, NFTs develop and accumulate in limbic structures such as the subiculum of the hippocampal formation (stage Ⅲ) and the amygdala, thalamus, and claustrum (stage Ⅳ). Finally, NFTs spread to all isocortical areas (isocortical stage), with the associative areas being affected prior and more severely (stage Ⅴ) than the primary sensory, motor, and visual areas (stage Ⅵ). A severe involvement of striatum and substantia nigra can occur during the late isocortical stage. Of note, this neurofibrillary degeneration follows a laminar pattern affecting preferentially the stellate neurons of layer Ⅱ, the superficial portion of layer Ⅲ, and the large multipolar neurons of layer Ⅳ within the entorhinal cortex; the stratum pyramidale of CA1 and subiculum within the hippocampal formation, and the pyramidal neurons of layers Ⅲ and Ⅴ within the isocortical areas.

Clinicopathological correlations

Multiple clinicopathological studies from different groups have established that the amount and distribution of NFTs correlate with the severity and the duration of dementia. Moreover, the selective rather than widespread topographical distribution of NFTs described above matches with the hierarchical neuropsychological profile typical of the AD-

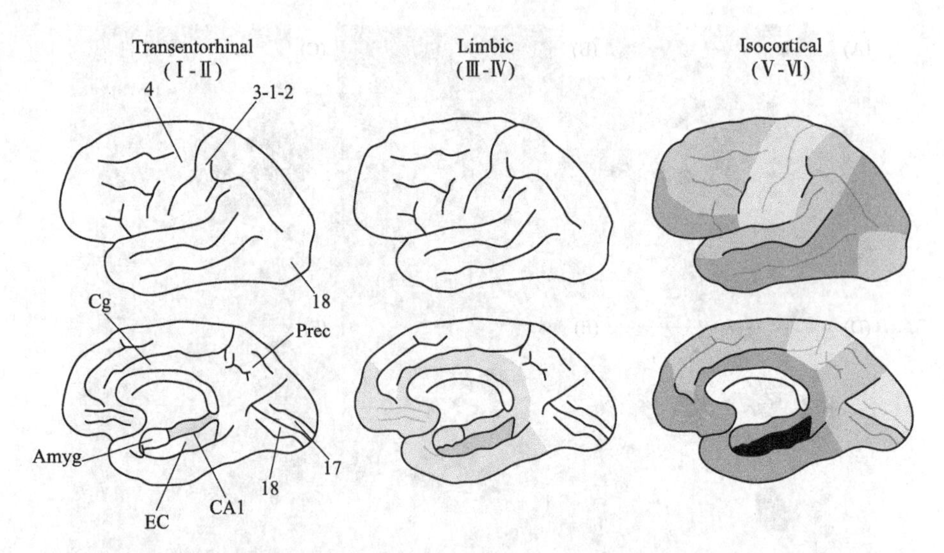

Fig. 1. 24 Spatiotemporal pattern of neurofibrillary degeneration. Shading indicates the distribution of NFTs with darker colors representing increasing densities. Amyg=Amygdala; EC=Entorhinal cortex; CA1=Cornus ammonis 1 hippocampal subfield; Cg=Cingulate cortex; Prec=Precuneus; 4=Primary motor cortex; 3-1-2=Primary sensory cortex; 17=Primary visual cortex; 18=Associative visual cortex.

type dementia syndrome. The prominent initial impairment of episodic memory characteristic of AD is explained by the isolation of the medial temporal lobe structures from the association isocortex and the subcortical nuclei because of the ongoing massive neurofibrillary degeneration. Next, the involvement of multimodal high-order association isocortical areas accounts for the progressive impairment of additional cognitive domains, including executive dysfunction (prefrontal cortex), apraxias (parietal cortex), visuospatial navigation deficits (occipitoparietal cortex), visuoperceptive deficits (occipitotemporal cortex), and semantic memory (anterior temporal cortex), giving rise to the full-blown dementia syndrome. By contrast, the late involvement of primary motor, sensory, and visual isocortical areas explains the sparing of motor, sensory, and primary visual functions. However, as discussed below, whether NFT formation is a necessary precursor of the neuronal death in AD or represents a protective response of damaged neurons (and thus more of a surrogate marker of the ongoing pathological process) is still controversial.

Amyloid plaques

Composition

The senile plaques described by Alois Alzheimer in his original case report result from the abnormal extracellular accumulation and deposition of the amyloid-β peptide (Aβ) with 40 or 42 amino acids (Aβ40 and Aβ42), two normal byproducts of the metabolism of the amyloid precursor protein (APP) after its sequential cleavage by the enzymes β- and γ-secretases in neurons. Because of its higher rate of fibrillization and insolubility, Aβ42 is more abundant than Aβ40 within the plaques.

Morphological characteristics

Attempts to understand the evolution of the amyloid plaque after its formation based on morphological criteria gave rise to a number of terms, including "primitive," "classical," and "burn-out" plaques. However, a more practical and widely used morphological classification distinguishes only two types of amyloid plaques—diffuse versus dense-core plaques-based on their staining with dyes specific for the β-pleated sheet conformation such as Congo Red and Thioflavin-S. This simpler categorization is relevant to the disease because, unlike diffuse Thioflavin-S negative plaques, Thioflavin-S positive dense-core plaques are associated with deleterious effects on the surrounding neuropil including increased neurite curvature and dystrophic neurites, synaptic loss, neuron loss, and recruitment and activation of both astrocytes and microglial cells. Indeed, diffuse amyloid plaques are commonly present in the brains of cognitively intact elderly people, whereas dense-core plaques, particularly those with neuritic dystrophies, are most often found in patients with AD dementia. However, the pathological boundaries between normal aging and AD dementia are not clear-cut and, as we will further discuss below, many cognitively normal elderly people have substantial amyloid burden in their brains.

Electron microscopy studies revealed that the ultrastructure of dense-core plaques is comprised of a central mass of extracellular filaments that radially extend toward the periphery, where they are intermingled with neuronal, astrocytic, and microglial processes. These neuronal processes, known as dystrophic neurites, often contain packets of paired helical filaments, as well as abundant abnormal mitochondria and dense bodies of probable mitochondrial and lysosomal origin. Plaque-associated neuritic dystrophies represent the most notorious evidence of Aβ-induced neurotoxicity and feature many of the pathophysiological processes downstream Aβ. Their origin can be axonal or dendritic and their morphology can be either elongated and distorted or bulbous. They can be argyrophylic [Fig. 1. 23 (c)] and Thioflavin-S positive because of the aggregation of β-sheet pleated tau fibrils, which can also be shown with many phosphotau and conformation-specific tau antibodies. Interestingly, dystrophic neurites can also be immunoreactive for APP. Cytoskeletal abnormalities in dystrophic neurites explain their immunoreactivity for neurofilament proteins. These cytoskeletal abnormalities can lead to a disruption of the normal axonal transport and, indeed, a subset of dystrophic neurites are positive for mitochondrial porin and chromogranin-A because of the abnormal accumulation of mitochondria and large synaptic vesicles, respectively. Moreover, some axonal dystrophic neurites contain either cholinergic, glutamatergic, or gabaergic markers, suggesting a plaque-induced aberrant sprouting. Finally, dystrophic neurites can be displayed with immunohistochemical studies for ubiquitin and lysosomal proteins, indicating that there is a compensatory attempt to degrade and clear the abnormal accumulation of proteins and organelles. A less-evident expression of the plaque-induced neuritic changes is the increase in the curvature of neurites located in the proximity of dense-core plaques.

Topographic distribution

Unlike NFTs, amyloid plaques accumulate mainly in the isocortex. Although the spatiotemporal pattern of progression of amyloid deposition is far less predictable than that of NFTs, in general the allocortex (including entorhinal cortex and hippocampal formation), the basal ganglia, relevant nuclei of the brainstem, and the cerebellum, are involved to a lesser extent and later than the associative isocortex. The dissociation between amyloid and NFT burdens in the medial temporal lobe is particularly noticeable. Among the isocortical areas, likewise NFTs, primary sensory, motor, and visual areas tend to be less affected as compared to association multimodal areas. Despite this poorer predictability of the progression of amyloid deposition, two staging systems have been proposed. Braak H and Braak E distinguished three stages: (1) Stage A, with amyloid deposits mainly found in the basal portions of the frontal, temporal, and occipital lobes; (2) Stage B, with all isocortical association areas affected while the hippocampal formation is only mildly involved, and the primary sensory, motor, and visual cortices are devoid of amyloid; and (3) Stage C, characterized by the deposition of amyloid in these primary isocortical areas and, in some cases, the appearance of amyloid deposits in the molecular layer of the cerebellum and subcortical nuclei such as striatum, thalamus, hypothalamus, subthalamic nucleus, and red nucleus. Thal et al. proposed a descendent progression of amyloid deposition in five stages: (1) Stage 1 or isocortical; (2) Stage 2, with additional allocortical deposits (entorhinal cortex, hippocampal formation, amygdala, insular, and cingulated cortices); (3) Stage 3, with additional involvement of subcortical nuclei including striatum, basal forebrain cholinergic nuclei, thalamus and hypothalamus, and white matter; (4) Stage 4, characterized by the involvement of brainstem structures, including red nucleus, substantia nigra, reticular formation of the medulla oblongata, superior and inferior colliculi; and (5) Stage 5, with additional amyloid deposits in the pons (reticular formation, raphe nuclei, locus ceruleus) and the molecular layer of the cerebellum. These five Thal stages can be summarized in three: stage 1 or isocortical; stage 2, allocortical or limbic, and stage 3 or subcortical (Fig. 1. 25).

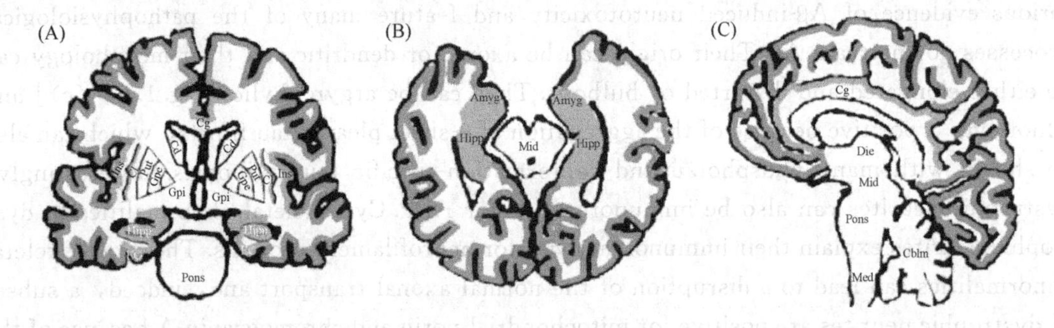

Fig. 1. 25 Spatiotemporal pattern of amyloid plaque deposition according to Thal et al. (2002). Coronal (A), axial (B), and sagittal (C) views of the brain. The five Thal stages of amyloid deposition are here summarized in three stages. Amyloid deposits accumulate first in isocortical areas (stage 1 or isocortical, in red), followed by limbic and allocortical structures (stage 2 or limbic, in orange), and in a later stage, by subcortical structures including basal ganglia, selected nuclei in diencephalon and brainstem, and the cerebellar cortex (stage 3 or subcortical, in yellow). Amyg=Amygdala; EC=Entorhinal cortex; Hipp=Hippocampus; Cg=Cingulate cortex; Cd=Caudate nucleus; Put=Putamen; Gpe=Globus pallidus externus; Gpi=Globus pallidus internus; Cl=Claustrum; Ins=Insular cortex; Die=Diencephalon; Mid=Midbrain; Med=Medulla oblongata; Cblm=Cerebellum.

Amyloid deposits usually involve the six layers of the isocortex, although layers Ⅰ and Ⅵ are usually relatively more spared than layers Ⅱ-Ⅴ. However, in advanced cases it is frequent to observe band-like diffuse amyloid deposits in the subpial surface of the cortex and even a few amyloid deposits in the white matter close to its transition with the cortical layer Ⅵ.

Clinicopathological correlations

Clinicopathological studies have established that the amyloid burden (either total amyloid plaques, dense-core plaques or only neuritic plaques) does not correlate with the severity or the duration of dementia. Indeed, in a region of early amyloid deposition such as the temporal associative isocortex, the amyloid burden reaches a plateau early after the onset of the cognitive symptoms or even in the preclinical phase of the disease and not even the size of the plaques grows significantly with the progression of the disease. However, it is possible that the amount of amyloid measured over the entire cortical mantle does increase during the clinical course of the disease as the distribution of amyloid deposits "spread" following the above stages. Preliminary data from longitudinal amyloid PET imaging studies in living patients have recently supported this possibility.

Cerebral amyloid angiopathy

Composition

The amyloid-β peptide not only deposits in the brain parenchyma in the form of amyloid plaques but also in the vessel walls in the form of cerebral amyloid angiopathy (CAA). Indeed, the more insoluble and aggregation-prone Aβ42 peptide tends to accumulate in the core of senile plaques, while the more soluble Aβ40 peptide is the major constituent of CAA, accumulating mainly in the interstitium between the smooth muscle cells of the tunica media. Although CAA can also appear in isolation (pure CAA), it is more common in the context of AD, with≈80% of AD patients showing some degree, usually mild, of CAA at autopsy.

Morphological characteristics

The same methods described for the examination of amyloid plaques are valid for CAA, that is Thioflavin-S or Congo red staining or immunohistochemical studies with anti-Aβ antibodies. A morphological staging system has been implemented to describe the severity of CAA within a single vessel: grade 0 or absence of staining; grade 1 or congophilic rim around an otherwise nomal-appearance vessel; grade 2 or complete replacement of the tunica media by congophilic material; grade 3 or cracking of ≥50% of the circumference of the vessel, giving a "vessel-within-vessel" or "double-barrel" appearance; and grade 4 or fibrinoid necrosis of the vessel wall, often accompanied by additional amyloid deposits in the surrounding neuropil. In this severe stage, Prussian blue (Perl's) staining is useful to show hemosiderin-laden macrophages in the parenchyma surrounding CAA-affected vessels, indicative of chronic microbleeds.

Topographic distribution

CAA usually affects cortical capillaries, small arterioles and middle-size arteries as well as leptomeningeal arteries, whereas venules, veins, and white-matter arteries are rarely involved. For unknown reasons, posterior parietal and occipital areas are usually more prominently affected than frontal and temporal lobes, and within the same area, leptomeningeal arteries usually show more severe CAA than cortical arteries. A semiquantitative scoring system has been proposed to characterize the severity of CAA within a region of the cortex: 0= no Thioflavin-S-stainedleptomeningeal or cortical vessels; 1= scattered positivity in either leptomeningeal or cortical vessels; 2=strong circumferential positivity in at least some vessels either leptomeningeal or cortical; 3=widespread circumferential staining in many leptomeningeal and cortical vessels, and 4=presence of "dysphoric" perivascular amyloid deposits in addition to score 3. A global severity score can be obtained by averaging the scores from several regions.

Clinicopathological correlations

According to the Boston criteria, CAA should be suspected after one or multiple major symptomatic lobar hemorrhages in an elderly patient. But in the context of AD, unless it becomes symptomatic because of this hemorrhagic complication, CAA is usually diagnosed at autopsy. However, three independent postmortem longitudinal studies have revealed that the otherwise apparently asymptomatic CAA can also be a synergistic contributor to cognitive decline in AD.

Granuovacuolar degeneration and Hirano bodies

Granulovacuolar degeneration (GVD) andHirano bodies are two poorly understood lesions present in the cytoplasm of hippocampal pyramidal neurons of AD patients. Although they are increasingly observed with aging in cognitively intact elderly people, these two lesions are more severe and frequent in age-matched AD patients.

GVD consists of the accumulation of large double-membrane bodies. Their origin and significance are uncertain. Early immunohistochemical studies reported immunoreactivity of GVD bodies for cytoskeletal proteins including tubulin, neurofilament proteins and tau. Because GVD bodies are also positive for some tau kinases, a role in tangle formation has been proposed. Other authors have postulated a role in the apoptotic cell death because of their immunoreactivity for activated caspase-3. More recent studies have suggested that these bodies might derive from the endoplasmic reticulum and represent the stress granules that feature the unfolded protein response, because they are positive for several stress kinases. Finally, based on their positivity for ubiquitin and autophagic markers, it has been proposed that these granules are late-stage autophagic vacuoles.

Hirano bodies are eosinophilic rod-like cytoplasmic inclusions relatively common in the stratum lacunosum of the hippocampal CA1 region in the elderly. However, in AD patients the number of Hirano bodies is abnormally high and they are translocated to the neurons of the stratumpyramidale. Although the significance of Hirano bodies in AD is not completely understood, they are recognized by antibodies against tau, neurofilament proteins, actin, and other cytoskeletal proteins. Other immunoreactivities associated with Hirano bodies are inducible nitric oxide synthase, advanced glycation endproducts, and the carboxy-terminal fragments of APP.

Glial responses

Reactive astrocytes and activated microglial cells are commonly associated to dense-core amyloid plaques, indicating that amyloid-β is a major trigger of this glial response. However, we have recently observed a linear increase in reactive astrocytes and activated microglial cells through the entire disease course despite an early plateau in amyloid deposition in the temporal associative isocortex. Indeed, we found a highly significant positive correlation between both astrocytosis and microgliosis and NFT burden but not between both reactive glial cell types and amyloid burden, suggesting that glial responses are also related to neurofibrillary degeneration.

Neuronal loss

Neuronal loss is the main pathological substrate of cortical atrophy and, although usually evident in sections stained with hematoxylin and eosin, it can be more readily shown with a Nissl staining or a NeuN immunohistochemistry. Nissl staining (for example with cresyl violet) reveals the negatively charged ribosomic RNA present in the ribosomes of the rough endoplasmic reticulum (Nissl substance or granules), giving a dark blue appearance to the perinuclear region of neurons. By contrast, NeuN is a neuronal-specific nuclear antigen, although NeuN immunohistochemistry also stains the perinuclear region and some proximal processes of neurons.

The regional and laminar pattern of neuronal loss matches that of NFTs, but, importantly, within the same region neuronal loss exceeds the numbers of NFTs, so that it is a better correlate of cognitive deficits than the number of NFTs. Indeed, quantitative stereology-based studies of neurons, iNFTs and eNFTs have concluded that iNFTs can last for up to two decades and that neurons bearing iNFTs might still be viable as evidenced by their positive Nissl staining. This dissociation between the extent of neuronal loss and that of NFTs suggests that there are least two mechanisms of neuronal death in AD: one affecting tangle-bearing neurons, that will lead to the appearance of ghost extracellular tangles, and another affecting tangle-free neurons. Although themechanisms of neuronal death in AD are beyond the scope of this article, it will be noted that postmortem studies on apoptosis have yielded controversial results, with some studies showing a widespread distribution of apoptotic markers, while others have only reported a scattered distribution.

Synapse loss

Besides neuronal loss, synapse loss is another contributor to the cortical atrophy of the AD brain. Synapse loss in AD was shown with immunohistochemical studies using antibodies against pre- or postsynaptic proteins—typically the presynaptic protein synaptophysin—and with electron microscopy studies.

The spatiotemporal and laminar pattern of synapse loss matches that of neuron loss. Synaptic loss is not only caused by neuronal loss but can exceed the existing neuronal loss within a particular cortical area. This indicates that synapse loss predates neuronal loss and that the remaining neurons become less well connected to their synaptic partners than expected just by the number of viable neurons surviving in a particular circuit. Likely this is why synaptic density is the best correlate of cognitive decline in AD. Interestingly, an inverse correlation has been observed between synaptic density and the size of remaining synapses as measured by the length of the postsynaptic density. This enlargement of remaining synapses has been interpreted as a compensatory response, rather than as selective loss of small synapses.

Criteria for the pathological diagnosis of Alzheimer disease

Of all pathological features described above, amyloid plaques and NFTs are the most characteristic of AD and, understandably, the criteria for the pathological diagnosis of AD rely on their amount and/or distribution.

The first pathological criteria for the diagnosis of AD were based on the highest density of total amyloid plaques (both diffuse and neuritic) in any cortical field, adjusted for age so that the older the patient at death, the greater the density required for diagnosis. The presence of NFTs was not required and diffuse plaques—relatively frequent in nondemented elderly people—had the same consideration as neuritic plaques. Although meritorious, these criteria were soon abandoned because, despite a very high sensitivity to diagnose AD dementia, they lacked sufficient specificity. In 1991, the Consortium to Establish a Registry for Alzheimer Disease (CERAD) proposed more specific diagnostic criteria by emphasizing the importance of neuritic plaques over diffuse plaques. CERAD criteria use a semiquantitative score of the density of neuritic plaques in the most severely affected region of the isocortex (frontal, temporal, or parietal) and the patient's age at death to obtain an age-related plaque score. This score is then integrated with clinical information regarding the presence or absence of dementia to establish one of three levels of certainty that dementia is explained by the AD pathological changes: possible, probable, and definite. A diagnosis of AD is made if the criteria for probable or definite AD are met. Although higher than that of *Khachaturian criteria*, the specifiy of CERAD criteria proved to be still insufficient because they did not incorporate the scoring of the severity of NFTs. By contrast, the use of Braak H and Braak E staging of NFTs alone—with the isocortical stages Ⅴ and Ⅵ as criteria of definite AD—showed a high specificity at the expense of a low sensitivity.

Current pathological criteria for AD were defined in 1997 by a workshop of the National Institute of Aging and the Reagan Institute (NIA-RI). The NIA-RI consensus recommendations combine the CERAD semiquantitative score of neuritic plaques and the Braak H and Braak E staging of NFTs to distinguish three probabilistic diagnostic categories: (1) high likelihood, if there are frequent neuritic plaques (CERAD definite) and abundant isocortical NFTs (Braak stage Ⅴ/Ⅵ); (2) intermediate likelihood, if there are moderate neuritic plaques (CERAD probable) and NFTs are restricted to limbic regions (Braak Ⅲ/Ⅳ), and (3) low likelihood, if there are infrequent neuritic plaques (CERAD possible) and NFTs are restricted to the entorhinal cortex and/or hippocampus (Braak Ⅰ/Ⅱ). A diagnosis of AD is made when the criteria for intermediate or high likelihood of AD are met and the patient had a clinical history of dementia. Because experience has revealed infrequent cases with many AD pathological lesions but no or few cognitive symptoms (and vice versa) and these circumstances were not addressed by the NIA-RI consensus workgroup, these diagnostic criteria are currently under review.

Both CERAD and NIA-RI criteria also incorporated the assessment of other pathologies, particularly vascular and Lewy body diseases, already recognizing the high prevalence of mixed pathologies underlying dementia in elderly people, a circumstance well documented by more recent longitudinal community-based clinicopathological studies. Thus, in many practical instances, the CERAD criteria for "possible AD" and the NIA-RI criteria for "intermediate probability of AD" are not only based on a moderate amount and distribution of AD pathology but also on the coexistence of vascular or Lewy body pathology with sufficient severity to contribute to the patient's dementia.

Neuropathology of mild cognitive impairment and early Alzheimer disease

Clinicopathological correlation studies have taught us that at the moment of the clinical diagnosis, patients with AD-type dementia often already have a Braak stage Ⅴ or Ⅵ of neurofibrillary degeneration and a substantial and widespread synaptic and neuronal loss. To anticipate the clinical diagnosis of AD before the stage of full-blown dementia, a new clinical construct was needed. Petersen et al. proposed the concept of "mild cognitive impairment" (MCI) as a new diagnostic entity for the transition between normal aging and AD dementia. Patients with MCI have already some cognitive complaints that are detectable with the appropriate cognitive tests and represent a decline from a previous higher baseline level but that, unlike the definition of dementia, do not interfere with their activities of daily life. Importantly, MCI patients have an increased risk of developing dementia, which has been reported between 10% and 15% per year.

Autopsy studies on MCI patients are scarce but they have reproducibly found a stage of AD pathology intermediate between cognitively intact subjects and demented patients, particularly regarding neurofibrillary degeneration, that is consistent with the idea of a transition phase between normal aging and definite AD. Specifically, MCI patients usually have a moderate number of neuritic plaques and a limbic stage of NFTs (Braak stage Ⅲ or Ⅳ), fitting into the NIA-RI category of intermediate likelihood of AD (sufficient to cause dementia) and providing a pathological validation for this clinical construct. Along the same lines, patients with a Clinical Dementia Rating score of 0.5 (equivalent to MCI or very mild AD) have already a≈30% of neuron loss in the entorhinal cortex compared to cognitively intact controls (CDR=0), but still no evident neuronal loss in the superior temporal sulcus. Moreover, electron microscopy studies have shown that MCI patients also have an intermediate number of synapses between nondemented controls and mild AD patients in the hippocampus, further indicating that many individuals with the clinical symptoms of MCI have early AD. Of note, a paradoxical, presumably compensatory, up-regulation in the density of presynaptic glutamatergic boutons has been reported in the frontal cortex of MCI patients compared to nondemented controls and mild AD patients.

Although AD was the most common pathological diagnosis underlying MCI in the above case series, it should be noted that there was a high degree of pathological heterogeneity underlying the clinical diagnosis of MCI, with vascular disease, Lewy body disease, argyrophilic grain disease, and hippocampal sclerosis as major concurrent or alternative pathologies. In addition, in the largest study a high proportion (up to 25%) of MCI patients had no pathology at autopsy. Finally, no significant pathological differences have been observed between the amnestic and the nonamnestic subtypes of MCI nor in their pathological outcome after conversion to dementia.

Alzheimer neuropathology in "normal aging"

Longitudinal prospective clinicopathological studies in nondemented elderly people have revealed that up to 45% of nondemented elderly would meet the NIA-RI criteria for AD had they been demented, usually the intermediate likelihood category of these criteria, and rarely the high likelihood category. Moreover, the pattern of regional distribution of pathological changes in nondemented controls matches that of AD patients. Thus, mounting evidence from clinicopathological studies support the view that AD is a continuous spectrum between asymptomatic lesions in cognitively normal elderly and dementia, with MCI as a transition phase between these two ends.

The apparent dissociation between AD pathology and cognitive status in some elderly people is remarkable because these so-called "high-pathology nondemented controls" or "individuals with asymptomatic AD" seem to be resilient to the neurotoxic effects of amyloid plaques and NFTs and to contradict the aforementioned positive correlation between NFT burden and cognitive decline. Understanding the biochemical and morphological substrates of this resilience to cognitive decline in the presence of abundant AD pathology might be crucial to discover new therapeutic targets for the disease. As expected from the highly significant clinicopathological correlations of synaptic and neuronal loss in AD, high-pathology controls have preserved synaptophysin levels compared to AD patients with a similar burden of plaques and NFTs, and they do not seem to have significant neuronal loss, not even in vulnerable regions such as the entorhinal cortex and the hippocampus. Moreover, they have lower levels of neuroinflammatory markers than pathology-matched AD patients. This resistance to AD pathology has also been related to a nucleolar, nuclear, and cell body hypertrophy of the hippocampal and cortical neurons, suggestive of a compensatory metabolic activation to face the neurotoxic effects of AD lesions. In keeping with these pathological reports, a MRI-neuropathological correlation study revealed larger brain and hippocampal volumes in high-pathology controls than in pathology-matched demented patients, further supporting the preservation of both neurons and synapses.

Overlap of ad with lewy body disease

Alzheimer disease and Parkinson's disease (PD) are the leading causes of dementia and movement disorders in the aging population. It is estimated that over 10 million people live with these devastating neurological conditions in the United States, and that this country alone will see a 50% annual increase of AD and PD by the year 2025.

PD and AD are two distinct clinicopathological entities. While in AD, abnormal accumulation of misfolded Aβ protein in the neocortex and limbic system is thought to be responsible for the neurodegenerative pathology, intracellular accumulation of α-synuclein has been centrally implicated in the pathogenesis of PD. In AD, Aβ protein accumulates in the intracellular and extracellular space, leading to the formation of plaques, whereas intracellular polymerization of phosphorylated cytoskeletal molecules such as tau results in the formation of neurofibrillary tangles. In PD, intracellular accumulation of α-synuclein—an abundant synaptic terminal protein—results in the formation of characteristic inclusions called Lewy bodies (LBs) [Fig. 1. 23 (F)] . The new consortium criteria for the classification of Lewy body diseases (LBD) recognizes two clinical entities, the first denominated dementia with LBs (DLB) and the second PD dementia (PDD) . While in patients with DLB, the clinical presentation is of dementia followed by parkinsonism, in patients with PDD the initial signs are of parkinsonism followed by dementia. Interestingly, the brains of patients with DLB and PDD display very similar pathology, with the exception that recent studies have shown extensive deposition of Aβ and α-synuclein in the striatum and hippocampus in DLB compared to only α-synuclein in PDD cases. Because of the implications for the management and treatment of parkinsonism and dementia in patients with PD and DLB, loss of dopaminergic neurons in the midbrain and cholinergic cells in the nucleus basalis of Meynert have been characterized in detail. Although the severity of the neuronal loss within these subcortical regions might explain some of the neurological deficits in patients with PD and DLB, the neuronal populations responsible for the more complex cognitive and psychiatric alterations have not been completely characterized. Abnormal accumulation of α-synuclein in the CA2-3 region of the hippocampus, insula, amygdala and cingulate cortex has been shown to be an important neuropathological feature.

Remarkably, despite being initially considered distinct clinicopathological conditions, several studies have now confirmed that the clinical features and the pathology of AD and PD can overlap. Approximately 25% of all patients with AD develop parkinsonism, and about 50% of all cases of PD develop AD-type dementia after 65 years of age. Moreover, 70% of patients with sporadic AD display the formation of α-synuclein-positive LB-like inclusions in the amygdala and limbic structures. Similarly, in patients with familial AD (FAD) and Down syndrome, LB-like pathology and parkinsonism have been reported. Last, as mentioned above, the single most important neuropathological finding that distinguishes PDD from DLB is the presence of Aβ deposits in the striatum and in the hippocampus.

A number of studies provide extensive support for an interaction between pathogenic pathways in AD and LBD and argue against a coincidental concurrence of both disorders (i. e., merely because of their high prevalence in the elderly). FAD cases with presenilin mutations that present with significant LB pathology strongly support an interaction between Aβ and α-synuclein. Although plaques, tangles and LBs are useful neuropathological and diagnostic markers of these disorders, the initial injury that results in the cognitive and movement alterations is likely the damage of the synaptic terminals in selected circuitries. Several lines of investigation support the notion that oligomeric forms of Aβ and α-synuclein, rather than the polymers and fibrils associated with plaques and LBs, accumulate in the neuronal membranes and lead to the characteristic synaptic pathology. Some studies have shown that underlying interactions between α-synuclein and Aβ play a fundamental role in the pathogenesis of LBD. Specifically, Aβ promotes the oligomerization and toxic conversion of α-synuclein, Aβ exacerbates the deficits associated with α-synuclein accumulation, Aβ and α-synuclein colocalize in membrane and caveolar fractions, and Aβ stabilizes α-synuclein multimers that might form channel-like structures in the membrane. Both lysosomal leakage and oxidative stress appear to be involved in the process of neurotoxicity and pathological interactions between Aβ and α-synuclein.

Therefore, it is possible that the combined effects of α-synuclein and Aβ might lead to synaptic damage and selective degeneration of neurons in the neocortical, limbic, and subcortical regions. A more precise mapping of the neuronal populations affected in these regions is needed to understand the cellular basis for the characteristic cognitive dysfunction in PDD and DLB and to develop new treatments for these conditions.

Conclusions

Classical neuropathological lesions including senile amyloid plaques and neurofibrillary tangles define AD but they likely represent the "tip of the iceberg" of the pathological alterations that cause the cognitive decline associated with AD. Indeed, the development of new biomarkers and imaging tools has made evident that these neuropathological stigmata of AD begin to accumulate a decade or more prior to a clinical diagnosis of dementia. Synaptic loss, plasticity changes, neuronal loss, and the presence of soluble microscopic oligomeric forms of Aβ and even of tau, likely contribute to the progressive neural system failure that occurs over decades. An understanding of this natural history of the disease is critical to design primary or secondary prevention strategies to halt the disease progression before the damage to the neural system becomes irreversible.

Glossary

Amyloid plaques: extracellular deposits of amyloid β abundant in the cortex of AD patients. Amyloid plaques are commonly classified in diffuse and dense-core based on their morphology and positive or negative staining with Thioflavin-S or Congo Red.

Dense-core plaques: fibrillar amyloid deposits with compact core that stains with Thioflavin-S and Congo Red. Dense-core plaques are typically surrounded by dystrophic neurites (neuritic plaques), reactive astrocytes and activated microglial cells, and associated with synaptic loss. A semiquantitative score of neuritic plaques is used for the pathological diagnosis of AD because their presence is generally associated with the presence of cognitive impairment.

Diffuse plaques: amorphous amyloid deposits with ill-defined contours that are Congo Red and Thioflavin S negative. Diffuse plaques are usually nonneuritic and not associated with glial responses or synaptic loss. This plaque type is not considered for the pathological diagnosis of AD because it is a relatively common finding in the brain of cognitively intact elderly people.

Cerebral amyloid angiopathy (CAA): deposits of amyloid b in the tunica media of leptomeningeal arteries and cortical capillaries, small arterioles and medium-size arteries, particularly in posteriorareas of the brain. Some degree of CAA, usually mild, is present in ≈ 80% of AD patients. If severe, CAA can weaken the vessel wall and cause life-threatening lobar hemorrhages.

Amyloid β: a 40 or 42 amino acid peptide derived from amyloid precursor protein (APP) after its sequential cleavage by β- and γ-secretases. Its physiological role is likely related to the modulation of synaptic activity although still controversial. In AD Aβ accumulates forming intermediate soluble oligomers that are synaptotoxic as well as insoluble β-sheet pleated amyloid fibrils that are the main constituent of dense-core plaques (mainly Aβ42) and cerebral amyloid angiopathy (primarily Aβ40) .

Neurofibrillary tangles (NFTs): intraneuronal aggregates of hyperphosphorylated and misfolded tau that become extraneuronal ("ghost" tangles) when tangle-bearing neurons die. NFTs have a stereotypical spatiotemporal progression that correlates with the severity of the cognitive decline. In fact, a topographic staging of NFTs is used for the pathological diagnosis of AD.

Neuropil threads: axonal and dendritic segments containing aggregated and hyperphosphorylated tau that invariably accompany neurofibrillary tangles in AD.

Tau: a microtubule-associated protein normally located to the axon, where it physiologically facilitates the axonal transport by binding and stabilizing the mictrotubules. In AD, tau is translocated to the somatodendritc compartment and undergoes hyperphosphorylation, misfolding, and aggregation, giving rise to neurofibrillary tangles and neuropil threads.

Glossary

aberrant [ə'ber(ə)nt] adj. 异常的，畸变的，脱离常轨的，迷乱的
actin ['æktin] n. [生化] 肌动蛋白
aggregate ['ægrigət;(for v.)'ægrigeit] n. 合计，集合体，总计；adj. 聚合的，集合的，合计的；vt. 集合，聚集，合计；vi. 集合，聚集，合计
allocortex 旧皮层，古皮层
Alzheimer ['ælz'emə:] n. 阿尔茨海默病（等于 Alzheimer's disease）
amnestic adj. 遗忘的，记忆缺失的（等于 amnesiac）
amygdala [ə'migdələ] n. [解剖] 杏仁核，扁桃腺，苦巴旦杏
amyloid ['æmilɔid] n. 淀粉体，淀粉质食物；adj. 含淀粉的，淀粉质的
angiopathy [ˌændʒi'ɔpəθi] n. 血管病，淋巴管病
apoptotic [æpɔp'tɔtik] 凋亡
apraxia [ə'præksiə] n. [医] 失用症；[医] 运用不能症
argyrophilic 嗜银的，嗜银性
arteriole [ɑ:'tiəriəul] n. [解剖] 小动脉，细动脉
artery ['ɑ:təri] n. 动脉，干道，主流
astrogliosis 星形胶质细胞增生
asymptomatic [əˌsimptə'mætik;ei-] adj. 无症状的
atrophy ['ætrəfi] n. 萎缩，萎缩症，发育停止；vi. 萎缩，虚脱
autophagic 自体吞噬的
axon ['æksɔn] n. 轴索，[解剖] 轴突（神经细胞）
bouton [bu:'tɔ:ŋ] n. 疖，小结（哑铃状脓肿），[解剖] 棒头
calbindin 钙结合蛋白
capillary [kə'piləri] n. 毛细管；adj. 毛细管的，毛状的
caspase n. 半胱天冬酶，切冬酶（胱冬肽酶），细胞凋亡蛋白酶
categorization [ˌkætigəri'zeiʃən] n. 分类，分门别类，编目方法
caveolae ['kæviəuli:] n. 小窝，小凹，细胞膜穴样内陷
cerebral amyloid angiopathy 大脑淀粉样血管病
cerebral ['seribr(ə)l;sə'ri:br(ə)l] adj. 大脑的，脑的
cerebrovascular [ˌseribrə(ʊ)'væskjʊlə] adj. 脑血管的
cholinergic [ˌkəʊli'nɜ:dʒik] adj. 类胆碱的
chromogranin n. 嗜铬粒蛋白
cingulate ['siŋgjulit,-leit] adj. 有色带的，（昆虫腹部）有色带环绕的
circuitry ['sɜ:kitri] n. 电路，电路系统，电路学，一环路
circumference [sə'kʌmf(ə)r(ə)ns] n. 圆周，周长，胸围
claustrum ['klɔ:strəm] n. [解剖]（大脑的）屏状核
coeruleus 天蓝色，蓝色的，青的
cognitive ['kɔgnitiv] adj. 认知的，认识的
colliculi n. 丘（colliculus 的变形）
colocalize vt. 共区域化，局部化
coloration [kʌlə'reiʃ(ə)n] n. 着色，染色

concomitant [kən'kɔmit(ə)nt] n. 伴随物；adj. 相伴的，共存的，附随的
concurrent [kən'kʌr(ə)nt] n. [数] 共点，同时发生的事件；adj. 并发的，一致的，同时发生的
Congo red 刚果红（一种染料）
Congophilic adj. 刚果红（染色）的
consortium [kən'sɔːtiəm] n. 财团，联合，合伙
continuum [kən'tinjuəm] n. [数] 连续统，[经] 连续统一体，闭联集
cortex ['kɔːteks] n. [解剖] 皮质，树皮，果皮
cortical ['kɔːtikəl] adj. 皮质的，[生物] 皮层的，外皮的
cortices ['kɔːti'siːz] n. 皮质（cortex 的复数）；adj. 外皮的
cresyl violet 甲酚紫
cross-sectional adj. 截面的，断面的，剖面的
CSF（cerebrospinal fluid） abbr. 脑脊髓液
cytoskeleton [ˌsaitəu'skelətən] n. [细胞] 细胞骨架，细胞支架
deleterious [ˌdeli'tiəriəs] adj. 有毒的，有害的
dementia [di'menʃə] n. [内科] 痴呆
demyelination [diːmaiəli'neiʃən] n. 髓鞘脱失，[病理] 脱髓鞘
dendritic [den'dritik] adj. 树枝状的，树状的
dense-core 致密核心
dentate ['denteit] n. 配位基；adj. 齿状的，有齿的，锯齿状的
dopaminergic [ˌdəupəmi'nəːdʒik] adj. 多巴胺能的
dysfunction [dis'fʌŋ(k)ʃ(ə)n] n. 功能紊乱，机能障碍，官能不良；vi. 功能失调，出现机能障碍，垮掉
dystrophic [dis'trəufik;-'trɔfik] adj. 营养不良的，营养障碍的
ementia with Lewy bodies 一种以路易小体为病理特征的神经变性疾病（痴呆）
endoplasmic reticulum [细胞] 内质网
entorhinal 内嗅的
eosinophilic [ˌiːəsinə'filik] adj. 嗜曙红的，嗜酸性的，嗜曙红细胞的
eosin ['iːə(ʊ)sin] n. 曙红（鲜红色的染料）
fibril ['faibril] n. [生物] 纤丝，[生物] 原纤维，须根
fibrinoid necrosis 纤维素样坏死
filamentous [filə'mentəs,ˌfilə'mentəs] adj. 纤维所成的，细丝状的，如丝的
frontal ['frʌnt(ə)l] n. 额骨，额部，房屋的正面；adj. 额的，正面的，前面的
gabaergic n. γ-氨基丁酸能的
ganglia ['gæŋgliə,'gæŋgliːə] n. 神经节，神经中枢
glial ['glaiəl] adj. 神经胶质的
granulovacuolar adj. 颗粒空泡的
gyrus ['dʒairəs] n. [解剖] 脑回（形成大脑半球的组织），回转
hematoxylin [ˌhiːmə'tɔksilin] n. [有化] 苏木精，苏木紫（等于 haematoxylin）
hemorrhage ['heməridʒ] n. [病理] 出血（等于 haemorrhage），番茄汁；vt. [病理] 出血
hemosiderin [ˌhiːməu'sidərin,ˌhem-] n. 血铁质，[生化] 血铁黄素
hierarchical [haiə'rɑːkik(ə)l] adj. 分层的，等级体系的
hippocampal [ˌhipə'kæmpəl] adj. 海马的，海马趾的
hippocampus [ˌhipə(ʊ)'kæmpəs] n. [解剖] [脊椎] 海马
Hirano body 平野小体（一种病理学中的结构体）
histopathological ['histəˌpæθə'lɔdʒikəl] adj. 组织病理学说的
hydrocephalus [ˌhaidrə'sef(ə)ləs;-'kef-] n. [内科] 脑积水
hyperphosphorylated adj. 过度磷酸化的
hypertension [haipə'tenʃ(ə)n] n. 高血压，过度紧张
hypertrophy [hai'pɜːtrəfi] n. [病理] 肥大，过度增大

immunohistochemical [ˈimjunəuˌhistəuˈkemikəl] adj. 免疫组织化学的
impairment [imˈpeəm(ə)nt] n. 损伤，损害
impregnation [ˌimpregˈneiʃən] n. 注入，受精，怀孕，受胎
infarct [ˈinfɑːkt;inˈfɑːkt] n. 梗死，梗塞
innervation [ˌinəːˈveiʃən] n. 神经分布，神经支配
insula [ˈinsjulə,ˈinsə-] n. [解剖] 脑岛
interstitiumn. 小间隙
intraneuronal 神经细胞内的
isocortex 同形皮质
lacunar infarct 腔隙性梗塞
lacunar [ləˈkjuːnə] n. 花格平顶，(花格平顶的) 嵌板；adj. 有花格平顶的
laminar [ˈlæminə] adj. 层状的，薄片状的，板状的
leptomeningeal 柔脑膜的
lesion [ˈliːʒ(ə)n] n. 损害，身体上的伤害，机能障碍
Lewy bodies 路易氏小体；[医] 向心性多层的圆形小体
limbic [ˈlimbik] adj. 边的，缘的
lobar cerebral hemorrhage 脑叶出血
lobar hemorrhage 脑叶出血，脑叶出血的
lobar intracerebral hemorrhage 脑叶出血的，脑叶出血
lobe [ləʊb] n. (脑、肺等的) 叶，裂片，耳垂，波瓣
locus coeruleus [ˈləukəs siˈruːliəs] 蓝斑
longitudinal [ˌlɔn(d)ʒiˈtjuːdin(ə)l;ˌlɔŋgi-] adj. 长度的，纵向的，经线的
medial temporal lobes [解剖] 内颞叶
medulla [meˈdʌlə] n. 髓质
medulla oblongata [解剖] 延髓
microglial 小神经胶质 (细胞) 的
microgliosis 小胶质细胞增生
microinfarct [内科] 微型心肌梗死，微梗死
microtubule [maikrə(ʊ)ˈtjuːbjuːl] n. [细胞] 微管
mitochondria [ˌmaitəʊˈkɔndriə] n. 线粒体 (mitochondrion 的复数)
morphological [ˌmɔːfəˈlɔdʒikəl] adj. 形态学的
morphology [mɔːˈfɔlədʒi] n. 形态学，形态论，[语] 词法，[语] 词态学
MRI (magnetic resonance imaging) abbr. 核磁共振成像
neocortex [ˌniːəuˈkɔːteks] n. 新 (大脑) 皮质
neurite [ˈnjuərait] n. [组织] 神经突
neuritic [ˌnjuəˈritik, njuəˈritik] adj. 神经炎的
neurofibrillaryn. 神经元纤维
neurofilament [ˌnuərəˈfiləmənt] n. [细胞] [组织] 神经丝，神经纤毛
neuronal adj. [解剖] 神经元的
neuropathological [njʊərəʊpæθəˈlɔdʒikl] adj. 神经病理学的
neuropil [ˈnjuərəpail] n. 神经纤维网
neuropsychology [ˌnjʊərə(ʊ)saiˈkɔlədʒi] n. 神经心理学
neurotoxic [ˌnjuərəuˈtɔksik] adj. 毒害神经的
neurotoxicity [ˌnjuərəutɔkˈsisəti] n. [农药] 神经中毒性，神经毒性
NIA-Reagan 一种 AD 诊断的病理标准
nigra [ˈnigrə] n. 黑质，黑人 (等于 niggra)
NPY abbr. 神经肽 Y (一个 36 个氨基酸的肽)
nucleolar [ˌnjuːkliˈəulə, njuːˈkliːələ] adj. [细胞] 核仁的

nucleus [ˈnjuːkliəs] n. 核，核心，原子核
oblongata [ˌɔblɔŋˈgɑːtə] n. 延髓
occipital [ikˈsipitəl] n. 枕骨；adj. 枕骨的，枕部的
occlusive [əˈkluːsiv,ɔ-] n. 闭塞音；adj. 咬合的，闭塞的
parenchyma [pəˈreŋkimə] n. [组织] 实质，软细胞组织
parietal [pəˈraiit(ə)l] n. 顶骨，头顶骨；adj. [解剖] 腔壁的，颅顶骨的，(美) 学院生活的
parkinsonism [ˈpɑːkins(ə)niz(ə)m] n. 震颤麻痹，帕金森症（等于 parkinson's disease）
perikaryon [ˌperiˈkæriən] n. 神经细胞的核周体
perivascular [ˌperiˈvæskjulə] adj. 血管周的
periventricular [ˌperivenˈtrikjulə] adj. 室周的
PET (Positron Emission Tomography) abbr. [生物物理] 正电子成像术，正电子放射断层造影术
petechial [pəˈtiːkiəl] adj. 瘀点的，瘀斑的
Phosphotau n. 磷酸化 tau 蛋白
plaque [plæk;plɑːk] n. 斑块，血小板，匾，饰板
pons [pɔnz] n. [解剖] 桥，[解剖] 脑桥
porin n. 孔蛋白
posterior parietal 后顶叶区，顶叶，后顶叶
postmortem [pəustˈmɔːtəm] n. 验尸，检视，尸体检查；adj. 死后的，死后发生的
postsynaptic [ˌpəustsiˈnæptik] adj. 突触后的
potentiation [pətenʃiˈeiʃən] n. 增强作用，势差现象
presenilin n. 早老素，早老蛋白
presynaptic [ˌpriːsiˈnæptik] adj. [解剖] 突触前的
Prussian blue [颜料] 铁蓝，[颜料] 普鲁士蓝
psychiatric [ˌsaikiˈætrik] adj. 精神病学的，精神病治疗的
pyramidal neuron 锥体细胞，锥体神经元
quantitative EM 电镜定量
sclerosis [skliəˈrəusis;sklə-] n. [病理] 硬化，[医] 硬化症，细胞壁硬化
soma [ˈsəumə] n. 体细胞，躯体，身体
spatiotemporal [ˌspeiʃiəuˈtemperəl] adj. 时空的，存在于时间与空间上的
spinesn. [解剖] 脊柱；棘状突起，体刺（spine 的复数）
stigmata [ˈstigmətə] n. 气孔，皮肤红斑
stratum lacunosum 腔隙层
stratum pyramidale 锥体细胞层
striatum n. 纹状体，终脑的皮层
subcortical [ˌsʌbˈkɔːtikəl] adj. 皮质下的，[心理] 皮层下的
subfield [ˈsʌbfiːld] n. [数] 子域，分区
subiculum n. [植] 菌丝层，脑下脚
subpial adj. 软膜下的
substantia nigra [解剖] 黑质
sulcus [ˈsʌlkəs] n. [动] 沟，槽，裂缝
symptom [ˈsim(p)təm] n. [临床] 症状，征兆
synaptic [siˈnæptik;sai-] adj. 突触的，(染色体) 联合的
synaptophysin n. 突触小泡蛋白，突触素
syndrome [ˈsindrəum] n. [临床] 综合症状，并发症状，校验子，并发位
synergistic [ˌsinəˈdʒistik] adj. 协同的，协作的，协同作用的
synuclein n. 突触核蛋白
temporal horn [解剖] 下角
thalamus [ˈθæləməs] n. [解剖] 丘脑，花托

thioflavin n. 硫黄素
topographical [ˌtɔpə'græfikəl] adj. 地志的，地形学的（等于 topographic）
transsynaptic adj. 跨突触的
tubulin ['tju:bjulin] n. [生化] 微管蛋白
tunica ['tju:nikə] n. [生物] 膜，被膜，[生物] 被囊
ubiquitin [ju:'bikwitin] n. 泛激素，泛素
ultrastructural adj. 超微的，超微结构的
vacuole ['vækjuəul] n. [细胞] 液泡，[地质] 空泡
vascular ['væskjulə] adj. [生物] 血管的
vein [vein] n. 血管，静脉，叶脉，[地质] 岩脉，纹理，翅脉，性情
venule ['venju:l] n. 小翅脉，[解剖] 小静脉
volumetric [ˌvɔlju'metrik] adj. [物] 体积的，[物] 容积的，[物] 测定体积的
ventricule n. 心室，脑室

难句分析

1. In addition，clinicopathological correlation studies have been crucial to generate hypotheses about the pathophysiology of the disease，by establishing that there is a continuum between "normal" aging and AD dementia，and that the amyloid plaque build-up occurs primarily before the onset of cognitive deficits，while neurofibrillary tangles，neuron loss，and particularly synaptic loss，parallel the progression of cognitive decline.

   译文：此外，临床病理相关性的研究已经表明，重要的是建立关于这种疾病的病理生理学的假说，即建立一个从"正常"老化到阿尔茨海默病连续的，以及以淀粉样斑块形成为主的（评价标准），（这些）发生在认知功能障碍、神经元纤维缠结、神经元丢失，特别是突触丧失、伴有认知衰退进程的发生之前。

2. The hierarchical pattern of neurofibrillary degeneration among brain regions is so consistent that a staging scheme based on early lesions in the entorhinal/perirhinal cortex，then hippocampal Ammon subfields，then association cortex，and finally primary neocortex is well accepted as part of the 1997 NIA-Reagan diagnostic criteria.

   译文：在大脑区域中的神经元纤维变性的分层模式是很稳定的，因此，基于在内嗅皮层/边缘皮层早期病变，然后是海马阿蒙区域，再然后是联合皮层，最后是初级皮质，这样一种分阶段变化进程是公认的 1997 NIA-里根诊断标准的一部分。

3. Although the gross visual examination of the AD brain is not diagnostic，a typical symmetric pattern of cortical atrophy predominantly affecting the medial temporal lobes and relatively sparing the primary motor，sensory，and visual cortices，is considered strongly suggestive of AD being the condition underlying the patient's dementia.

   译文：虽然阿尔茨海默病人大脑的粗略视觉检查不能作为诊断（标准），但是，主要影响内侧颞叶，波及初级运动、感觉和视觉皮层的一个典型的对称性皮质萎缩模式被强烈建议作为引起病人痴呆的阿尔茨海默病的条件。

4. Next，the involvement of multimodal high-order association isocortical areas accounts for the progressive impairment of additional cognitive domains，including executive dysfunction (prefrontal cortex)，apraxias (parietal cortex)，visuospatial navigation deficits (occipitoparietal cortex)，visuoperceptive deficits (occipitotemporal cortex)，and semantic memory (anterior temporal cortex)，giving rise to the full-blown dementia syndrome.

   译文：其次，多种模式的高度有序的关联同形皮质区的参与通过引起额外的认知区域的

进行性损害，包括执行功能障碍（前额叶皮质）、运用不能症（顶叶皮质）、视觉空间导航缺陷（枕顶叶皮质）、视觉障碍（枕颞皮质）和语义记忆（颞前皮质），导致了全面的痴呆综合征。

5. This indicates that synapse loss predates neuronal loss and that the remaining neurons become less well connected to their synaptic partners than expected just by the number of viable neurons surviving in a particular circuit.

 译文：这表明神经突触损失早于神经元损失，并且，剩余的神经元变得比预期的可见的在特定回路中存活的神经元更不易与神经传递有关的参与者相接触。

6. Longitudinal prospective clinicopathological studies in nondemented elderly people have revealed that up to 45% of nondemented elderly would meet the NIA-RI criteria for AD had they been demented, usually the intermediate likelihood category of these criteria, and rarely the high likelihood category.

 译文：纵观在非痴呆老年人的临床病理学研究中发现，达到45%的非痴呆老年人满足已经发展成阿尔茨海默病人的 NIA-RI 标准，（但他们）通常满足标准分类中的中度可能性，很少满足标准分类中的高度可能性。

（选自：Serrano-Pozo A，Frosch MP，Masliah E，Hyman BT. Neuropathological alterations in Alzheimer disease. Cold Spring Harb Perspect Med. 2011；1（1）：a006189.）

Section Ⅱ How to Write a Scientific Paper (英语学术论文写作)

1 Introduction（序论）

撰写科技论文是科学研究工作中必不可少的组成部分。不论实验研究，还是理论研究所获得的结果如何重要，在其没以论文的形式公开发表之前，这项研究工作都不算最后完成。因此，撰写科技论文是完成某项研究课题中的阶段性的或最后的程序。

一位大学本科生，一位研究生，甚至一位科学家，其科研能力不但由其在实验室中操作的熟练程度来衡量，也要由他所从事的科研领域的宽和窄来衡量，还要由其智力和潜力来衡量，更要由其发表论文的水平和知名度来衡量，而只有将科研成果公之于世，才能变成知识，才能为社会所共享。所以本科生、研究生在大学学习期间就必须学会撰写科技论文。

1.1 科技论文的基本特征

所谓科技论文，就是指在科学研究的过程中，通过足够的、可以重复的研究结果，用同行得以评价和信服的素材论证，揭示事物发展的规律，正式发表在科技期刊上的论述性文章。因此完备的科技论文应该具有科学性、创新性、逻辑性、通达性。

所谓科学性，一是指论文表达的内容科学，即表达的内容是客观存在的事实，或可被实践检验的理论，探讨和论述的问题必须符合客观事物的发展规律，符合被实践规律证明了的法则，经得起别人的检验，无论何时、何地，只要按相同的条件进行实验就能得到相同的结果，绝不是那些伪科学或假科学的内容。二是指内容表达科学，即立论客观、论据充足、论述严密。立论必须以足够的和可靠的实验数据以及观察到的现象作为基础，不能认为观察到的某些现象或得到的某些数据和结果与自己的预测相反就随意取舍，不能凭个人的好恶主观臆造。论据必须翔实、充分和可靠；论证要从事实出发，以严格周密的逻辑推理得出恰当的结论。

所谓创新性就是指论文揭示事物的现象、属性及运动规律是前所未见的，而不是堆砌材料、罗列现象或重述、模仿别人的成果和结论。创新是科技论文的灵魂，那种囿于别人的见解，无所发现、无所发明、无所创造、无所前进的文章是不能称其为科技论文的。

所谓逻辑性指的是论文篇章结构脉络清晰、结构严密、前提完备、演算正确、符号规范、图表精致、推断合理、前后照应、自成体系，通过对客观事实的分析和推理，提高到理论的高度，而不是一堆堆无序的数据和一个个观察到的现象的无规律的堆砌。

所谓通达性是指论文文字表述流畅通顺，通理达意，要符合大多数读者的思考习惯和阅读习惯，要让人看懂。切忌行文呆板，语句晦涩，结构松散，词不达意。

1.2 科技论文的分类

严格并科学地对科技论文进行分类是一件很不容易的事情。因为从不同的角度去分析，就会有不同的分类结果。通常在高等院校攻读学士、硕士、博士学位的大学本科生和研究生为取得相应的学位，依据科学研究获得的结果所撰写的论文称作学位论文（dissertation）。而科研人员总结的在科研中所获得的新成果发表在科技期刊上的论文称作学术论文。就学术论文而言，其写作题材亦多种多样，常见的有研究论文（paper or article）、研究简报（note）、研究快报（letter or communication）和综合评述（review）四种体裁类型。

1.2.1 研究论文

研究论文是指对实验和观察所得到的结果从一定的理论高度所进行的分析和总结，形成一定的科学见解，并对自己提出的见解，用事实和理论进行周密且符合逻辑的论证。所以一篇研究论文不但要有创新性，而且还要强调理论性。

1.2.2 研究简报

研究简报是指对课题研究所取得的有特殊意义的结果所作的初步报道，其写作体裁的鲜明特点是文章短小精悍。所谓短小就是文章篇幅短。所谓精悍就是指文章主要表达作者从事某项研究所采用的最主要的研究方法、所得到的重要结果和得出的结论。从文章中虽见不到作者为支持结论而进行论证过程，但其主要观点和独到的研究方法以及有价值的结果却会使人一目了然。通常一项科研课题在某个研究阶段已得到某些突破性的结果，但离全面完成这项研究尚有一段距离。这时作者需要认真总结写成研究简报形式予以发表，以便及时与同行进行研究动态交流。但研究简报一旦在科技期刊上发表后，其内容在任何时候和任何地方都不能重复发表。

1.2.3 研究快报

研究快报（letter）也称作研究通讯（communication 或 correspondence），是指对一项科研课题所取得的重大突破性进展所作的快速报道，这种论文体裁不仅要求篇幅短，内容精悍，而且必须快速发表。文章内容务须论述支持所得结论的独特证据，而不必涉及有关的实验细节。在研究快报发表之后，还可撰写详细的研究论文在发表快报的刊物或在其他刊物上申请全文发表。

1.2.4 综合评述

综合评述（review）是一类较为特殊的科技论文，是指在广泛查阅某一学科、某一研究领域科技文献资料之后，运用分析和综合的方法，进行鉴别、分类、归并，进行整体上的深入研究和重新组合所形成的一种更有价值的文献论文。一篇好的综述文章既要包括已有的文献内容的回顾和某些尚未公开发表的新资料和新思想，更要阐明某一课题的发展演变规律并对其发展趋势进行展望，提出合乎逻辑的、具有启迪性的观点和建议。这类文章要求较高，具有权威性，对所讨论的研究领域的进一步发展具有引导作用，是科研工作者在选题、立题和开题前必须查阅的文献资料。

1.3 科技论文的基本框架

在科技期刊上公开发表的科技论文常见的写作格式按从开头到结尾的顺序包括文题、作者及其工作单位、摘要、关键词、前言（或引言）、正文（其中包括实验部分或理论基础、实验结果或理论结果、讨论、结论）、致谢、参考文献和附录等。

1.3.1 文题

文题（title or headline）即论文的题目或篇名。文题位于论文开篇之首，是论文中简洁、恰当描述并反映文章主要内容，对读者具有启迪作用的部分。

1.3.2 作者及其工作单位

作者（authors）是指参与论文的选题、从事具体的研究工作，并对论文主题内容进行构思及具体执笔撰稿等方面的全部或部分工作的重要贡献者，是能对文章内容负责的法定责任人。工作单位（addresses）指的是作者完成该研究论文的地点（如高等学校或研究所等）或作者的工作单位或作者的通信地址，是作者与同行之间进行学术交流所必需的通讯信息。作者及其工作单位在论文中位于论文首页的文题之下。

1.3.3 摘要

摘要（abstract or summary）又称提要（synopsis）。摘要是科技论文主要内容和基本思想的缩影或概述，具有独立性和自明性。摘要不但可以使读者通过阅读就可能了解到论文中的主要创新信息，还能为科技文献数据库和检索类刊物提供转载的信息。通常，长篇科技论文都需要提供摘要，并编排在作者及其工作单位之后。研究快报、通讯和简报，因其短小精悍，亦可不加摘要。

1.3.4 关键词

关键词（key words）也常称主题词（subject words），是指从文题、摘要和正文中选出的最能反映科技论文主要内容和特征并具有实在意义的词或词组。它可使读者通过查阅关键词即可追溯到原文，同时也给文献数据库和检索系统提供了编制索引的信息。目前国内外有些科技期刊在录用论文时都要求作者提供几个能代表论文内容主题方面的关键词列于论文摘要之后。

1.3.5 引言

引言（introduction）亦称导言、前言或序言，科技论文的开篇，主要介绍论文的研究背景，与此文相关的领域中前人的研究历史和现状，作者的研究目的和预期达到的目标以及该目标在理论和实践上的意义和价值等。它类似于一部交响乐的序曲，一部长剧的序幕，一篇宣言的总纲。通过阅读前言可使读者对所进行的这项研究一目了然。

1.3.6 正文

正文（main body）是科技论文的主体或核心组成部分，在整篇论文中占有较大的篇幅。正文是作者通过对科研实践中所获得的数据、结果、现象这些感性资料进行去粗取精，去伪存真，由表及里的整理、综合、分析、逻辑推理，并上升到理论认识的文字表达。通常实验性的学术论文中应包括实验部分、实验结果、讨论和结论等几个部分。理论性的论文亦包括理论基础、理论方法、理论结果、讨论和结论等几个部分。

① 实验部分（experimental section）主要介绍实验所用材料或试剂和方法。

② 实验结果（results）主要是总结实验中所得到的各种现象和数据，并对这些现象和数据进行定性或定量的分析，得出规律性的东西。

③ 结果与讨论（results and discussion）主要是对实验结果做出合理解释，回答引言中所提出的问题，将实验结果由感性认识上升到理性认识的文字表述。

④ 结论（conclusion）是论文的结束语或整篇论文的归结。它是在理论分析和讨论实验结果的基础上通过严密的推理而形成的富有创新性和指导性的并与引言相互呼应的概括总结。

1.3.7 致谢

致谢（acknowledgments）是指对此论文的完成做出贡献（如给予指导或提出重要建议，或提供实验材料或仪器，或做过某些测试）的人，给予研究经费资助的单位和团体所表达的必不可少的诚挚谢意。通常置于正文部分的结论之后。

1.3.8 参考文献

参考文献（references）指的是撰写或编辑论文引用的有关文献信息资源，通常以文后参考文献表形式出现。参考文献表就是附在论文之后论文中引用或参考他人或本人已公开发表的论文、图书、专利等出处的一览表。参考文献是论文对前人研究成果的继承和发展的标志，也是统计影响因子和总被引频次、考察论文影响力的基本资料，是论文的重要组成部分之一。

1.3.9 附录

附录（appendix）也称为论文的附件。它不是论文的必要组成部分。它旨在向读者介绍一些不便于列入正文的具有参考价值的资料，如详尽的推导、演算过程等或补充提供一些辅助资料等。

1.4 科技论文的撰写语言

不同的国家、不同的民族都可以用自己的母语来撰写科技论文，但是各国科技工作者都希望将自己的论文放到国际同行中进行比较，以增大学术交流的效果和扩大论文在国际学术界的影响。要想实现这一目标，采用国际上大多数科技工作者都熟悉的和使用得最为广泛的语言，即英语，撰写科技论文，则是最佳途径。

随着我国改革开放事业的不断深入和科技事业的迅猛发展，我国科技工作者完成的科技论文的数量大幅度增加。据统计，几乎4篇文章中就有一篇是用英语撰写的，并且用英语撰写论文数量的增加趋势越来越显著。因此我国高等学校的本科生和研究生在校学习期间很有必要学习一些科技英语写作知识，尤其是学习如何用英语撰写科技论文的基础知识，以达到能够正确使用英语语言来报道自己在学校学习期间或毕业走上工作岗位后所完成的科研成果，使国际同行能读懂，以便取得较好的学术交流的效果。正是基于这一目的，在本教科书中专门编写了“英语学术论文写作”一章，希望对在高等学校学习的本科生和研究生能有所裨益。

2 How to Write the Title（论文题目的写法）

2.1 文题的重要性

论文题目（简称文题）是科技论文的重要组成部分之一，是将一篇论文与其他论文相区别的重要标志，也是文章主题内容的直接反映。

科学引文索引（Scientific Citation Index，SCI）、工程索引（Engineering Index，EI）、化学文摘（Chemical Abstracts，CA）等著名的权威数据库和文献检索系统都是依据论文题目来进行摘引和编制索引的。而成千上万的读者往往都是先从期刊的目录上、文献检索刊物上或文献数据库中先看到文题，然后再决定是否看原文。尤其是在20世纪80年代之后，全世界每年的科技文献量已远远超出1亿册，面对数量如此巨大且迅速激增的科技信息，人们即使夜以继日地查阅，所接触到的科技信息也只不过是沧海一粟，因此科技人员查阅论文题目是他们获得所需要的科技信息的重要渠道。鉴于此，在撰写科技论文的过程中首先要写好文题。如果文题写得不好，读者可能看不懂，文献检索系统和数据库很可能不予摘录。这篇论文就会因失去读者而被束之高阁，文章虽然发表了，但由于无人问津而失去其应有的价值。

2.2 题目务需精悍

好的论文题目应该是用词精练、语法正确，且能准确、完整地表达论文的核心内容。所谓精练就是指用词要尽可能地少。少的标准是应该让读者看懂，包含能反映论文主题内容的几个关键词和其他必要的词。而不能令读者莫名其妙，丈二和尚摸不着头脑。

例如有一篇论文的题目为“Studies on Brucella”（关于布鲁菌的研究）。这个题目给人以含混不清的感觉。从题目中很难判断论文究竟是属于Brucella的分类学（taxonomic）、遗传学（genetic）、生物化学（biochemical）、医学（medical）中的哪一方面研究，这样的文题尽管精练得只有3个词，却由于太笼统而毫无价值。

但在大多数科技英文论文中，文题长是经常可见的。长的文题中常常含有多余的字，往往这些多余的字看上去似乎是正确的，并大多出现在文题的开头，如“Studies on…”、“A Study of…”、“Investigation on…”、“Research on…”、“Report on…”、“Observation on…”等是司空见惯的。实际上，在大多数情况下，这些词都是多余的，完全可以略去。例如“Studies on Synthesis of Analogues of Marine Anti-tumour Agent Curacin A”中的Studies on应略去。在大多数情况下，题目中应尽量不用冠词“the”，对那些无定量意义的修饰词如“rapid”和“new”等也尽可能少用或不用。

2.3 用词务需具体明确

文题中涉及的术语不能用一般性的术语涵盖具体的术语，选择专业术语应该恰当反映正文中所研究的具体内容。例如“Action of Antibiotics on Bacteria”（抗生素对细菌的作用）这一文题，看上去没有多余的词，但仔细分析起来，antibiotic（抗生素）是多种多样的，bacteria（细菌）不仅种类繁多，而且数量亦相当大，该文题所涉及的正文绝不可能研究所有的抗生素对各种各样细菌的作用，充其量也只不过是研究了一种或几种抗生素对一种或几种细菌的作用。另外，action（作用）一词也未具体说明是属于哪一种作用。根据论文的主体内容，若将文题写成“Inhibition of Growth of *Mycobacterium tuberculosis* by Streptomycin”（链霉素对结核分枝杆菌生长的抑制作用），其含义就会更为明确。再如“New Color Standard for Biology”和“New Color Standard for Biologists”两个文题中，虽然只有Biology和Biologists的差别，但表达的内容却是完全不同的。前者旨在描述动植物样本的颜色指标，而后者旨在告诉人们通过新指标可将不同颜色的生物相互区分开来。

2.4 语法务需正确

文题中涉及的语法现象虽然较少，但必须使用得正确，否则将产生歧义。例如“Mechanism of Suppression of Nontransmissible Pneumonia in Mice Induced by Newcastle Disease Virus”这一文题中，由于将Induced放错了位置，因而很容易使人误解为Pneumonia（肺

炎）是由鸡瘟病毒感染而引起的，而文章的核心内容是通过鸡瘟病毒来抑制小鼠感染的肺炎。因此将论文题目写成“Mechanism of Suppression of Nontransmissible Pneumonia Induced in Mice by Newcastle Disease Virus”（鸡瘟病毒对小鼠感染的非传播性肺炎的抑制机理）才能恰当地反映论文的内容。

在科技论文的题目中，常常见到 using、employing 等组成的短语，在使用这样的短语时，务必考虑其在文题中的作用。例如文题“Isolation of Antigens from Monkeys Using Complement-Fixation Techniques”中的 Using 引起的短语是状语。但易将 Using 理解成现在分词当作修饰 Monkeys 的定语，而实际上 Monkeys 并不是 Using 的逻辑主语，而是动名词，该动名词与介词搭配构成动名词短语才能起到状语作用，即将文题写成“Isolation of Antigens from Monkeys by Using Complement-Fixation Techniques”（采用补体固定技术从猴子中分离抗原）。

2.5 尽可能不用句子

文题像一条标签，通常是通过将词汇或术语按语法规则规范地安排成序而写成的。切忌使用主、谓、宾语结构构成的句子来写文题。例如“β-Endorphin is Associated with Overeating in Genetically Obese Mice and Rats”（内啡肽在遗传上与肥胖老鼠过度饮食的关系）这一文题是用句子表述的，实际论文题中不用“is”并不影响对文题的理解。

2.6 避免使用化学式、公式、缩略语等

文题中应尽量避免使用化学式（chemical formulas）、数学公式（mathematical formulas）、读者不太熟悉的符号（symbols）、缩写（abbreviations）、非规范性的术语（jargon）和商标名（proprietary names）等。这是因为大多数文献数据库和检索系统中其索引都是以规范的词汇编制的。如在索引中，盐酸用的是 hydrochloric acid，而不用 HCl；脱氧核糖核酸用的是 deoxyribonucleic acid，而不用 DNA。

2.7 系列文题

所谓系列文题指的是在一个总的文题中，撰写若干篇论文，每一篇论文都有一个研究主题，为了对各篇论文加以区别，通常对总文题中的各篇文章按照完成或撰写的先后用罗马数字编序，然后再写上该篇论文的具体题目。例如 Studies on Bacteria Ⅳ. Cell Wall of *Staphylococcus aureus*（关于细菌的研究Ⅳ. 金黄色葡萄球菌的细胞壁）这一文题是细菌研究中的第四篇论文，而这篇论文是有关金黄色葡萄球菌细胞壁的研究结果。

系列论文往往不是同时完成的，也不是同时连续发表的，在多数情况下是在不同的刊物上发表的，所以显得文章与文章之间的联系松散，给人以一种零打碎敲的感觉，给读者查阅带来困难。此外，系列论文中的各篇文章都有自己的主旨内容，都有独立的研究结果和结论，因而总文题在此篇文章中不但无大用途，而且还为文题精练造成累赘。过去，系列文题在科技期刊中是很常见的。目前，这类文题用得越来越少了。鉴于此，在学习写论文时，亦尽量避免使用系列文题。

2.8 主副文题

所谓论文的主副文题指的是一篇论文中有两个文题，其中一个为主要的，称为主文题，另一个是进一步补充说明主文题的，称作副文题，总称为主副文题或悬挂式文题（hanging title）。例如“the Fragile Site in Somatic Cell Hybrids：an Approach for Molecular Cloning of Fragile Sites”这一文题中，冒号之前的部分为主文题，其后为副文题。这种文题虽可帮助读者理解全文，但由于多加了文字和标点符号，有时较难抓住重点，也不利于编制索引，若用一个文题“an Approach for Molecular Cloning of Fragile Site in Somatic Cell Hybrids”可能效果更好。

2.9 典型论文题目举例

① Acetic Acid Bacteria as Enantioselective Biocatalysts（Journal of Molecular Catalysis B：Enzymatic，2002，Vol 17：235.）

② Methods to Increase Enantioselectivity of Lipases and Esterases（Current Opinion in Biotechnology，2002，Vol 13：543.）

③ Antibody Multispecificity Mediated by Conformational Diversity（Science，2003，Vol 299：1362.）

④ Distinctive Roles of HPAP Proteins and Prothymosin-α in a Death Regulatory Pathway（Science，2003，Vol 299：223.）

⑤ Genome Sequence of the Human Malaria Parasite Plasmodium Falciparum（Nature，2003，Vol 419：498.）

⑥ DNA Sequence and Analysis of Human Chromosome 14（Nature，2003，Vol 421，601.）

⑦ Merotelic Kinetochore Orientation versus Chromosome Mono-orientation in the Origin of Lagging Chromosomes in Human Primary Cells（Journal of Cell Science，2002，Vol 115：507.）

⑧ Nglyl，a Mouse Gene Encoding a Deglycosylating Enzyme Implicated in Proteasomal Degradation：Expression，Genomic Organization，and Chromosomal Mapping（Boichemical and Biophysical Research Communications，2003，Vol 304：326.）

3 How to List the Authors and Addresses（作者及其工作单位的写法）

3.1 英语人名的写法

在英国、美国等国，人的姓名通常由三部分组成，即首名（first name）、中间名（middle name）和末名（last name），其末名即相当于姓氏（surname 或 family name）。通常首名和末名用全称，词的首字母用大写，中间名用该词的首字母大写形式，如 Christopher

J. Easton。也可以将首名和中间名都用缩写，如 C. J. Easton。也有作者的姓名由首名和末名两个词组成，都用全称且首字母大写，如 John Mann；也可以将首名用缩写字，如 J. Mann。

3.2 汉语人名的写法

按照中华人民共和国国家标准（GB 3259—92，GB/T 16759—1996）中的规定，汉语人名采用汉语拼音，按姓和名分写，姓在前，名在后，首字母均用大写。如 Wang Baoguo（王保国）和 Zhao Dong（赵东）等。当我国科学工作者在国外英文版刊物上发表论文时，为了使汉语人名符合英语人名的习惯，或防止外国读者对我国姓和名的误解，也可采用名在前、姓在后的方式书写，如 Baoguo Wang。

在论文后面所附的参考文献表中，著者的姓名一项往往采用名用首字母缩写，姓用全称的方式。这样在处理姓名为三个字的时候，往往名的部分就被缩写成一个字母，如 Baoguo Wang 被缩写成 B. Wang。为了解决这个问题，有的刊物将姓名中名部分的两个字之间加连字号，且这两个字的首字母用大写字母，如 Bao-Guo Wang，但这种写法尚未得到国家认可。

3.3 多个作者姓名的写法

当论文的作者为两位时，则在两个作者之间加连词“and”。如“Andrew S. Balnaves and Mark E. Light”；“Bao-Guo Wang and Dong Zhao”。

当论文的作者为三位或三位以上时，作者姓名之间加逗号“，”，最后两位作者之间加连词“and”。如“Andrew S. Balnaves，Mark E. Light and Christopher J. Easton”；“Bao-Guo Wang，Dong Zhao and Wan-Long Li”。

3.4 作者姓名的顺序

凡是对整个论文的完成（包括设计、实验、总结、撰写论文等）做出贡献的课题组成员都可以是论文的作者。当作者为两个或两个以上时，则按照对完成该论文做出贡献的大小来排序。排在前面者，其贡献大，同时他对该论文承担的责任也最大。

但目前也有这样的倾向，课题组负责人往往是学术带头人或导师，他们对该论文所做出的贡献最大，应负的责任也最大。但他们从鼓励其学生或年轻一代致力于科学研究的角度出发，往往将年轻人放在前面，而将他们自己放在最后，这种排法也是普遍的，但他们大多都是文章的联系人（correspondence）。

3.5 作者工作单位的写法

在一篇论文中，在作者姓名的后面注明其工作单位，目的在于辨认作者和提供作者的通讯地址，以便与同行之间进行学术交流，也可以用于与论文相关的一些通讯联系。

通常，论文的每位作者都有各自的工作单位，工作单位相同的作者可共用一个工作单位。工作单位不同的作者，作者姓名及其工作单位都要用英语字母或阿拉伯数字或其他符号按照作者姓名的先后顺序依次标记，且地址的先后顺序应与作者姓名的顺序一致。举例如下。

作者属于同一工作单位的情况如下。

David C. Sherrington，Alexander Swann and Lan M. Huxham

Department of Pure & Applied Chemistry，University of Strathclyde，259 Cathedral

Street，Glasgow，UK G 11XL

Tao Sun，Zhu-De Xu and Zhi-Sheng Jia

College of Life Science，Qianwei Campus，Jilin University，Changchun 130023，P. R. China

作者的工作单位不同的情况如下。

Christopher J. Easton[a]，Stephen F. Lincoln[b]，Adam G. Meger[a] and Hideki Onagi[b]

a. Research School of Chemistry，Australian National University，Canberra，ACT 0020，Australia.

b. Department of Chemistry，University of Adelaide，Adelaide，SA 5005，Australia

Pei-Yong Cao[1]，Li-Fan Chen[1] and Yu-Hang Song[2]

1. Department of Chemistry，Qianwei Campus，Jilin University，Changchun 130023，China

2. Changchun Institute of Applied Chemistry，Chinese Academy of Sciences，Changchun 130022，China

4 How to Write the Abstract（摘要的写法）

摘要置于论文正文之前，在科技期刊中有的在其前冠有 Abstract，有的冠有 Summary。还有的两个词都不用，只用与正文不同的字体或字号与正文相区别。

4.1 摘要的重要性

首先，一篇论文在投稿到科技期刊编辑部或期刊社之后，需经过编辑人员和同行专家审查后，才能确定是否录用并发表。他们在审稿时通常是先阅读摘要看其是否有价值，然后再决定是否审读全文。其次，论文在期刊上发表后，文献数据库和文摘刊物的文摘员亦是先看摘要的价值，再决定是否将其摘录。再次，科技工作者在查阅科技文献时，也是首先阅读摘要，不是阅读一次文献中的摘要，就是阅读如化学文摘（Chemical Abstract）、生物学文摘（Biological Abstract）等二次文献中的摘要，通过阅读摘要内容来了解他们感兴趣的信息，以便决定是否再看原文。由此看来，摘要对编者和审者，对于情报机构，对于读者都是十分重要的。因此，一篇质量高的论文必须有一篇高水平的摘要，要通过摘要引起编者、审者、文摘员和读者对论文的注意，引起他们的兴趣，使他们有一个总的概念，不得不看全文。即使看不到全文，也能对正文的主要内容有所了解。如果摘要写得不好，即使论文的内容再好，也有遭到拒稿的可能。

4.2 摘要的特点

① 要写得简明，应将一篇摘要中的实词数量控制在 250 个左右。

② 内容应客观、真实，不能涉及论文中没有出现的内容，不能加入作者的主观见解、解释和评论。

③ 犹如一篇完整的短文，要有独立性和自明性。

④ 不分章节、段落，不列图表，不注参考文献。

⑤ 结构严谨、语言文字确切，表述简明、一气呵成。

4.3 摘要的类型

科技论文的摘要通常有两种类型。一种是信息性摘要（informative abstract），一种是指示性摘要（indicative abstract）。

信息性摘要中需要囊括论文正文中的主要内容，要阐明论文研究的目的和目标，给出达到此目的和实现此目标所用的方法和手段；总结所获得的主要数据、观察到的现象和研究结果；阐明得出的主要结论。举例如下。

Summary

Defects in chromosome segregation play a critical role in producing genomic instability and aneuploidy, which are associated with congenital diseases and carcinogenesis. We recently provided evidence from immunofluorescence and electron microscopy studies that merotelic kinetochore orientation is a major mechanism for lagging chromosomes during mitosis in PtK1 cells. Here we investigate whether human primary fibroblasts exhibit similar errors in chromosomes segregation and if at least part of lagging chromosomes may arise in cells entering anaphase in the presence of mono-oriented chromosomes. By using in site hybridization with alphoid probes to chromosome 7 and chromosome 11 we showed that loss of singe sister is much more frequent than loss of both sisters from the same chromosome in anatelophases from human primary fibroblasts released from a nocodazote-induced mitotic arrest, as predicted from merotelic orientation of single kinetochores. Furthermore, the lagging of pairs of separated sisters was higher than expected from nandom chance indicating that merotelic orientation of one sister may promote merotelic orientation of the other. Kinetocheores of lagging chromosomes in anaphase human cells were found to be devoid of the mitotic checkpoint phosphoepitops recognized by the 3F3/2 antibody, suggesting that they attached kinetochore microtubules prior to anaphase onset. Living cell imaging of H2B histone-GFP-transfected cells showed that cells with mono-oriented chromosomes never enter anaphase and that lagging chromosomes appear during anaphase after chromosome alignment occurs during metaphase. Thus, our results demonstrate that the mitotic checkpoint efficiently prevents the possible aneuploid burden due to mono-oriented chromosomes and that merotelic kinetochore orientation is a major limitation for accurate chromosome segregation and a potentially important mechanism of aneuploidy in human cells.

(Journal of Cell Science, 2002, Vol 115: 507.)

指示性的摘要只要求介绍和阐明论文的主题和概括性的结果及结论，给读者一个指示性概括了解，根据需要再阅读原文。举例如下。

Abstract

A concise, multigram synthesis of (4*R*)-2-[(1′-*R*, 2′-*S*)-1′, 2′-methano-3′- (*tert*-butyldiphenylsiloxy) propyl] -4-hydroxymethyl-4, 5-dihydrothiazole has been achieved, and this compound has been used for the production of a range analogues of the anti-tumour agent curacin A. (J Chem Soc Perkin Trans Ⅰ, 1999: 2455.)

通常，研究论文普遍采用信息性摘要，综合评述性或评论性文章采用指示性摘要。

4.4 摘要中的人称

摘要中的人称指的是用作句子主语的人称代词。由于摘要主要是客观、概括地介绍正文的内容，所以每个句子的主语大多为名词。但有时在需要的时候，也可以使用代词。其中用得比较多的为“we”。例如，In this study, we report the genomic organization and mRNA distribution of the mouse Ngly1.（在本研究中，我们报道了老鼠聚糖酶中的基因组织及信使核糖核酸的分布。）(Biochemical and Biophysical Research Communications, 2003, Vol

34：326.）

再如，From an analysis of the CpG island occurrences，we estimate that 70% of these annotated genes are complete at their 5′end.（从对出现的 CpG 岛的分析中，我们估计在它们的 5′端，70%已注释的基因都是完整的。）（Nature，2003，Vol 421：601.）

此外，非人称代词“it”作为形式主语（实际主语为主语从句或不定式短语）时在摘要的结论部分中也时常可见。如 It is still not well understood why a certain lipase or esterase shows a certain enantioselectivity in a given reaction.（在给定的反应中为什么脂酶或酯酶表现出对映选择性尚不十分清楚。）（Chem. Biotech，2002，Vol 13：543.）

又如，It is noteworthy that nonanoic lactone was produced in one step by oxidation of 1,4-nondiol.（值得注意的是，壬内酯是由 1,4-壬二醇经一步氧化生成的。）（J Molec Catal B：Enzymatic，2002，Vol 17：235.）

4.5 摘要中的语态

凡谓语动词和非谓语动词（如动词不定式、动名词、现在分词）为及物动词时，都可以有主动语态和被动语态两种形式，具体使用哪种语态取决于其主语或逻辑主语是该动词的执行者还是承受者，若是前者，则采用主动语态，若是后者，则采用被动语态。

Example 1（主动语态谓语）

Analysis by both X-ray crystallography and pre-steady-state kinetics reveals an equilibrium between different preexisting isomers.（X 射线结晶学和前稳态动力学分析揭示在不同的预先已有的异构体之间存在一个平衡。）(Science，2003，Vol 299：1362.）

Example 2（谓语为被动语态）

These chromosomes are characterized by a hetero-chromatic short arm containing RNA genes and a euchromatic long arm containing protein coding genes.（这些染色体是以一个含 RNA 基因的异染色体短臂和含蛋白质编码基因的常染色体长臂为特征的。）（Nature，2003，Vol 421：601.）

Example 3（现在分词短语的主动语态）

The thermal stability of complexes involving these oligonucleotide analogues has been evaluated towards complementary single-stranded DNA and RNA.（针对辅助的单股 DNA 和 RNA，估算了含有这些低聚核苷配合物的热稳定性。）（J Chem Soc Perkin Trans Ⅰ，1999：2543.）

Example 4（现在分词短语的被动语态）

The rate of acetylation is influenced by the relative stereochemistry of C(3)-C(4)β-lactam carbon atoms，the *trans* isomers being transformed much faster than the *cis* ones. [乙酰化的速度受到 C(3)-C(4)β-内酯碳原子的有关立体化学的影响，反式异构体乙酰化速度比顺式异构体快得多。]（J Chem Soc Perkin Trans Ⅰ，1999：2489.）

Example 5（不定式短语的主语语态）

The evidence presented here suggests that thymosin β4，a protein involved in cell migration and survival during cardiac morphogenesis，may be redeployed to minimize cardiomyocyte loss after cardiac infarction.（这里提供的证据表明，在强心形态形成期间，细胞迁移或残留中所涉及的一种蛋白质——胸腺素 β4 在心肌梗死后可以重新调配以降低心肌细胞的损失。）（Nature，2004，Vol 432：467.）

Example 6（不定式短语的被动态）

In fact，recognition of the cleavage site has been believed to be considered as the key step in termination.（事实上已经确信可以将断裂点作为终止的关键步骤考虑。）（Nature，2004，Vol 432：456.）

4.6 摘要中的时态

摘要行文中常用的时态主要有 3 种，即一般现在时，一般过去时和现在完成时。

一般现在时态主要用于介绍研究论文的背景，其中包括目的和目标、描述研究的思路、方案、方法和过程；表述被作者的研究证明是正确的结果和结论或普遍存在的真理；提出新的理论或建议。

一般过去时态主要用于表述在写论文之前所做的工作（只注重于谓语动词本身发生在过去某一时间）。

现在完成时态主要用于表述写论文之前所做的工作到写论文时所产生的结果和影响。

Example 1

Abstract

In species as diverse as yeast and mammals，peptide：*N*-glycanase（PNG1 in yeast；Ngly1 in mouse）is believed to play a key role in the degradation of misfolded glycoproteins by the proteasome. In this study，we report the genomic organization and mRNA distribution of the mouse Ngly1. Mouse Ngly1 spans 61kb and is composed of 12 exons，the organization of which is conserved throughout vertebrates. Comparison of the mouse and human genomic sequence identifies a conserved gene structure with significant sequence similarity extending into introns. A 2. 6kb Ngly1 message was detected in all mouse tissues examined，with the highest abundance in the testis. In addition，a lower molecular weight transcript of 2. 4kb was detected in the testis. From analysis of dbESTs the alternative transcript of Ngly1 is predicted to be present in the human placenta. Given the key role Ngly1 plays in glycoprotein degradation，we predict that Ngly1 may be a contributing factor in "disease" susceptibility. To begin to address this question，we used radiation hybrid mapping to localize mouse Ngly1 to chromosome 14 and the human orthologue to chromosome 3 with a strong link with known genes.（Biochemical and Biophysical Research Communication，2003，Vol 304：326.）

该摘要中用一般现在时态介绍了论文的背景、内容和结果，用一般过去时态描述了实验中所做的工作。

Example 2

Abstract

Acetic acid bacteria（five strains of *Acetobacter* and five strains of *Gluconobacter*）were used for the biotransformation of different primary alcohols（2-chloropropanol and 2-phenylpropanol）and diols（1，3-butandiol，1，4-nonandiol and 2，3-butandiol）. Most of the tested strains efficiently oxidized the substrates. 2-Chloropropanol and 1，3-butandiol were oxidized with good rates and low enantioselectivity（enantiomeric excess＝18％-46％ of the *S*-acid），while microbial oxidation of 2-phenylpropanol furnished（*S*）-2-phenyl-1-propionic acid with enantiomeric excess（*e. e.*）＞90％ with 10 strains. The dehydrogenation of 2，3-butandiol was strongly dependent on the stereochemistry of the substrate；the meso form gave *S*-acetoin with all the tested strains，the only exception being a *Gluconobacter* strain. The formation of diacetyl was observed only by using *R*，*R*-2，3-butandiol with *Acetobacter* strains. Oxidation of 1，4-nonandiol gave γ-nonanoic lactone in one step，although with moderate enantioselectivity.

(Journal of Molecular Catalysis B：Enzymatic，2002，Vol 17：235.)

该摘要中只用了一般过去时态，着重描述了论文所做的工作，观察到的现象和得到的结果。

Example 3

A small molecule，α-(trichloromethyl)-4-pyridineethanol (PETCM)，was identified by high-throughput screening as an activator of caspase-3 in extracts of a panel of cancer cells. PETCM was used in combination with biochemical fractionation to identify a pathway that regulates mitochondria-initiated caspase activation. This pathway consists of tumor suppressor putative HLA-DR-associated proteins (PHAP) and oncoprotein prothymosin-α (ProT). PHAP proteins promoted caspase-9 activation after apoptosome formation, whereas ProT negatively regulated caspase-9 activation by inhibiting apoptosome formation. PETCM relieved ProT inhibition and allowed apoptosome formation at a physiological concentration of deoxyadenosine triphosphate. Elimination of ProT expression by RNA interference sensitized cells to ultraviolet irradiation-induced apoptosis and negated the requirement of PETCM for caspase activation. Thus，this chemical-biological combinatory approach has revealed the regulatory roles of oncoprotein ProT and tumor suppressor PHAP in apoptosis. (Science，2003，Vol 299：223.)

该摘要中用了一般过去时态介绍了论文的具体做法；用一般现在时态介绍研究背景；用现在完成时态介绍了所用方法已显示的作用。

5 How to Choose the Key Words（关键词的选择）

为了便于文献检索，许多科技刊物都要求提供5～8个关键词。

关键词一般是指在弄清论文的主题概念和中心内容基础上，从论文题名、摘要、层次标题以及论文的其他内容中挑选出来的能反映论文主题的名词或名词性词组，是文献检索的标识。因此，合理恰当地选择关键词可以增加论文被检索和引用的概率。

关键词具有专指性，即一个词或词组只能表达一个概念。首先应该选择论文的研究对象，如学科、理论、定律、现象、物质、性质、化学反应；其次应选择采用的技术、方法和手段等。

如 *Acetobacter*（醋酸杆菌属），peptide（肽），glycophorin（血型糖蛋白），oxidation（氧化），resolution（拆解），electron paramagnetic resonance（电子顺磁共振），Langmuir-Blodgett film（LB膜），gas sensor（气体传感器），chemical modification（化学修饰）等。

选择关键词必须规范，不能凭经验或主观想象去选择。通常可从大型检索类刊物如美国化学文摘（CA）中的关键词索引（keyword index）或主题索引（subject index）中选出适合所写论文中心内容的词和词组。如果某些词或词组在这两种索引中查不到，它们又确实能体现论文的主题内容，则可参照这两种索引中的选词规律来进行选择。

选择关键词必须具体贴题，例如有一篇论文报道的内容是有关老鼠基因的，选“Gene”作为关键词，初看起来似乎有道理，仔细研究一下，则不贴题，这时应选择“mouse gene”为关键词。

应避免将空泛的词选做关键词，如 synthesis，determination，property，method，characterization，analysis 等。为了适应检索系统的要求合理地逻辑组合，可将多词构成的关键词按照逻辑规律进行拆解。如可将 enzymolysis kinetics 改为 enzymolysis 和 kinetics 两个关键词。

6 How to Write the Introduction（引言的写法）

6.1 引言包括的内容

引言是论文的开头。常言道：万事开头难，写科技论文也是如此。但只要了解引言应涵盖的内容和写作要求及英语的语言表述规律，先从“读书破万卷”开始，就会获得“下笔如有神”的效果。

一篇好的引言至少应该包括以下五个部分：①提出研究的主题，即说明开展这项课题研究的理由和目的；②提供该领域的背景资料，即对该领域的文献资料进行回顾，阐明其研究现状和存在的问题，明确该文要解决的问题及其作用和目的；③说明该研究课题的理论依据和主要采用的研究方法和手段；④简述研究中所取得的主要结果；⑤给出所得到的主要结论。

6.2 引言的写作要求

①引言的表述应言简意赅，突出重点。要表述的内容应该一语道破，常识性的或人所共知的知识不必赘述。②背景材料主要来源于参考文献，首先应仔细查阅大量的参考文献，从中选取那些在该领域有重要影响的与本文所述观点相似或相悖的直接相关文献进行综合性的表述。③在引言中如需介绍自己的工作，对其评价务必客观和实事求是，像“达到国际先进水平”的说法切忌使用，对前人和别人的工作也需客观评价，不应蓄意贬低或夸大。④引言和摘要不同，不要把引言写得与摘要雷同，变成摘要的扩充或注释。总之，引言的写法并无统一的格式，其长短也不一，主要视论文背景和内容而定。如果能用规范的英语重点写好背景、理由、目的和目标，就会为论文开了一个好头，从而使读者饶有兴趣地读下去。

6.3 引言中常用的人称、语态和时态

引言表述中常见的人称、语态和时态与摘要中的类似，这里不再赘述。就时态而言，通常一般现在时态和一般过去时态用得较多。

6.4 引言常见写法实例

Example 1

The draft sequences of the human genome[1,2] have provided an unprecedented wealth of information on our genome, and have facilitated the identification of genes involved in human diseases. These drafts, however, contain a number of inconsistencies and gaps that have to be resolved to obtain a reliable molecular infrastructure on which we can anchor the entire set of human genes including their transcriptional start and stop signals, exons, splicing variants and regulatory elements, as well as the sequence variations found in various human populations. Nearly complete sequences of chromosomes 22 (ref. 3), 21 (ref. 4) and 20 (ref. 5) have already been achieved in the last three years. As an additional contribution to this goal, we present here the sequence of the euchromatic region of human chromosome 14. This chromosome contains two megabase (Mb)-long regions of prime importance for the immune system—the α/δ T cell receptor（TCR）located close to the centromere，and the immunoglobulin heavy chain（IGH）locus adjacent to the telomere—as well as about 60 genes which，when defective，are known to lead to genetic diseases，including spastic paraplegia，Niemann-Pick disease，early onset Alzheimer's disease and a severe form of Usher syndrome.

The sequence quality reaches the internationally adopted standard of 99. 99%. We estimate that we cover more than 99. 9% of the euchromatic portion of chromosome 14，which，as in the other acrocentric chromosomes，namely 13，15，21 and 22，is restricted to its long arm. Particular care has been devoted to evaluating the integrity and accuracy of clone coverage. For its annotation，this sequence has benefited from the availability of the continuously increasing amount of both genomic and expressed sequence data from humans and other vertebrates.（Nature，2003，Vol 421：601.）

Example 2

Despite more than a century of effort to eradicate or control malaria，the disease remains a major and growing threat to the public health and economic development of countries in the tropical and subtropical regions of the world. Approximately 40% of the world's population lives in areas where malaria is transmitted. There are an estimated 300-500 million cases and up to 2. 7 million death from malaria each year. The mortality levels are the greatest in sub-Saharan Africa，where children under 5 years of age account for 90% of all death due to malaria[1]. Human malaria is caused by infection with intracellular parasites of the genus Plasmodium that is transmitted by Anopheles mosquitoes. Of the four species of Plasmodium that infect humans，Plasmodium falciparum is the most lethal. Resistance to anti-malarial drugs and insecticides，the decay of public health infrastructure，population movements，political unrest，and environmental changes are contributing to the spread of malaria[1]. In countries with endemic malaria，the annual economic growth rates over a 25-year period were 1. 5% lower than in other countries. This implies that the cumulative effect of the lower annual economic output in a malaria-endemic country was a 50% reduction in the per capita GDP compared to a non-malarious country[2]. Recent studies suggest that the number of malaria cases may double in 20 years if new methods of control are not devised and implemented[3].

An international effort[4] was launched in 1996 to sequence the *P. falciparum* genome with the expectation that the genome sequence would open new avenues for research. The sequences of two of the 14 chromosomes，representing 8% of the nuclear genome，were published previously[5,6] and the accompanying letters in this issue describe the sequences of chromosomes 1,3-9 and 13（ref. 7），2，10，11 and 14（ref. 8），and 12（ref. 9）. Here we report an analysis of the genome sequence of *P. falciparum* clone 3D7，including descriptions of chromosome structure，gene content，functional classification of proteins，metabolism and transport，and other features of parasite biology.（Nature，2002，Vol 419：498.）

Example 3

Cytochrome c release from mitochondria to the cytosol marks a defined moment in a mammalian cell's response to a variety of apoptotic stimuli，in which the normal electron transfer chain is disrupted and caspases become active [1,2]. The released cytochrome c readily binds to apoptotic protease activating factor 1（Apaf-l）and induces a conformational change that allows stable binding of deoxyadenosine triphosphate/adenosine triphosphate (dATP/ATP) to Apaf-1, an event that drives the formation of a heptamer Apaf-1-cytochrome c complex called the apoptosome [3,4]. The apoptosome recruits and activates procaspase-9, which in turn activates the downstream caspases such as caspase-3, caspase-6, and caspase-7[5,6]. These caspases cleave many intracellular substrates, ultimately leading to cell death [7].

The mitochondrial caspase activation pathway is tightly regulated. One major regulatory step is at the release of cytochrome c from mitochondria, a process controlled by the Bcl-2 family of proteins [8,9]. The inhibitors of apoptosis (IAP) also regulate this pathway by directly inhibiting caspase activity [2,10]. IAP proteins are antagonized by mitochondrial proteins such as Smac/Diablo and Omi/HtrA2 after they are released to cytoplasm[11-15].

We have identified a death regulatory pathway by using a combined high-throughput chemical screen and biochemical fractionation approach. The pathway consists of tumor suppressor PHAP proteins and the oncoprotein ProT, each playing a distinctive role in regulating apoptosome formation and activity. (Science, 2003, Vol 299: 223.)

Example 4

Since the identification of antibodies at the end of the 19th century scientists have sought to explain how a limited repertoire of antibodies can bind and thereby protect against an almost infinite diversity of invading antigens. The discovery of clonal selection revealed that the combinatorial arrangement of a rather small number of different antibody gene segments has the potential to generate a highly diverse repertoire of antibodies. However, the number of antibodies in the primary response is finite, whereas antigen space is effectively limitless. A possible explanation is that each antibody is capable of binding more than one antigen. Such a scenario was envisaged in the 1940s. Pauling proposed that specific binding sites were selected out of an ensemble of preexisting antibody conformations[1]. Indeed, antibodies have been shown to cross-react with multiple antigens ever since they were discovered [2-6]. The first immunologists also realized that cross-reactivity, in addition to expanding the antibody repertoire, could result in the immune system turning against the organism it is meant to defend, or in Ehrlich's chilling words, horror autotoxicus [3]. Cross-reactivity is now known to play a central role in autoimmunity and allergy [7,8]. The ability to distinguish between invading antigens and self-proteins can be bypassed when antibodies raised against pathogenic antigens promiscuously bind self-antigens or innocuous environmental molecules[9,10].

Antibodies are renowned for their exquisite specificity, so how can they be both promiscuous and specific? Structures of antibody fragments that have been cocrystallized with different steroid molecules demonstrate that cross-reactivity can be accomplished through shared ligand chemistry or molecular mimicry [11,12]. The D1. 3 antibody to lysozyme binds both lysozyme and an antiidiotype antibody through the rearrangement of several side chains[13]. An antibody to the HIV-1 protein p24 also binds several other peptides that fit into the same binding site[6]. These studies demonstrate that a single antibody-binding site can accommodate difference, if related, ligands. However, Pauling had proposed an equilibrium between different preexisting structural isomers (pre-equilibrium), each of which provides a different binding site and binding specificity. Many antibodies exhibit different crystal structures in their free and bound states [14,15]. In the absence of solution kinetics, however, structural differences between free and complexed antibodies are generally ascribed to an induced-fit mechanism in which the ligand binds the free structure and induces a conformational change leading to the complexed structure[14] rather than to pre-equilibrium. Although there is no structural evidence for pre-equilibrium，there are kinetic studies that unambiguously demonstrate its existence. Pecht [16]，followed by Foote and Milstein，revealed the operation of both pre-equilibrium and induced-fit in antibodies；the latter also postulated that conformational diversity might have fundamental implications for the immune response [17]. To date，however，there has been no structural evidence for equilibrium between preexisting antibody isomers or that different isomers of the same antibody can bind different antigens.

Here we describe our studies of antibody SPE7，a monoclonal immunoglobulin E（IgE）raised against a 2,4-dinitrophenyl（DNP）hapten[18,19]. SPE7 exhibits high affinity [a dissociation constant（K_d）of 20 nmol/L] and specificity toward DNP；its affinity for close analogs of DNP，such as 2-nitrophenol（NP）and 2-nitro-4-iodophenol，is negligible（$K_d >$ 100 μmol/L）[20]. However，screening has revealed that SPE7 also binds several unrelated compounds with a broad range of affinities [5]，such as alizarin red（Az，K_d=40 nmol/L）and furazolidone（Fur，K_d=1. 2 μmol/L）. We have also used repertoire selections to identify a protein antigen of SPE7（Trx-Shear3）. Data obtained by X-ray crystallography and pre-steady-state kinetics show that SPE7 adopted at least two different preexisting conformations that were independent of antigen，each conferring a different antigen-binding function.（Science，2003，Vol 299：1362.）

6.5 引言中部分常用词、短语和句型

6.5.1 常用词

科技论文的前言中在表述研究的课题时常用的名词：paper，article，work，report，investigation，research，contribution，project 等。

例 1. As an additional contribution to this goal，we present here the sequence of the euchromatic region of human chromosome.

例 2. In this work，we have employed acetic acid and bacteria for the oxidation of racemic primary alcohols and chiral and prochiral diols with strains belonging to the genus *Acetobacter* and *Gluconobacter*.

表达论文的研究目的时常用的名词：aim，goal，purpose，approach，attempt，object，objective 等。

例：The main purpose of the present study is to obtain some information about neuronal genes.

前言表述中常用的动词：do，get，make，report，study，investigate，determine，obtain，observe，elucidate，focus，examine，measure，record 等。

例：Considerable attention has been focused on the crystallization kinetics of high polymers from the molten state.

6.5.2 常用短语

（1）表示时间

as early as the 1990s；as for back as the 1990s；as long ago as the 1950s（早在……年代）；since 1980（自从 1980 年以来）；during the last 20 years of the last century（在上个世

纪最后的20年里）；in the past decade（在过去的10年中）；in recent years（近年来）；until quite recently（直到最近）。

（2）表示大量，许多

a large number of…，a great number of…，a great deal of…，a considerable number of…，a lot of…，a great variety of…，a majority of…，the bulk of …。

（3）表示范围、领域

in the field of …，in the area of …。

（4）表示对比、对照

as opposed to…，as contrasted to…，as compared with，in contrast to…，in comparison with …。

6.5.3 常用句型

（1）某问题受到重视，感兴趣……

…have attracted many scientists in…；…have fascinated many workers in …；there has been much interest in…。

（2）某项研究的主要目的

The main aim（purpose，object，objective）of the present study（work，research，investigation，research，paper，article）is to obtain（assess，find out，establish，reveal）…。

（3）文献背景

there are no published data for…，owing to the lack of …，a survey of literature indicated that …，on a preliminary study of…，an earlier paper reported the studies of…，HPLC is a suitable（valuable）method of…for …。

6.5.4 常见例句

Example 1

During the past a few years，a number of techniques have been introduced in biology and applied to biological macromolecule analysis，combining chromatographic separation with spectroscopic detection，including gas chromatography-mass spectrometry.［在最近几年中，在生物学中已引进了几种技术，将色谱分离与光谱检测（其中包括气相色谱-质谱方法）结合起来用于生物大分子的分析。］

Example 2

On contrast to a number of experimental studies there have been only a few theoretical investigations.（与大量的实验研究相反，只有很少有关理论方面的研究。）

Example 3

In the present study，we report on investigations of the self-diffusion of PEP-PDMS block copolymer forming different morphologies using NMR.（本文中，我们报道了有关采用NMR技术对形成不同形态的PEP-PDMS嵌段共聚物的自扩散行为的研究。）

Example 4

The results obtained by the proposed method are compared with the self-diffusivities of the corresponding homopolymers.（将用本文提出的方法所获得的结果与相应的均聚物的自扩散系数做比较。）

Example 5

The aim of this work is to show the complex shear modulus of semi-infinite viscous layers can be determined by impedance measurements at thickness-shear mode resonators.（本文的目的是要说明半无限黏性膜的复杂的剪切模量可通过厚度-剪切模型共振腔上的阻抗测定结果来确定。）

Example 6

In this paper we apply the technique X-ray absorption spectroscopy using synchrotron radiation to examine and characterize structurally the nature of pure and brominated Cu-BTA complex.（本文中，我们采用带有同步辐射的 X 射线吸收光谱技术从结构上考察和表征纯的和溴代的 Cu-BTA 配合物的本质。）

Example 7

Although there are a few reports of using artificial neural networks（ANN）in the prediction of physicochemical properties，these reports have generally been restricted to equilibrium properties.（用人工神经网络来预测物理化学性质虽然已有一些报道，但这些报道仅局限于平衡性质方面。）

Example 8

This report will demonstrate that molecular dynamics simulations can give insights into the mode of action of incrustation inhibitors in atomistic detail.（本报道将会证明用分子动力学模拟可以对表面缓蚀剂的作用模型在原子水平上有一个深入的了解。）

7 How to Write the Main Body（正文的写法）

7.1 实验部分

7.1.1 实验部分所包含的内容

在实验性科技论文中，实验部分（experimental section）是正文中的第一部分，它包括实验所用的试剂或材料、实验仪器或设备以及实验的方法和过程等。

（1）试剂或材料（reagents or materials）

要求给出足够的信息，如化学名称（chemical name）[勿用商品名称、分子式（formulas）]、纯度（purity）、规格（specification）、生产厂家（company name）。用前是否经过提纯或特殊处理。配制的溶液需指明浓度、配制方法及标定方法、储存方式等。如果使用生物样品，则需注明年龄、性别、种属、生理状况和来源等。

（2）仪器或设备（apparatus or equipment）

要求给出标准名称、型号、技术规格、附件、使用条件和生产厂家等。若为自制仪器，需给出装置图或示意图，指明操作条件。

（3）实验方法及过程（method or procedure）

若使用自己新建立的方法则需详细描述，若使用常规或标准的方法，可只给出本文所需要的细节，并注明其来源的文献。

7.1.2 实验部分的写作要求

为了使同行能按照上述内容将实验重复出来，并得到与作者相同或相近的结果，此部分内容须充分翔实、准确、客观。若结果不能被别人重复，此篇论文便毫无价值，故写作时务必仔细。

各篇论文的实验部分因其具体内容不同，内容多少不同，故文章的长短亦不同，写作风格也不同。如果内容不多，则可直接叙述；若是内容较多，则可以分层次叙述或列出几个层次标题，并在层次标题下分别叙述具体内容。

7.1.3 实验部分中所用的时态

实验部分中的具体工作都是在写论文之前完成的，故英文表述中句子的谓语动词应采用一般过去时态。

7.1.4 实验部分写作实例

Example 1

materials and methods

(1) construction of the caspase-8 probes

The bacterial version of the caspase-8 probes were generated by site-directed mutagenesis of a C_3 probe that we made earlier (for details, see Luo et al[7]). The DNA sequence encoding the FRET sensor was then sub-cloned into a mammalian vector of pECF-Cl (clontech). The linker sequence between the two GFP molecules are shown in the first part of the results section.

(2) in vitro caspase assay using purified fusion proteins

The polyhistidine-tagged FRET sensor protein was expressed in bacteria and purified by using a Ni-NTA affinity column [7]. The purified sensor proteins were incubated with specific caspases and subjected to SDS-PAGE and Western blot analysis by using anti-GFP and anti-polyhistidine antibodies. The fluorescence emission spectrum of each sample was obtained by using a spectrofluorometer (SPEX 1681).

(3) living-cell imaging analysis

Plasmid DNA of different sensors was introduced into Hela cell by electroporation[11]. After expression for 20-24h, cells were treated with TNFα (10 ng/mL) and CHX (10μg/mL) and observed by using a fluorescence microscope (Axiovert 35 zeiss). The cells were maintained in 7.5mmol/L Hepes-buffered MEM culture medium containing 10% fetal bovine serum at 37℃. To measure the FRET effects, cells were excited by using excitation filter (440nm±10nm). The emission images of YFP (535nm±13nm) and CFP (480nm±15nm) were recorded by using a computer-controlled cooled CCD camera. The digital fluorescence images were then processed by using Meta Morphor 4.6 software. (Biochemical Biophysical Research Communications, 2003, Vol 304: 217-218.)

Example 2

2 Materials and Methods

2.1 protein expression and purification

Cloning of the hisC gene encoding HPAT and overexpression and purification of the recombinant enzyme were as described in ref.[7]. The mutant plasmids were constructed by using Quick Change Site-directed Mutagenesis kit (stratagane, La Jolla, CA, USA). The hisC-deficient *E. coli* strain used for overexpression of the mutant enzymes was constructed by the P′phage-mediated homologous recombination method as described in ref.[8].

2.2 Spectroscopic analysis

All spectroscopic measurements were carried out at 25℃. Absorption spectra were measured by using Hitachi U-3300 spectrophotometer (Tokyo, Japan). The buffer solution contained 50mmol/L buffer components and 0.1 mol/L KCl. The buffer components used were MES-NaOH, HEPES-NaOH, and TAPS-NAOH. The concentration of the HPAT subunit in solution was determined spectrophotometrically. The apparent molar extinction coefficient at 280nm for the PLP-forms of HPAT was 45680, which was calculated based on the number of tryptophan and tyrosine residues in HPAT. (Biochimica at Biophysica Acta, 2003, Vol 1647: 322.)

7.1.5 常用短语

（1）描述实验所用溶液的术语

culture solution	培养液	standard solution	标准溶液
stock solution	储备液	internal standard reference	内标
working solution	工作液	external standard reference	外标
sample solution	样品溶液	concentrated solution	浓溶液
buffer solution	缓冲溶液	dilute solution	稀溶液

（2）描述不确定量的术语

a stoichiometric amount	化学计量	a small amount	少量
a given (known) amount	已知量	a slight excess of amount	稍过量
an appropriate amount	适量	a large excess of amount	大过量
a desired amount	要求量	a large amount	大量
a trace amount	微量		

（3）描述实验过程的常用短语

be equipped with···	装配有……	be removed from	从……中除去
be incorporated into···	将……插入（装入、混入）	be evaporated to (near) dryness	蒸发至（近干）
be fitted with···	匹配有……	be slowly added dropwise	缓慢滴加
be coupled with···	连有……	heat gently to···	缓慢加热至……
be connected with···	连接有……	cool back to···	冷却至……
be located at···	固定在……	with constant stirring	不断搅拌下
be fixed to···	固定到……	under reduced pressure	在减压下
be soluble in···	溶于……	under an argon atmosphere	在氩气气氛中
be slightly (very) soluble in···	微（易）溶于……	without further purification	未进一步纯化
be insoluble in···	不溶于……	in the absence of···	在无……情况下
be warmed to···	温热至……	in the presence of···	在……存在下
be left at room temperature	于室温放置	at the head of···	在……起始处
be heated under reflux	加热回流	at the end of···	在……末端
be in agreement with···	与一致（吻合）……	at the top of···	在……顶部
		at the bottom of···	在……底部
be described elsewhere	另文报道	at the foot of······	在……下部
be kept at···	保持在……	in vitro	在体外
be maintained at···	维持在……	in vivo	在体内

7.2 结果部分

实验结果或理论结果是科技论文的核心，是评价论文是否有价值的关键部分。

7.2.1 结果部分涵盖内容

结果部分（result section）主要是总结实验中所获得的数据和观察到的现象，并对这些数据和现象进行定性或定量分析，给出规律性的东西。

7.2.2 结果部分写作要求

由于观察到的现象和得到的数据都来自研究工作的记录中，所以写入论文中的那些现象和数据必须具有代表性。也就是说能说明在论文前言部分中提出的要解决问题的那些结果。如前所述，前言部分中立了题，并阐明了立题的理由，实验部分中给出了完成这一课题的方法、手段和途径，而在结果部分及后面的讨论部分中则需要将研究结果由实践上升到理论。由此可以清楚地看到，前言和实验部分都旨在得到理想的结果，而讨论部分则是以结果为依据来实现认识的飞跃，所以结果部分必须写得真实、客观、可靠，并具有再现性和普遍性，要为科研同行重复该项研究或对论文结果进行验证，提供充分和必要的条件。

因此写入结果中的内容要从研究记录中进行精心选择，达到去伪存真的理想效果。

7.2.3 结果部分的具体写法

最常见的写法是将从结果中选出的有规律性的、最能说明问题的数据列成表格，将主要的实验现象或规律性的数据绘成图，然后再用语言进行必要的且不重复的说明，亦即采用表格、图和行文并茂的方式将研究结果表述出来。

在结果的语言描述中，谓语动词常见的时态为一般过去时态和一般现在时态。如实验中发现的情况常用一般过去时态，图表的说明常用一般现在时态。总之，在科技论文中，这两种时态的使用场合与常规使用场合是一致的。

（1）表格的制作

表格是按照统计学原理，将研究中观察和计算所得到的各种结果和数据，按一定的方式集中地排列和组合在一起的一种有效的表达研究结果的方式。由于表格不仅具有简洁、清晰和准确的特点，而且其逻辑性、规律性和准确性很强，因此表格作为文字的翅膀已成为科技论文中的重要表达手段。

一篇论文中不是表格越多越好，而是要根据其必要性进行精选，凡是能用一两个句子可说明的内容就不必列表。当要表述相同研究结果时，可以用表格表述，也可以用图描述，但只能选择其中之一。通常，如果重点在于描述参量变动的总体趋势，则可以选择插图形式；如果重点在于比较量与量之间的关系，则应选择列表形式。

设计表格时应重点突出，与要说明的主题无关的内容则不列入表内；表格表述应简洁，凡分析或运算的过程不应列于表内；表格设计应科学，要明确目的性，把背景条件、比较前提、使用方法、实测数据或计算结果都逐个分列清楚，有条理，逻辑性强，使同行一目了然。

表格的种类常见的有 3 种，即无线表、系统表和卡线表。凡项目和数据较少，内容简短，则可选用无线表（见 Table 2.1）。

Table 2.1 Mass fraction of amino acids in fish powder

Amino acid	Mass fraction/%
Histidine	23
Lysine	80
Arginine	52
Threonine	40

当表述隶属关系多层次事项时，应采用系统表的形式（见 Table 2. 2）。

Table 2. 2 Classes of organic reactions

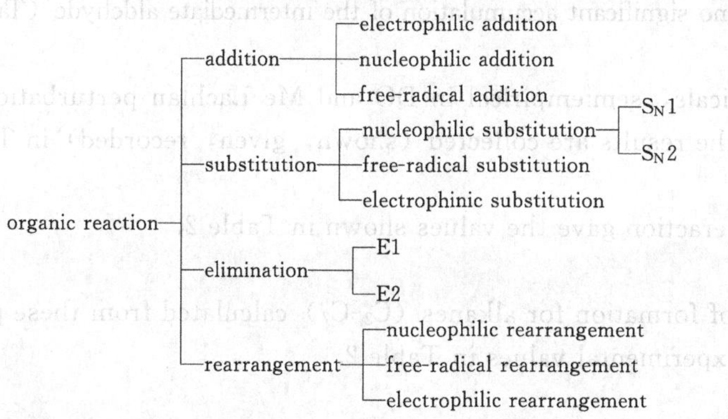

当表述的内容较多时，通常用卡线表。卡线表中，横向各栏之间用栏线隔开，纵向各栏之间用行线隔开，各项数据之间分隔得很清楚，隶属关系一一对立。但由于横线和竖线太多，显得繁杂，故国内外科技期刊普遍采用卡线表的简化形式之一，即三线表。其组成要素如下。

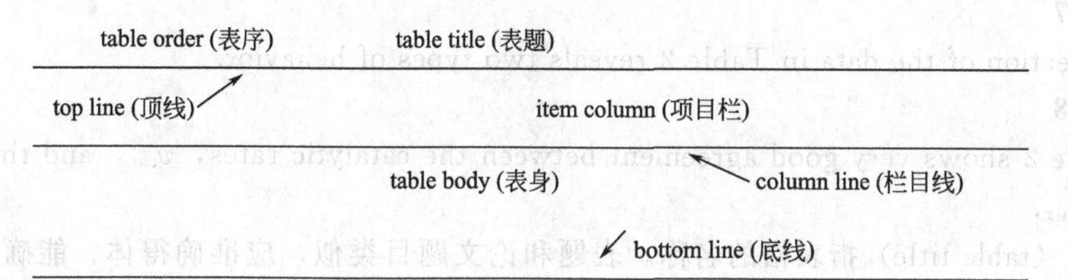

三线表实例，见 Table 2. 3。

Table 2. 3 Molar conversions and *e. e.* obtaned in the oxidation of 2-chloropropanol and 2-phenylpropanol（2. 5g/L）with acetic acid bacteria

strains	2-chloropropanol			2-phenylpropanol		
	conversion	*e. e.*/%	time/h	conversion	*e. e.*/%	time/h
A. aceti DSM 3508	35	40(*R*)	1	40	95(*S*)	24
A. aceti MIM 2000/28	20	35(*R*)	1	25	92(*S*)	24
A. aceti MIM 2000/50	37	43(*R*)	2	43	97(*S*)	24
A. hansenii MIM 2000/5	50	39(*R*)	4	38	91(*S*)	24
A. xylinus MIM 2000/13	25	45(*R*)	1	34	90(*S*)	24
G. asaii MIM 1000/14	23	46(*R*)	4	33	>97(*S*)	24
G. cerimts DSM 9534	27	44(*R*)	4	20	95(*S*)	24
G. frateurii DSM 7146	35	28(*R*)	3	22	91(*S*)	24
G. oxydans DSM 2343	20	45(*R*)	4	<5		24
G. oxydans DSM 1000/9	45	18(*R*)	4	45	97(*S*)	24

表序（table order）指的是表格序号。在科技论文中的表格按出现的先后顺序用阿拉伯数字或希腊数字依次连续编号，若全文只有一个表，则编序为 Table 1。表格一般位于正文中出现处之后。在正文中表述表格出现的常见行文形式举例如下。

Example 1

Two strains failed to oxidize the substrates, while the strains furnished the correspondent carboxylic acid with no significant accumulation of the intermediate aldehyde (Table 2).

Example 2

For the radicals, semiempirical in DO and Me Lachlan perturbation calculations were carried out and the results are collected (shown, given, recorded) in Table 1.

Example 3

An EPR interaction gave the values shown in Table 2.

Example 4

Enthalpies of formation for alkanes (C_2-C_7) calculated from these parameters are compared with the experimental values in Table 2.

Example 5

From Table 2, it can be seen that most of the physical properties used in this work are highly correlated.

Example 6

Also given in Table 2 are the mass changes observed on reduction of the adsorbed complex.

Example 7

Inspection of the data in Table 2 reveals two types of behavior.

Example 8

Table 2 shows very good agreement between the catalytic rates, v_{cat}, and the mixture rates, v_{mix}.

表题（table title）指表格的名称。表题和论文题目类似，应准确得体，能确切表达表格中的特定内容，简短精练，勿用缺乏专指性的行文，如“Data Table，Comparison Table，Parameter Table”等作表题。

项目栏（item column）指的是表格顶线（top line）和栏目线（column line）之间的部分。所谓栏目就是该栏名称，它必须表达出表身中该栏内信息的特征和属性。栏目如果由量的名称（或其符号）和单位组成，应写成“量名称/单位”的形式，如 velocity/(m/s) 或 v/(m/s)等。若栏目为多层次的，彼此之间应加辅助线（见上述所给三线表实例 Table 2.3）。

表身（table body）指栏目线和底线（bottom line）之间的部分，为表格的主体内容。表身内容如果是数据，其后不能附单位或（%），单位或%应该放在栏名称＋斜线的后面。同一栏中的各行数据都在同一栏目名称之下，故应以个位数或小数点或范围号“-”为准上下对齐，表身中的各个数据的有效数字应相同。表身中上下左右相邻栏内的数据即使相同，也应分别列出，切不可用“ibid”（同上）表示。表身中无数据处，应视具体情况分别处理，若属无此项数据或未测此项数据则用空白表示；如果为未发现此项内容，则用“-”表示；若实测结果为零，则用“0”表示。

如果表身中某项内容需要用文字加以说明时，则应在表的底线下加表注。表身中需说明的那项内容应在右上角加符号标记，并将该标记写在表注之前。若表注为2条及其以上时，该标记应该使用阿拉伯数字或英文字母，并按从左至右和先上后下的原则依次编序。

若一个表格太长，一页排不下时，则可采用续表形式。先将该表在适当处断开，并用细线封底，在下页顶头续排该表。为便于阅读，表的栏目项需要重排，然后再排续表表身，在续表栏目项之上应加“continued”（续表）字样。

如果表格纵向栏目过多，而横向栏目较少时，可以转栏编排，转栏处需加“双细线”分隔开来（见 Table 2.4）。

Table 2.4 History of the discovery of amino acids

amino acid	time discovery	discover	source	amino acid	time discovery	discover	source
glycine	1820	Braconnot	gelatin	histidine	1896	Kossel	sturin
leucine	1820	Braconnot	wool, muscle	cystine	1899	Morner	horn
tyrosine	1849	Bopp	casein	valine	1901	Fischer	casein
serine	1865	Gramer	silk	proline	1901	Fischer	casein
glutamic acid	1866	Ritthausen	gliadin	trytophan	1901	Hopkins	casein
aspartic acid	1866	Ritthausen	conglutin, legumin	hydroxyproline	1902	Fischer	gelatin
phenylalanine	1881	Schultze	lupine sprouts	isoleucine	1904	Ehrlich	fibrin
alanine	1881	Weyl	silk fibroin	thyroxine	1915	Kendall	thyroid
lysine	1889	Drechsel	coral	methionine	1922	Mueller	casein
arginine	1895	Hedin	horn	hydroxylysine	1925	Schryver	isinglass
diiodotyrosine	1896	Drechsel	coral	threonine	1935	Mccoy, Meyer	casein

卡线表的另一种形式为二线表。二线表与三线表的差别在于其栏目项居纵向左侧，全表只有顶线和底线两条线（见 Table 2.5）。

Table 2.5 Recovery of amino acid afer acid hydrolysis for different time and temperatures

time/h	24	96	24	96
temperature/℃	100	100	106	106
recovery of serine/%	96	84	95	65
recovery of threonine/%	97	92	97	83
recovery of isoleucine/%	58	86	72	88

（2）插图的设计

插图是形象化的语言。它可以作为论文行文的另一只翅膀，形象、直观、简洁、准确地表达有关内容。因此，它是科技论文的重要组成部分，是科技论文中不可缺少的辅助表述手段。

插图与其他绘画或摄影作品一样都要求有鲜明的主题和高超的表现技巧，追求内容与形式的完美统一。但科技论文中的插图和艺术作品中插图之间的差别在于重点表达事物的结构、各组成部分的内在联系或相互位置关系以及量化关系。

插图作为科技论文中的文字表述的辅助表达形式，尤其是表达那些用文字很难叙述清楚的内容，其重点在于通过示意突出主题，在于真实地反映事物的形态、运动的规律或量的关系，因此必须依据研究的结果科学地设计和绘制，绝不能随意取舍，主观臆造和虚构。

另外，科技论文中的插图和论文本身一样，是用于学术交流的，因此插图设计和绘制必须规范，所谓规范就是作者、科技刊物的编者和读者之间的共同约定，也就是说，论文中给出的插图应是专业人员能一目了然的。如果不遵守约定，自行其是，给出的图无人能看得懂，这样的插图也就没有任何意义了。

科技论文的插图按印刷工艺可分为墨线图和照片图。墨线图是指用墨线绘制出来的图形，按其表达功能可分为示意图、流程图、构造图、线路图、程序框图、函数图（包括分析测试仪器绘制出来的各种谱线图）。照片图是指用摄像机拍照下来的真实物体照片或用特殊设备拍摄下来的反映物体内部真实结构的 X 射线照片和扫描电子显微镜（scanning electron microscopy）照片、扫描隧道显微镜（scanning tunnel microscopy）和原子力显微镜

(atomic force microscopy)照片等。

科技论文中的插图表达的内容必须服务于论文的主题，应同行文、表格有机地构成一体，共同论证论文的中心。插图应该有自明性，即不看行文也可以理解图意。插图应该精心选择，可以用简单文字表达清楚的问题不用插图，不要与表格重复表述。图形要简明，所用各种符号、计量单位和名词术语必须符合国际标准。图内文字说明要简明，可用代号或数字标注，再辅以注释加以说明。绘制插图时，要求用黑色墨水清绘，线条粗浅均匀，比例协调。

在科技论文的研究结果中最常见的有函数图、结构图和照片图。

函数图常见的有条形图、线形图、点形图、圆形图和谱线图，举例如下。

Example 1　条形图（Fig. 2. 1）

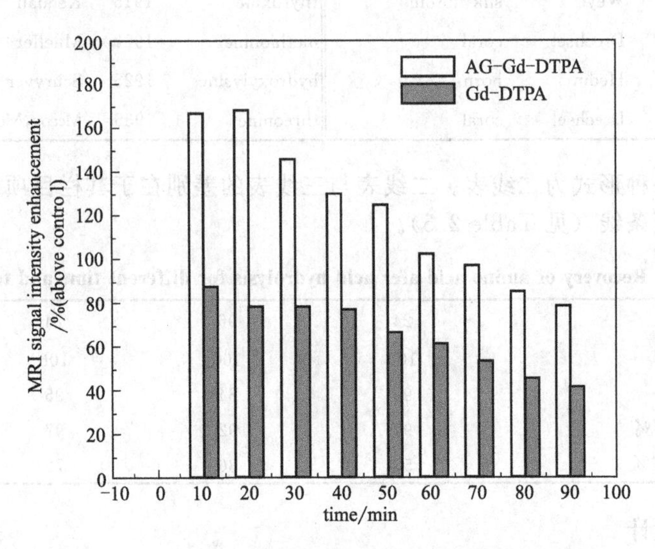

Fig. 2. 1　The mean percentage enhancement of kidney at different time intervals following intravenous administration of AG-Gd-DTPA and Gd-DTPA

Example 2　线形图（Fig. 2. 2）

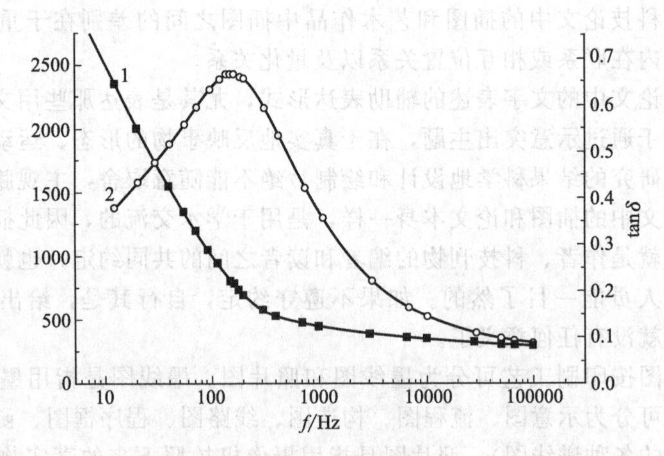

Fig. 2. 2　Dependence of ε and tan δ on f for samples calcined at 400℃

1— ε；2—tanδ

Example 3　点形图（Fig. 2. 3）

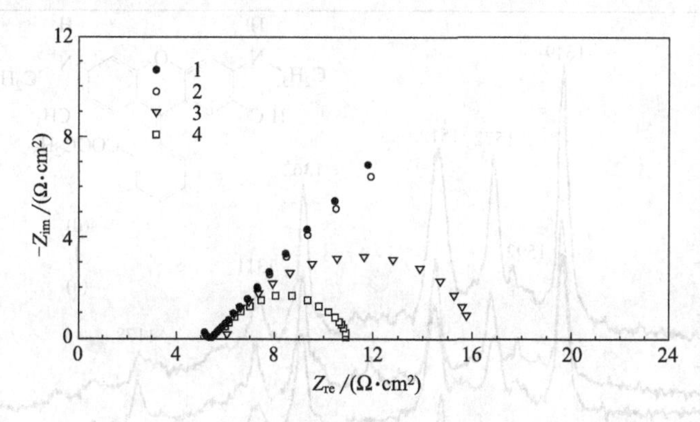

Fig. 2. 3　EIS of air electrode using purified carbon nanotubes as a catalyst layer materials
DC bias（vs. open-circuit potential）：1—0mV；2——50mV；3——100mV；4——150mV

Example 4　圆形图（Fig. 2. 4）

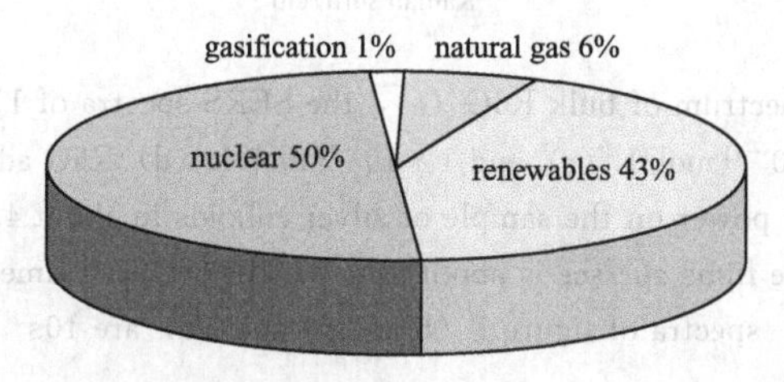

Fig. 2. 4

用各种分析测试仪器记录的谱图大多属于函数图。

Example 5　拉曼光谱图（Fig. 2. 5）

科技论文中的插图需要按文中出现的先后顺序用阿拉伯数字、希腊数字或罗马数字编序号，即图序。若全文只有一个插图，编序为 Fig. 1。插图一般紧随正文中出现处之后。

在正文中表述插图出现时的常见句式举例如下。

Example 1

Fig. 2 displays（shows）the SANS intensity distribution obtained with redial and tangential beam configuration.

Example 2

Also shown in Fig. 2 is the birefringence of the sample with the shorter block copolymer.

Example 3

The vesicles can also be seen with polarization microscopy（see Fig. 2).

Example 4

The particle size distributions are shown in Fig. 2.

Example 5

The reason for the apparent agglomeration of the two distinct population shown in Fig. 2 is uncertain at this time.

Example 6

Fig. 2 presents the evolution of the cell parameter of the quenched δ-phase versus the annealing temperature for sample 8.

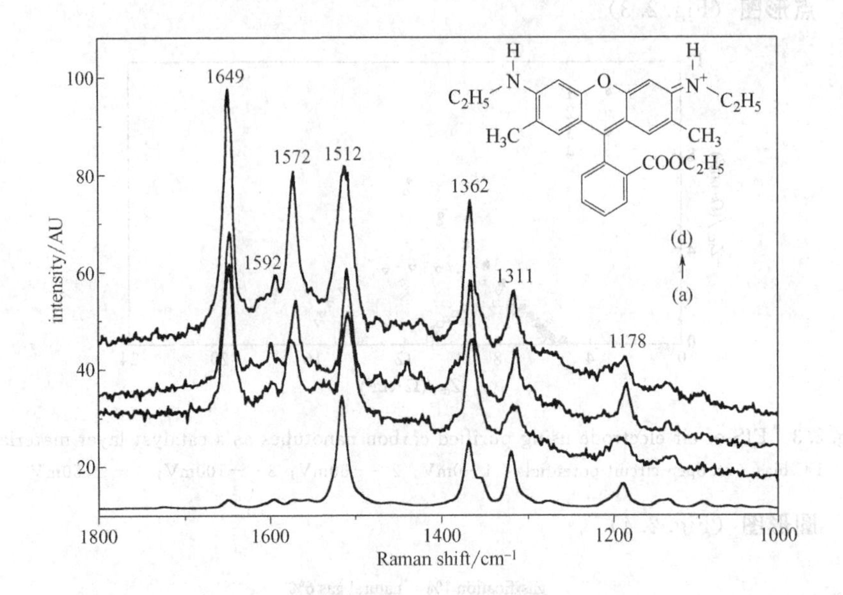

Fig. 2. 5 The FT-Raman spectrum of bulk R6G (a), the SERS spectra of 1×10^{-8} mol/L R6G adsorbed on silver colloid (b), 1×10^{-11} mol/L (c) and 1×10^{-12} mol/L (d) R6G adsorbed on aggregated silver particle films. The laser power on the sample of silver colloids in about 4. 2 mW, on aggregated silver particle films surface is about 40 μW. The acquired times for Raman spectra of figure 1 (b), (c) and (d) are 10s

Example 7

Fig. 2 demonstrates (illustrates) the fit between the experimental data and the theoretical data generated from eqn. (10).

Example 8

Fig. 2 summarizes the phase-transformation temperature as a function of the amount of silica.

每个插图的图序后面要给出图的标题，即图题。图题应简短、准确，能恰当地概括插图所表达的核心内容。图题的写法与文题类似，此处不再赘述。

对于函数图和谱线图，在横坐标轴下方，纵坐标轴的旁侧需给出标目。标目是说明坐标轴物理意义的必要项目。标目由物理量的名称或其符号和相应的单位组成，按国际单位制(SI，international system of units) 的规定，物理量名称的符号用斜体，单位符号用正体，中间用斜线“/”隔开，如 I/A、p/Pa，c/(mol/L) 等，其中斜线意为相除，量和单位的比值即为坐标轴上的标值。

标值是坐标轴定量表述的尺度，标记在坐标轴的外侧，紧靠标值线的位置上。为防止标值过分密集，要认真选取标目中的物理量，通常标值的数字应不超过 3 位数。如 l=30000m 可改为 l=30 km，标目为 l/km；m=0. 00005 g，可改 m=5μg，标目为 m/μg。

插图中的代号或符号的含义需用图注加以说明。图注通常放在图题的下方。图注的文字表述力求简洁明确。

7. 2. 4 结果部分的写法实例

RESULTS

Partial Amino Acid Sequence of Bovine LHPPase—To investigate whether humans have

LHPPase, we analyzed the primary structure of the bovine LHPPase and used the data to isolate a cDNA encoding the human counterpart. Bovine LHPPase was purified from the liver as previously described [11] and subjected to amino acid sequencing. However, no substantial sequence was obtained, suggesting that the bovine LHPPase N-terminus was blocked. To obtain its internal sequence, the protein was digested with V8 protease and lysylendopeptidase. Two peptide fragments were isolated and their partial amino acid sequences were determined. A homology search indicated that the two peptides were similar to those encoded in cDNA clone H75601 analyzed in the human EST project [14].

A cDNA Encoding the Human Homologue—Although the homology search suggested that the cDNA encoded a human counterpart of the bovine enzyme, the human EST cDNA sequence was incomplete and too short to encode the full length of a corresponding human enzyme. Thus, we tried to isolate a cDNA clone that encodes a full-length human homologue. A cDNA clone encoding an insert of 1814 bp was isolated from a cDNA library of HeLa cells (Fig. 1) . The translation initiation site was predicted from the Kozak's sequence in vertebrates [20,21]. The cDNA contained an in-frame stop codon sequence between 976 bp and 978 bp. There was a stop codon (145-147 bp) upstream of the predicted in-frame initiation codon and no other intervening initiation codon. This indicated that the sequence upstream of the predicted initiation codon did not encode any amino acids. Thus, the open reading frame encoded 270 amino acids, and the molecular mass of the protein was predicted to be 29192D. Because the sequence was similar to the partial amino acid sequences of the bovine LHPPase as shown in Fig. 2, we designated the gene as lhpp.

The human LHPPase sequence had no similarity to the consensus sequences of known inorganic pyrophosphatases [22,23]. Thus, we compared the human LHPPase sequence with that of other phosphatases. Proteintyrosine phosphatase, dual-specific phosphatase and low molecular weight acid phosphatase are known to have a consensus sequence of XCX_5RS (T) at their active sites [24]. We compared the sequences around the five cysteine residues of human LHPPase with the consensus sequence and found two homologous regions, as shown in Fig. 1. There is a candidate sequence of XCX_5KS at residues 52-60, although lysine is substituted for arginine, another basic amino acid. Another candidate is XCX_7RT at residues 225-235, although there are seven amino acids between cysteine and threonine instead of five. Motif analysis also indicated that human LHPPase had a leucine-zipper—like sequence (i. e. three leucines accounted for every seventh residue at positions 8, 15, and 22) .

Expression of Human LHPPase—LHPPase was originally purified from bovine liver, but its presence in other tissues has not been reported. We analyzed the expression of human LHPPase by Northern blot. Although the mRNA was expressed at low level in other tested tissues, it was highly expressed in the liver and kidney, and moderately in the brain (Fig. 3). The expression patterns were the same in both fetal and adult tissues. The results indicated that LHPPase plays an important role in these tissues. The RNA detected in the Northern blot was 1. 8 kb in length.

Subunit Interaction of Human LHPPase—Because the purified bovine enzyme is a homodimer, we analyzed the molecular mass of human LHPPase. We used the His-hLHPPase because it was possible to purify the proteins containing His • Tag sequence in one step by immobilized metal affinity chromatography. The His-hLHPPase was expressed in *E. coli* and purified on a Ni^{2+} chelating affinity column as a single band protein on SDS-PAGE (Fig. 4). The fusion protein was composed of 322 amino acids, and the molecular mass was predicted to be 34900D. The migration of the fusion protein on SDS-PAGE was the same in the presence or absence of reducing reagents. The fusion protein was fractionated from Sephadex G-100 superfine gel and the molecular mass was estimated with standard proteins (Fig. 5). The fusion protein was eluted in the 80kD fraction, and its molecular mass was almost double the 35kD molecular mass of the fusion protein estimated from the sequence and migration on SDS-PAGE. These results suggest that the human LHPPase is also a homodimer, at least under this buffer condition, and the disulfide bonds are not used for LHPPase dimerization.

Optimal pH for the Phosphatase Activities—To confirm that the cDNA encodes an active enzyme, the phosphatase activities of the His-hLHPPase were assayed as previously described [11]. The hydrolysis of nitrogenphosphorus bonds in PNP and oxygen-phosphorus bonds in PPi by human LHPPase was tested through assays in various pH ranges of GTA buffers (Fig. 6). The optimal pH for the reaction was 7.0 and 5.5, respectively. Accordingly, the activity hydrolyzing PPi appeared to function in a lower pH range than the activity hydrolyzing PNP. These enzymatic characteristics were similar to those of the purified bovine enzyme. The enzyme activity to hydrolyze nitrogen-phosphorus bonds was higher than that for oxygen-phosphorus bonds at the optimal pH.

Effect of a Thiol Reagent—The PPase and PNPase activities of the His-hLHPPase were each measured at their optimal pH (Fig. 7A). High concentration of PNP inhibited the PNPase activity of human LHPPase. Similar substrate inhibition has been reported for the bovine enzyme [12]. Because *p*-CMPS inhibited both PPase and PNPase activities of the purified bovine LHPPase, we analyzed the effect of this strong thiol reagent on the human LHPPase. Both PPase and PNPase activities were inhibited by *p*-CMPS as predicted (Fig. 7B). More than half of the full activity of PPase and PNPase were inhibited by 0.1μmol/L and 1μmol/L *p*-CMPS respectively. The results indicate that the hydrolytic activity on PPi was slightly more sensitive to *p*-CMPS than that on PNP. These enzymatic characteristics were similar to those of the purified bovine enzyme (J Biochem, 2003, 133 (5): 610-612).

7.3 讨论部分

讨论部分是在研究结果基础上进行的，其目的在于阐述结果的意义和作者自己对研究结果的见解和观点。讨论部分通常是论文中最难写的部分。许多论文由于讨论部分写得很差而未被有关期刊录用而退回，即使其得到的结果是正确的和有意义的。

7.3.1 讨论部分中所包含的内容

对研究结果做出的解释，阐述所研究事物的内在关系，揭示所研究课题的内在本质及一般法则和规律，并上升到理论逐一加以论证。要阐明自己的研究成果与他人的研究成果及其观点的异同之处，并指出所获得结果与论文引言中提出的目标是否相符。要说明自己研究成果的理论价值和实际应用前景。

讨论部分中的最为关键之处是突出新的发现、新的发明、解释因果关系，说明研究结果的必然性或偶然性。如果该论文仍有尚不能定论的问题或有与常规相反的情况，也应做出解释，某些观点虽然还不能得到充分证明，但也可提出讨论并给出进一步研究的设想。

7.3.2 讨论部分写作的基本要求

讨论的表述不能重复研究成果，而是要以研究成果为依据，客观地、实事求是地进行推理，并阐明自己的观点，要由感性认识上升到理性认识，产生一个认识的飞跃。

在讨论中如需对比别人的工作，必须以文献形式注明出处以示对他人成果的尊重和避免剽窃之嫌。当遇到与已发表的结果或观点不同的情况时，应首先认真核对自己的结果，判断自己所得结果的准确性、推理和观点的正确性，切忌不要轻易武断地否定别人的结论或观点。

7.3.3 讨论部分的写法及时态

讨论部分作为一个相对独立的内容可以单独成文，这种写法在科技刊物上是一种常见的成文方式。将结果与讨论（results and discussion）两部分结合在一起撰写，边叙边议，一气呵成，是另一种常见的成文方式。后一种行文方法将研究成果一个一个地描述出来，同时指出变化趋势，揭示各种规律，阐明自己的观点，具有直观性强、条理清晰、精练等优点。具体采用哪种方法，应视需要和作者写作习惯和风格而定。

讨论部分中，谓语动词主要采用一般现在时和一般过去时这两种时态，通常自己的研究结果采用一般过去时态，别人的工作采用一般现在时态，图表的说明采用一般现在时态，一般性的结论、已承认和公认的原理也都要采用一般现在时态，要继续做的工作和预期的结果采用一般将来时态。

7.3.4 讨论部分实例

DISCUSSION

LHPPase was found as a novel inorganic pyrophosphatase in bovine liver in 1997[11]. Because the enzymatic characteristics of the purified bovine enzyme were unique，it was expected that LHPPase sequence would be quite different from those of typical inorganic pyrophosphatases. We confirmed this hypothesis by sequencing. To investigate whether humans have the same enzyme，we isolated a cDNA clone encoding the human homologue. We registered the human cDNA sequence as LHPP to DDBJ/EMBL/GeneBank database in 2000（AB049629）. After our registration，the human genome projects reported the draft sequence of the whole human genome in 2001[25,26]. In this paper，we detail the isolation of the human LHPPase cDNA and the enzymatic activities of its translated product. A homology search against the genome sequence database indicate that LHPPase is encoded by seven exons on chromosome 10 q26.13. The official gene name of LHPP is now identified as LHPP in the human genome database.

We suggested that human LHPPase is a dimer，but the disulfide bonds are not used for dimerization. There is a leucine-zipper-like sequence（three leucines at intervals of seven residues at positions 8，15，and 22）. This suggests a possible dimerization domain of LHPPase to form a homodimer. Mutation analysis of these residues will determine the dimerization domain.

The purified bovine enzyme has two different catalytic sites for hydrolysis of imidodiphosphate and *N*-phosphorylated amino acids，respectively[12]. Two candidates for the active site may function to promote hydrolysis of the different substrates. The actual active site may be determined by point mutation analysis in further studies. A similar sequence was also found in phosphoarginine phosphatase purified from rat liver [27-30]. Although human LHP-Pase hydrolyzes O-P bonds in inorganic pyrophosphate, it hydrolyzes N-P bonds more effectively. This suggests that another, more suitable substrate than inorganic pyrophosphate exists *in vivo*. One possibility is that LHPPase works as a protein phosphatase which hydrolyzes phospho-proteins containing phospholysine and/or phosphohistidine. The similarity of the putative active site to the consensus sequence of other phosphatases functioning as signal transduction components suggests that LHPPase may transduce signals in a novel signal transduction pathway.

It was suggested that the N-terminal amino acid of the bovine enzyme was blocked. In the case of a phosphoarginine phosphatase, the first methionine was removed and the second alanine was acetylated [30]. LHPPase may also have such a modification. Tandem mass analysis of the modified N-terminal peptide fragment will determine the type of modification. Although the meaning of the N-terminal amino acid modification is not known at present, the N-terminus of LHPPase does not seem important for the enzymatic activity itself. In fact, the fusion protein containing a large extension at its N-terminal region of this protein has an activity to hydrolyze N-P and O-P bonds at the same level as that reported for the purified bovine enzyme.

The nucleotide sequence reported in this paper has been submitted to the DDBJ/EMBL/Gene Bank databases with an accession number of AB049629.

[J Biochem, 2003, 133(5): 612.]

7.3.5 结果与讨论写在一起的实例

2 Results and discussion

2.1 *Oxidation of 2-phenylpropanol and 2-chloropropanol*

Twelve strains belonging to different species of acetic acid bacteria were used for gaining information about the enantioselective potential of this microbial group. Acetic acid bacteria were grown in submerged cultures using GlyY medium and tested for oxidation at 28℃ (Scheme 1). Biotransformations were performed with cells grown for 24 h; Table 1 reports the average dry weights obtained.

The oxidation of racemic 2-phenyl-1-propanol (1) and 2-chloropropanol (2) was firstly investigated. These compounds were chosen as model substrates due to the importance of optically pure 2-arylpropionic acids [14] and 2-chloropropionic acid [15]. Two strains failed to oxidize the substrates [*Acetobacter liquefaciens* Deutsche Sammlung von Mikroorganismen (DSM) 5603 and *A. pasteurianus* DSM 3509], while the other strains furnished the correspondent carboxylic acid with no significant accumulation of the intermediate aldehyde (Table 2).

The oxidation of (*R*, *S*)-2-phenyl-1-propanol generally proceeded with high enantioselectivity, while the oxidation of the racemic mixture of 2-chloropropanol occurred with much lower enantioselectivity, the highest enantiomeric excess (*e.e.*) being below 50%. Higher rates were generally observed with 2-chloropropanol, which may indicate that less bulky substituents (e.g., chloro versus phenyl) have a positive effect on velocity and a negative effect on enantioselectivity.

Low *e.e.* observed with whole cells may be due to a number of factors, such as the presence of various dehydrogenases with different enantioselectivities, the action of a single enzyme with a low enantioselectivity or racemization of the formed product. Racemization was ruled out by carrying out experiments in which optically pure *S*-chloropropanoic acid was added to the cells and observing that no significant change in the enantiomeric composition took place. Biotransformations were also carried out at various substrate concentrations using cells resuspended in different buffers with pH values from 4 to 9. The variation of these parameters did not influence significantly the stereochemical outcome with *e. e.* ranging between 45% and 55%. The action of the involved dehydrogenases seems, therefore, stereo-convergent, whatever, the conditions of the biotransformations.（Journal of Molecular Catalysis B：Enzymatic，2002，17：237.）

7.4 结论部分

多数科技论文的正文末尾都以结论结束，也有些论文将结论融入讨论中。无论采用哪种形式，结论都不应该是正文内容尤其是结果与讨论部分的简单总结，而应该是在研究结果和推理分析讨论的基础上经过判断、推理、归纳的过程，完整、准确、简洁地阐述如下内容。

① 由对研究对象进行考察和对研究结果进行分析、推理所揭示出来的原理及普遍规律。

② 与他人的研究结果或作者先前的研究成果对比后得到的差异和发展。

③ 对研究中所发现的例外结果的分析和解释。

④ 该论文在理论上的建树和应用方面的价值和前景。

⑤ 对遗留的尚未解决的问题提出解决和深入研究的设想。

在写作格式上，如果结论的内容较多，为了使条理清晰，可以按内容的不同分条表述；若内容较少，可以用几句话阐明。

在文字表述方面，要注意逻辑严密，每个句子都要归结为一个认识、一个概念、一条规律或一个结论。用准确鲜明、非含糊其辞、模棱两可的语气来阐述新的发现和新的见解。在否定或肯定某种观点时，用词要确切，切忌用大概、可能、或许等词，以免造成似是而非的结果。

如果论文未得出明确的结论，结论部分也可以不写。现将结论的具体写法举例如下。

4 Conclusions

Acetic acid bacteria can be advantageously used as enantioselective biocatalysts for the oxidation of primary alcohols or diols. The stereobias of acetic acid bacteria mediated oxidations appears to be mainly dependent on the substrate and not on the stain employed. Complete chemoselectivity was observed when 1, 3 or 1, 4-diols were employed, with oxidation only of the primary hydroxy group; it is noteworthy that nonanoic lactone was produced in one step by oxidation of 1, 4-nonandiol.

（Journal of Molecular Catalysis B：Enzymatic，2002，17：240.）

8 Acknowledgments（致谢）

致谢部分是科技论文的附属部分。在从事和完成该项科研工作的过程中，如果得到各个方面的帮助和指导，在论文正式发表时，必须用书面形式郑重表示致谢，以示对他们劳动的尊重和肯定。

致谢部分可分为两类，一类是对此项研究资金和经费资助的单位（如国家自然科学基金委员会，国家科技部、教育部等，各省科技厅，合同单位）、团体（如有关学会等）和个人等，这种致谢也为基金资助单位提供了检查完成基金情况和所得效益提供依据。另一类是感谢对论文的选题、构思或撰写、修改给予指导或提出有价值的意见的人；感谢为该项研究提供样品、试剂材料、仪器设备的单位或个人；感谢对研究过程中给予指导或帮助完成样品测试的单位和个人；感谢为论文的完成提供过某种重要信息、数据、图表、照片而非为论文作者的那些单位或个人。

对已采用其他形式表示过感谢的单位或个人（如已支付过劳动报酬）可不再用书面形式致谢。致谢应该用恳切、淳朴、实事求是、恰如其分的简短语气来表述，切忌浮夸或客套。

现将致谢部分的常见写法举例如下：

Example 1

We thank the staff of the Electron Microscopy Unit and Department of Chemistry at Australian National University for their assistance in FE-SEM and XRD observation of some samples.

Example 2

This research was supported by the National Natural Science Foundation of People's Republic of China.

Example 3

Financial support by the Ministry of Education of China is gratefully acknowledged.

Example 4

We gratefully acknowledged professor D. M. Taylar for his advice during the final preparation of this paper.

Example 5

We are grateful to Dr. Jack McCoubrey for his many helpful discussions.

Example 6

It is a pleasure to thank professor Wang Gang for fruitful cooperation.

Example 7

We are thankful to State Key Laboratory of Theoretical and Computational Chemistry for the use of various programs and computing facilities.

Example 8

All Authors would like to thank prof. Luo Guangming from the Key Laboratory of Enzyme Engineering of Educational Ministry of China for providing data of ref. [12] prior to publication.

9 Reference Citation and Reference Lists（参考文献及其著录）

科技论文正文后面所附参考文献，指的是撰写论文时引用的有关文献资料。在科技论文中，凡是引用了他人或前人以及本人已公开发表的观点、数据和资料等都需要在论文中出现的地方按出现的先后用序号予以标明，并在文后列出参考文献表，这一过程称参考文献著录。

9.1 著录参考文献的意义

科学研究是逐步深化的，某个时期的研究成果大多都是对前人和他人成果的继承和发展，某项研究在开题之前和研究之中都需要查阅大量的相关资料，以取其有益的东西而用之，避免不必要的重复。因此，撰写科技论文时，著录参考文献是必不可少的。除了根据著作权法规定，在合理引用他人作品后应该著录参考文献外，其意义还在于：①著录参考文献可以充分反映作者的科学态度和为论文提供广泛的、真实的和确凿的科学依据；②著录参考文献可充分说明本课题的研究背景、研究理由和目的，通过对前人和他人的工作的介绍和评价，可为证实作者的论点提供客观的依据；③著录参考文献可以表明作者的起点和高度；④著录参考文献可便于读者阅读和理解以及追索；⑤著录参考文献有利于大型文献数据库的存储、检索和输出，提高学术交流的效果。

在著录参考文献时，应杜绝暗引、崇引和转引。所谓暗引指的是作者在论文中引用了前人或他人已发表过的内容，但未作为参考文献给予标注，也未列入文献表中；所谓崇引指的是生搬硬套权威和知名人物的文章和论著，而与作者自己所写的论文无太大关系的文献；所谓转引是指的是作者并未查原始文献，而是从他人论文的文献中照搬过来的文献。其中"暗引"有抄袭或剽窃之嫌；崇引有拉大旗作虎皮，故意提高论文知名度之嫌；转引有可能造成引文不准确、引文不真实，或有悖于论文主题的可能。

因此科技论文中引用参考文献必须实事求是，要与论文主题贴切，否则，该论文很可能遭到科技期刊评审专家和编辑的拒绝。

9.2 参考文献的主要类型

可作为科技论文引用的参考文献主要有公开出版发行的印刷型出版物和电子型出版物。

印刷型出版物包括图书（books）（如专著、译著、手册、字典等）、会议论文集（proceedings）、期刊（journals）、报纸（newspapers）、学位论文（dissertation）、技术报告（technical report）、技术标准（technical standard）和专利（patent）等。

电子出版物包括网上数据库（database online）、磁带数据库（database on magnetic tape）、光盘图书（monograph on CD-ROM）、计算机程序（computer program）、网上期刊（serial online）和电子公告（electronic bulletin）等。

9.3 科技期刊参考文献著录体系及其格式

国际上科技期刊参考文献著录的体系有多种，其中常见的有顺序编码制和著者-出版年制。在我国，参考文献著录应符合国家标准 GB/T 7714—2005《文后参考文献著录规则》要求。各种不同的著录体系其著录的内容和格式亦不同。其中顺序编码制为国际上大多数科技期刊所采用，而著者-出版年制也有一些科技期刊采用。所以本书重点介绍顺序编码制，对著者-出版年制亦作简明介绍。

9.3.1 顺序编码制

（1）文内著录格式

采用顺序编码制时，在引用前人或他人已发表的有关内容的位置，应按照在论文中出现的先后顺序用阿拉伯数字连续编号，将序号置于方括号内，并视引文的具体情况将序号放在引出处的右上角或作为句子的一部分。举例如下。

Specific strategies mediating gene repression and gene silencing are required to generate cell-type diversity and promote inheritable cell-type identity（reviewed in reference [1]). For example, the transcriptional repression of neuronal-specific genes is necessary to maintain functions unique to nonneuronal systems. Although the precise mechanisms responsible for this tissue-specific transcriptional inactivation remain unclear, it has been shown that repressor element RE-1 silencing transcription factor/neuronal restricted silencing factor（REST/NRSF）is a negative regulator that restricts expression of neuronal genes to neurons in a variety of genetic contexts [2-4]. About 35 neuronal target genes have been identified for REST/NRSF(reviewed in reference [4]). REST/NRSF is a 116kD protein that contains a DNA binding domain with eight zinc fingers and two repressor domains [4-6] and binds to a 21-to 23-base pair (bp) conserved DNA response element, RE-1/NRSE [2-4]. It has been shown that REST/NRSF can mediated repression, in part, through the association of its NH_2-terminal repression domain with the mSin3/histone deacethylase 1，2(HDAC 1，2) complex and with the nuclear receptor corepressor (N-CoR) participating in the context of certain genes [7,8]. The REST/NRSF COOH-terminal repression domain associates with at least one other factor, the transcriptional corepressor CoREST, characterized by two SWI3，ADA2，N-Cor，TFIIIB (SANT) domains [9], which may serve as a platform protein for assembly of specialized repressor machinery[10-12] (Fig. S1).

（2）文后参考文献表的编写格式

采用顺序编码制时，大多刊物在文后列出参考文献表（有的科技刊物参考文献表分列在引用页的地脚处）。各条文献按在论文中所编的序号依次排列。举例如下。

References

[1] Vasic-Racki D. History of industrial biotransformations-dreams and reality//Liese A，Seelbach K，Wandrey C，Eds. Weinheim，Germaney：Wiley，2000.

[2] Holland H L. Enzymes as bioconversion catalyst//Organic Synthesis With Oxidative Enzymes. New York：VCH Publishers，1992.

[3] Asai T. Biochemical Activities of Acetic Acid Bacteria//Acetic Acid Bacteria. University of Tokyo，1968：103-314.

[4] Gatfield I，Sand T. European Patent Application，1988：289，822.

[5] Svitel J，Sturdik E. Enzym Microb Technol，1995，17：546.

[6] Molinari F，Villa R，Aragozzini F，et al. Technol Biotechnol，1997，70：294.

[7] Gandolfi R，Ferrara N，Molinari F. Tetrahedron Lett，2001，42：513-514.

[8] Geelof A，Jongejan J A，Van Dooren T J G M，et al. Enzym Microb Technol，1994，16：1059.

（3）各类文献的著录格式及示例

对于文献表中各条文献的著录项目及其著录顺序和著录符号，各种科技期刊的要求亦不统一，在为不同科技期刊撰稿时，应事先熟悉该刊参考文献的著录格式，其中包括著录项目及其著录顺序和著录符号。现将大多数科技期刊中普遍采用的格式分述示例如下。

（Ⅰ）专著（monograph）

著录项目、顺序及符号：[序号] 著者姓名（西文姓在前，名在后并用缩写）．书名：其他题名信息．其他责任者．版本．出版地：出版社，出版年：参考页码．

举例如下。

[1] Eliel E L，Wilen S H，Memder L N. Stereochemistry of Organic Compounds. New York：Weiley，1994：235.

[2] Gettes L S，Caccio W E. The Heart and Cardiovascular System. 2nd ed. New York：Raven press，1992：2021

（Ⅱ）文集（collected works）

著录项目、顺序及符号：著者姓名．题名//责任者．书名．版本．出版地：出版社，出版年：页码．

举例如下。

Barresteros A，Boross L. Biocatalyst Performance//Straathof J J，Adlercreutz P. Applied Biocatalysis. 2nd ed. Amsterdam：Horword，2000：P287.

（Ⅲ）会议论文集（Proceedings）

著录项目、顺序及符号：著者姓名．题名//编者．文集名．出版地：出版者，出版年：页码．

举例如下。

Attila E P，George H R. How can Agricultural Help the Environment//Proceeding of CCS Congress. Beijing：Chinese Chemical Society，2000：1.

（Ⅳ）期刊论文（papers in journals）

著录项目、顺序及符号：作者姓名．题名（有些刊物不著论文题名）．期刊名，年，卷（期）（连续编号的期刊，如有卷号，期号也可不著）：起止页码（有的刊物只著起始页码）.

举例如下。

Chen C S，Fujimoto Y，Girdaukas G，et al. Quantitative Analysis of Biochemical Kinetic Resolution of Enantiomers. J Am Chem Soc，1982，104（16）：7294-7299.

（Ⅴ）专利文献（Patent）

著录项目、顺序及符号：专利发明者姓名．专利题名：专利国别，专利号．公告日期或公开日期．

举例如下。

Lelands S E. Aldehyde Condensation Products of Fluoroaliphatic Phenols：US，3832409. 1974-08-27.

（Ⅵ）学位论文（dissertation）

著录项目、顺序及符号：作者姓名．论文题名．保存地：保存者，年份．

举例如下。

Juaristi E. Enantioselective Synthesis of β-Amino acids［D］. Beijing：College of Chemistry and Molecular Engineering，Peking University，2002.

(4) 文献著录项目说明

a. 文献序号大多使用阿拉伯数字（有些科技期刊要求将该序号置于方括号内），该序号与其后面著录项目之间不用任何标点符号。

b. 作者姓名大多采用姓在前、名在后的著录格式（有些科技期刊则采用名在前、姓在后），名可以缩写为首字母。

c. 题目或书名按其来源的形式著录。

d. 专著的版本第一版不著录，其他各版均需注明，如 Fifth edition 或 5th ed.，Revised edition 或 Rev. ed.（再版）。

e. 外文科技期刊刊名通常采用标准缩写，如 J Agric Food Chem、J Bacterial、J Biol Chem、J Pharm Sci、Biochem、Biophys、Res commun 等。若无标准缩写，则用原刊名，如 Tetrahedron。

f. 出版者按文献来源所著形式著录。如 Wiley、Blackwell Scientific、Elsevier、Pergamon、Springer、Science Press、Jilin University Press 等。

g. 出版地指出版者所在城市名称，如 New York、Shanghai、Amsterdam 等。

h. 出版年，卷（期），起止页码，各种期刊要求的格式不统一，常见的有：

1999，64（10）：1464-1470.［年，卷（期）：起止页码．］

1989，85：977.（年，卷：起始页．）

139，(1974) 77.［卷，（年）起始页．］

9.3.2 著者-出版年制

这种引文体系亦称 Harvard 体系。该体系规定，在论文正文中引出参考文献时，于被引用的著者姓名之后的圆括号内标注该文献的出版年。举例如下。

We have examined a digital method of spread spectrum modulation for use with Smith's (1988) development of multiple-access communication and with Brown's (1989) technique of digitalmobile radiotelephony.

若正文中未出现著者姓名，只提及其发表的成果内容，则在其后圆括号内标注出著者姓名和出版年，举例如下。

These proteins can also be detected on unattached kinetochore of mono-oriented chromosomes at late prometaphase (Taylor and Mckcon，1997).

引用 3 个或 3 个以上著者的同一文献时，只标注第一作者姓，其后用 et al，举例如下。

The same behavior has been observed for the kinetochore phosphoepitopes recognized by the antibody (Gorbsky et al，1993；Campbell and Gorbsky，1995)．

采用著者-出版年制时，在文后所附的参考文献表中，各条文献均按著者姓的首字母排列，同一作者有多篇文献时，则再按出版年的先后排序。例如：

Klibanov A M. 1979. Enzyme Stabilization by Immobilization. Anal Biochem，93：1-25.

Malcata F X. 1992. Kinetics and Mechanism of Reactions catalysed by Immobilized Lipases. Enzyme Microb Technol，14：426-446.

10 Appendix（附录）

附录是科技论文的附属部分。凡插入论文正文后不但会增加论文的篇幅，而且还会影响论文主题内容的条理性、连续性和完整性，但对同行又有参考价值的资料，一般可以作为附录内容安排在参考文献之后。如正文中某个内容的详尽推导、演算、证明或解释以及不宜列入正文的有关数据、图表、照片、计算机程序、软件等辅助性资料。

附录一般在有特殊需要时，才附到正文后面。如果不属于上述这些情况，可不设附录。

11 Grammar（英语科技论文写作中的几个语法问题）

所谓科技英文论文语法指的是科技英语语言的结构方式。科技英语论文是一种严肃的英语书面文体，严谨周密是其写作风格，所以必须做到行文简练、语法正确、重点突出。

11.1 句子主语常用的人称代词

科技英语学术论文的语言主要是叙述和推理，句子中的主语多为名词，涉及人称代词的并不是很多，常用的人称代词有：第一人称（I，we）和第三人称（it，they），第二人称用得较少。

（1）第一人称

当作者表述自己的观点和自己所做的工作，且看重谓语的执行者时，可以用第一人称（I，we）作主语，若作者只有一位，可用 I，也可用 we；若作者有多位，则需用 we。例如著名的科学家爱因斯坦在他的论文“相对论”中就采用了“I”作为主语。

I am standing in front of a gas range. Standing alongside of each other on the range are two pans. So much alike that one may be mistaken for the other. Both are half full of water. I notice that steam is being emitted continuously from one pan，but not from the other. I am surprised at this，even if I have never seen either a gas range or a pan before.

下面例子中的主语使用了 we。

In this paper we describe the process of immobilized imprinting of the EH resulting in a so-called cross-linked imprinted epoxide hydrolase.

In the work reported herein，we have used different immobilization protocols to produce immobilization preparation of lipase from *Candida antarctica*.

We have demonstrated the utility of the CAL-B catalyzed enantioselective acylation of racemic amines in the resolution of some pharmacologically interesting β-substituted isopropylamines.

（2）第三人称

科技英语论文与写作，用 it 作形式主语具有普遍性。尤其是在作者重视的是自己论文中的内容和观点，而不是作者本人时，更是愿意用 it 来作形式主语代替主语从句、不定式短语和动名词短语。

例如：

It was observed that the butyrylation of （±5)-5 in toluene at 25 ℃ did not stop at 50% conversion .（代替主语从句）

It was shown that the enzymatic resolution of substrate（±5)-1 was effective.（代替主语从句）

It is also necessary to have a biocatalyst resistant under these drastic conditions.（代替不定式短语）

It is necessary measuring temperature with a calibrated thermocouple or thermometer in this work.（代替动名词短语）

用 it 来代替上面提到的单数名词或不可数名词，举例如下。

To achieve this，certain ligands are added to an enzyme solution and then the enzyme is precipitated or lyophilized before it is used in a non-aqueous environment.

用 it 来代替前面提到的事情，举例如下。

We can change a liquid into a solid by decreasing its temperature，or vice versa，it may be easily verified with heating.

用 it 构成强调句型，举例如下。

It was until the discovery of X-ray that a powerful means for the analysis of matter structure and medical treatment can be developed.

用 they 代替前面提到的复数名词，举例如下。

In order to assign the（*S*)-configuration for the remaining amines 1 and 4，they were transformed into the corresponding hydrochloride salts.

11.2 英语科技论文表述中常用的时态

英语句子中作谓语的动词发生的时间和存在状态不同，其表达形式亦不同，谓语动词的这种不同的形式称作时态。在写句子的时候，准确地把握句子的时态是科技论文写作的关键之一。英语句子中的谓语动词总共有 16 种时态，但在科技论文中常用的只有 5 种，即一般现在时、一般过去时、一般将来时、过去完成时和现在完成时。而用得最为频繁的只有一般过去时和一般现在时。

（1）一般现在时态

用于引用在科技刊物上正式发表的内容，举例如下。

Optically active amines bearing the stereogenic center at the α-position are important compounds because of their broad range of applications and their pharmacological properties[1].（引言中的引用）

用于引言中一般性背景叙述，举例如下。

In most cases，the pharmacological activities of these amines are related to the configuration of the stereogenic center.

用于图表的引出，举例如下。

Fig. 2 represents the binding mode of the substituents of the *R*-amines 1-6.

The results are collected in Table 1.

用于对实验结果的讨论，举例如下。

Amphetamine shows a lower *E* values than those of its methoxy analogous，which is due to the fact that（*R*)-methoxyamphetamines react faster than（*R*)-amphetamine and that（*S*)-methoxyamphetamines. In the same way，the higher *E* value for mexiletine is mainly a consequence of the slower reactivity of its（*S*)-enantiomer. These results can be explained to a large extent on the basis of the size of the substituents attached to the stereocenter and size of the small pocket.

用于一般性真理的描述，举例如下。

Streptomycin inhibits the growth of *tuberculosis*.

（2）一般过去时态

用于描述试剂和材料的使用，举例如下。

Lipase B from *C. antartica*，Novozym 435，was a gift from Novo Nordisk Co. and was employed without any further treatment. Solvents were of spectrophotometric grade and they were stored with 4 Å molecular sieves under nitrogen prior to use.（±)-Mexiletine was purchased as its chlorohydrate from Aldrich. The other racemic amines were obtained by reductive amination of the corresponding ketone.

用于测试仪器的使用，举例如下。

Melting points were determined on a Kofler hot-stage apparatus. Ultra-violet spectra were recorded by using Pye-Unicam SP800 and Philips PU8720 spectrometers. Infra-red spectra were recorded on a Perkin-Elmer PE1710 Fourier transform spectrometer，HNMR spectra were recorded on a Bruker WH360（360MHz) spectrometer and tetramethylsilane was used as internal standard. Mass spectra were recorded on Kratos MS80 or MS25 spectrometers.

用于实验全过程的描述，举例如下。

Imprinting of the derivatized protein

Imprinting molecules were added to the combined eluates. As imprinters were used，the racemic substrates，（*R*)-enantiomer of the substrate，or the（*S*)-enantiomer of the substrate，the samples were mixed and kept at room temperature for 20 min，then put on ice for 5 min. The imprinted proteins were precipitated with 30% isopropanol. After mixing for 10 min，the samples were put on ice for 60 min. After centrifugation，the precipitates were washed once with ice-cold isopropanol and after a second centrifugation step the precipitates were lyophilized.

用于实验结果的描述，举例如下。

The parameters k_1，k_2 and a_1 in eq.（2） were determined by a nonlinear regression program. Values of a_1 in immobilized lipases were similar to and greater than those of free lipase. The solid curves in Fig. 1 and 2 were calculated by using the parameters as estimated.

用于结论的描述，举例如下。

The octadecyl-sepabeads preparation was the most efficient among those examined. This preparation presented the highest enantioselectivity，good activity and high stability under the harsh reaction conditions.

The thermostability of lipase was enhanced by immobilization. Deactivation of the lipases followed a two-step series model. When incubation temperature was below 60℃，lipase immobilized in alginate gels was found to have very high thermostability compared with free and other immobilized lipases. This was because of the strong affinity between lipases and alginate. Likewise，thermostabilization was found for the lipase immobilized in inorganic microcapsules when incubation temperature was above 60 ℃. This was due to the stabilization of the intermediate state of the lipase.

（3）一般将来时态

用于打算要做的工作和预期结果的描述，举例如下。

We will next use the conversion values obtained after a reaction time of 7 h.

In some areas of the world，cancer has become or shortly will become the leading disease-related cause of death of the human population.

Studies on gene therapy will be carried out in the near future.

（4）现在完成时态

用于到写论文时为止的一段时间里，将结果与现在联系起来的情况描述，举例如下。

Though immobilization does not necessarily lead to stabilization，there have been many reports on enzyme stabilization by immobilization.

In most cases，including the present study，it has not been possible to clearly elucidate the nature of the interaction between the enzyme and the support.

Although a few chemotherapeutic regiments have yielded lasting remissions or cures，it is clear that new therapeutic options are necessary.

用于论文结论中对全文的概括性总结的描述，强调动作到现在所产生的结果和影响，举例如下。

The effect of enzyme immobilization on the catalytic properties of different CAL-B preparations in the resolution of several cyclopyrrolone compounds in aqueous media has been studied. This work has demonstrated the strong influence of the enzyme preparation and the nature of the substrate.

（5）过去完成时态

用来追述或补叙相对于过去某一时间之前所发生的动作，举例如下。

When the genetic information could be translated into therapeutics that could selectively ablate tumors without the systemic side effect often associated with cancer drugs，some of the first oncogene defects had been discovered.

时态是科技英语论文表述中一个关键性问题。在论文的各个不同章节中，尽管表述的重点是不同的，但谓语动词的时态的使用都是有规律的。通常描述作者自己的研究工作应该用一般过去时态，如论文使用的研究方法和过程，所得到的研究结果等。因为这些章节中主要描述作者自己在写论文之前做的工作、所发现的现象、所得到的数据和结果。通常在前言中叙述已经公开发表了的内容和作者在论文中对自己所得结果进行讨论时，大多使用一般现在时态，因为这些内容都是被确认了的知识。

11.3 英语科技论文表述中常用的语态

语态指的是句子中的主语和谓语之间的关系。当句子的主语是谓语的执行者时，其谓语形式为主动语态（active voice），当主语为谓语的承受者时，其谓语的形式为被动语态（passive voice）。在被动语态的句子中，其主语应该是主动语态句子中谓语的宾语。因此，被动语态句子中的谓语动词应是及物动词。

试比较如下。

We also determined four crystal structures of SPE 7 complexed with four different antigens.（主动语态）

Four crystal structures of SPE 7 complexed with four different antigens were also determined.（被动语态）

语态是强调句子中某一部分的一种表达形式。句子强调的是谓语动作的执行者时，常采用主动语态。举例如下。

To overcome these problems we have studied the biotransformation of foreign substrate by plant-cultured suspension cells.

当句子强调的是谓语动作的承受者时，通常采用被动语态。举例如下。

β-Lactams can be transformed into valuable β-amino acids, which are of immense interest from both pharmacological and chemical viewpoints.

在英语科技论文的表述中，被动语态的句子用得较多，但被动语态用词较多，使句子冗长或头重脚轻。故通常能够用主动语态表述的内容尽量不用被动语态表述。举例如下。

A facile transformation of 3, 4-disubstituted 2-azetidinones to chiral 5, 6-dihydro-2-pyridones, which can serve as valuable chiral intermediates for different piperidine and indolizidine alkaloids and azasugars, was reported by Lee et al.

此句子明显头重脚轻，头大尾小，给人一种句子不稳的感觉，若改成如下的主动句子，则可克服这一毛病。

Lee et al reported a facile transformation of 3, 4-disubstituted 2-azetidinones to chiral 5, 6-dihydro-2-pyridones, which can serves as valuable chiral intermediates for different piperidines and indolizidine alkaloids and azasugars.

在被动语态的句子中，凡是有动词时，则不用该动词相应的名词，以使句子简练。举例如下。

“Separation of the compounds was accomplished”，可改成“The compounds were separated”。

在科技英文论文的表述中，常见的这类名词及其相应的动词如下。

determination→determine, identification→identify, speculation→speculate, discussion→discuss, description→describe, treatment→treat, assignment→assign, interpretation→interpret, explanation→explain, calculation→calculate, differentiation→differentiate, synthesis→synthesize, preparation→prepare, dissolution→dissolve, dilution→dilute, concentration→concentrate, evaporation→evaporate, extraction→extract, analysis→analyze。

由形式主语“it”代替主语从句先行引出的被动句，在不影响句子表述的内容的情况下，可以不用这种形式。举例如下。

“It has recently been reported by Mateo et al that some immobilization methods can enhance the thermostability of an enzyme”可改成为“Mateo et al has recently reported that…”。

11.4 主语和谓语动词在数和人称上的一致

在科技英语表述中，主语的人称和数必须和谓语动词保持一致。这一语法现象看来似较简单，但用起来却不那么容易，其关键在于对主语的数做出正确的判断，以此为依据确定谓语动词的形式。

① 在主语中含并列连词“and”时，应正确区分是单个主语还是并列主语。若属前者，谓语动词为单数形式，若属后者，谓语动词则为复数形式。举例如下。

Description of the experimental conditions and the results of the present investigation is made in detail in this paper.

此句中的主语为 description，the experimental conditions and the results 是并列的，与 of 构成介词短语作定语，故谓语动词用“is made”。

The experimental conditions and the results of the present investigation are described in detail in this paper.

此句子主语为 the experimental conditions and the results，故谓语用了“are described”。

由 and 连接名词看起来像并列主语，但实际上却是一个事物，应按单数主语处理。举例如下。

The formation and accumulation（成聚）of several secondary metabolites does not normally occur in the plant.

Ice and water（冰水）was used as the coolant in these experiment.

② 主语中含 as well as 时，谓语动词应与其前面的名词在人称和数上保持一致。举例如下。

The composition，as well as molecular weight was confirmed by elemental analysis and mass spectrometry.

③ 主语中含 or，neither…nor，either…or，not only…but also 等时，谓语动词与其最近的名词在人称和数上保持一致。举例如下。

Ether ethylene chlorohydrin or bromohydrin is treated with strong alkali to give ethylene oxide.（与 bromohydrin 一致）

Coagulation of AgCl precipitate occurs at the equivalence point，where neither chloride nor silver ions are in excess.（与 ions 一致）

④ 由 there be（exist，remain，stand，seem，appear 等）作谓语时，应与其后面的第一个主语在人称和数上保持一致。举例如下。

There remains a test and an objective appraisal to be carried out before putting the device into operation.

⑤ 量和单位构成的主语，谓语动词用单数形式，举例如下。

Five grams of NaCl was added to the solution.

⑥ 不定代词 each，every 和 everyone 等作主语时，谓语用单数形式，举例如下。

Every rat injected and dosed orally was included.

⑦ 抽象名词作主语时，谓语动词用单数形式。举例如下。

The formation of acetaldehyde and propylene oxide is favoured in the case of inactive SiC.

⑧ 物质名词作主语时，谓语动词用单数形式。举例如下。

Lithium carbonate decomposes on heating in a stream of hydrogen.

⑨ 定语从句中关系代词作主语时，谓语动词的数应与该从句所修饰的名词（即先行词）保持一致，举例如下。

The current-voltage curves，which are shown in Fig. 5，clearly demonstrate the reversibility of all four processes.

Despite much research，there are still certain elements in the life cycle of the insect that are not fully understood.

⑩ 集体名词（collective nouns）作主语时，当其作为一个整体考虑时，谓语动词用单数形式。举例如下。

The series is arranged in order of decreasing size.

当作为整体中各个个体考虑时，谓语动词用复数形式。举例如下。

A series of compounds were tested.

The amount of pressure，which the materials are subject to，affects the quality of the products.

⑪ 动词不定式短语、动名词短语和从句作主语时，其谓语动词用单数形式。

Thorough removing the water-soluble proteins from the mitochondria is required before further fractionation is attempted.

⑫ 某些特殊变化的复数名词切不可误认为是单数，这样的词（如 analyses，bacteria，criteria，formulae，phenomena，fungi，mice，data 等）作主语时，谓语动词用复数形式。例如：The mice were cloned from three distinct libraries，including two from testis and one from spermatogenesis cells.

Extensive data on the thermodynamically properties of the elements are available.

⑬ 某些以“s”为结尾的名词不是复数名词，这样的词（如 statistics，kinetics，thermodynamics，mumps，apparatus 等）作主语时，谓语动词用单数，举例如下。

Mumps is a very common disease which usually affects children.

⑭ a number of＋复数名词作主语，谓语动词用复数形式；the number of＋复数名词，谓语动词用单数形式，其中 number 表示数目。举例如下。

A number of di-and trivalent cations have been successfully titrated with this reagent.

The number of times an object is magnified by a telescope was determined by someone who knows the focal length of the objective lens and of the eyepiece.

⑮ an average of＋复数名词作主语时，谓语动词用复数形式；the average of＋复数名词作主语，谓语动词用单数形式。举例如下。

The average of a series of values is obtained simply by adding all the individual values and dividing the sum by the number of values.

An average of 100 kinds of reagents were in regular supply from the company each year.

⑯ a lot of（most of，plenty of）＋复数名词作主语时，谓语动词用复数形式；a lot of（most of，plenty of）＋不可数名词作主语时，谓语动词用单数，举例如下。

Most of the better-known organic solvates are all suitable for these tests.

There is plenty of distilled water in the bottle.

There are plenty of test tubes on test table.

12 Quantities and Units（科技英语表述中的物理量及其单位）

物理量（简称量）及其单位是科技英语表述中必不可少的部分。在国内执行中华人民共和国法定计量单位在制。量是现象，是物体与物体可定性区别和定量确定的一种属性。一切量都可以与其他量建立数学关系，进行数学运算。例如，做匀速运动的质点的速度为 $v=l/t$，式中 l 为质点运动的距离，t 为时间。为了对量进行定量确定，在同一类量中选出一个称为单位的参考量，而这一量中的其他量都可以用这个单位与一个数的乘积表示，这个数叫做该量的数值。如 $v=15\mathrm{m/s}$，v 为某质点运动的速度，m/s 是速度的单位（其中 m 为米，s 为秒，“/” 斜线表示除），15 则是以 m/s 为单位时某质点运动速度的数值。

按照量和单位的正规表达式，上述关系可以写成：

$$A=\{A\}\cdot[\mathrm{A}]$$

式中，A 为某一物理量的符号；[A] 为某一单位的符号；$\{A\}$ 为以单位 [A] 表示的量 A 的数值。

20 世纪 60 年代以前，所使用的量和单位制比较混乱，各种不同的单位制，如公制、市制及混杂制到处可见。如英制（English system）中，表示距离用 yard，表示质量用 pound；米制（metric system）中表示距离用 meter，表示质量用 kilogram；市制（Chinese system）中表示距离用里，表示质量用斤等。

为了使量和单位在国际上得到统一，1960 年在第 11 届国际计量大会上制定并通过了国际单位制（the International System of Units），简称 SI 单位制（System International）。目前世界各国科技工作者在撰写科技论文时，普遍采用国际单位制。

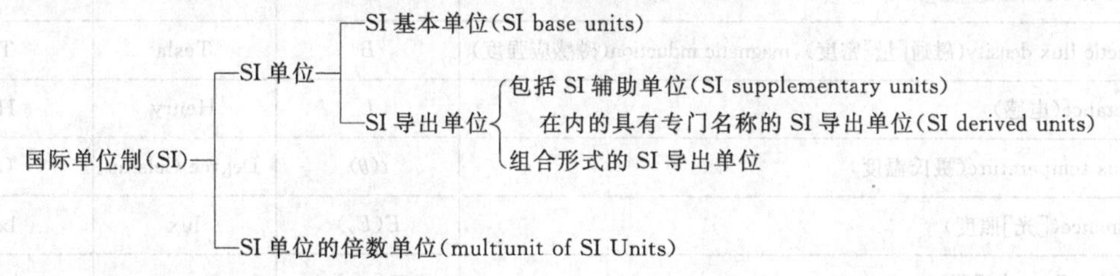

（1）SI 基本单位和辅助单位

SI 基本单位共有 7 个（详见 Table 2.6）。

Table 2.6 SI base units

physical quantity(物理量)	symbol(符号)	unit(单位)	symbol(符号)
electric current(电流)	I	Ampere	A
luminous intensity(发光强度)	$I(I_V)$	Candela	cd
thermodynamic temperature(热力学温度)	$T(\Theta)$	Kelvin	K
mass(质量)	m	kilogram	kg
length(长度)	L,l	meter	m
amount of substance(物质的量)	$n(v)$	mole	mol
time(时间)	t	second	s

SI 辅助单位共 2 个（详见 Table 2.7）。

Table 2.7 SI Supplementary units

physical quantity(物理量)	symbol(符号)	unit(单位)	symbol(符号)
plane angle(平面角)	$\alpha,\beta,\gamma,\theta,$	Radian(弧度)	rad
solid angle(立体角)	Ω	Steradian(球面度)	sr

（2）SI导出单位

SI导出单位是用基本单位以代数形式表示的单位。在SI中，具有专门名称的导出单位共有19个（见Table 2.8）

Table 2.8 SI derived units

physical quantity(物理量)	symbol(符号)	unit(单位)	symbol(符号)
frequency(频率)	F,υ	Hertz	Hz
force(力)	F	Newton	N
pressure(压力),stress(应力),压强	p	Pascal	Pa
energy(能[量]),work(功),quantity of heat(热量)	E,W,Q	Joule	J
power(功率),radiant flux(辐[射能]通量)	$P,\Phi(\Phi_e)$	Watt	W
electric charge(电荷),quantity of electricity(电量)	Q	Coulomb	C
voltage(电压),electric potential(电位),electromotive force(电动势),(电势)	$V,\varphi,U(v),E$	Volt	V
capacitance(电容)	C	Farad	F
electric resistance(电阻)	R	Ohm	Ω
conductance(电导)	G	Siemens	S
magnetic flux(磁通[量])	Φ	Weber	Wb
magnetic flux density(磁通[量]密度),magnetic induction(磁感应强度)	B	Tesla	T
inductance(电感)	L	Henry	H
Celsius temperature(摄氏温度)	$t(\theta)$	Degree Celsius	℃
iluminance([光]照度)	$E(E_v)$	lux	lx
luminous flux(光通量)	$\Phi(\varphi_v)$	Lumen	lm
activity of a radionuclide([放射性]活度)	A	Becquerel	Bq
absorbed dose(吸收剂量),比授[予]能,比释动能	D	Gray	Gy
dose equivalent(剂量当量)	H	Sielvert	Sv

（3）SI单位的组合单位

由两个或两个以上单位通过乘或除构成的单位叫做组合单位。相乘组合单位将各单位连写，也可以在单位之间用点乘号，如动力黏度（dynamic viscosity）的单位可写成Pas或Pa·s。对于相除组合单位，各单位之间可用分数线，也可用斜线，还可用点乘该单位负指数，例如运动黏度（kinematic viscosity）的单位可以写成m^2/s、$m^2 \cdot s^{-1}$。相除组合单位符号中的斜线“/”只能用1条，如传热系数（coefficient of heat transfer）的单位，只能写成$W/(m^2 \cdot K)$，不能写成$W/m^2 \cdot K$，也不能写成$W/m^2/K$。

（4）SI单位的倍数词头及倍数单位

指SI单位及SI导出单位的十进制倍数和分数单位。SI单位倍数单位是由SI词头和SI单位构成的。例如1401Pa可以写成1.401kPa，kPa为SI倍数单位，其k表示10^3。SI词头共有24个（详见Table 2.9）。

Table 2.9 SI multiplying prefixes

name of prefix	symbol	factor	name of prefix	symbol	factor
vendeca	V	10^{30}	deci(分)	d	10^{-1}
xenna	X	10^{27}	centi(厘)	c	10^{-2}
yotta(尧)	Y	10^{24}	milli(毫)	m	10^{-3}
zetta(泽)	Z	10^{21}	micro(微)	μ	10^{-6}
exa(艾)	E	10^{18}	nano(纳)	n	10^{-9}
peta(拍)	P	10^{15}	pico(皮)	p	10^{-12}
tera(太)	T	10^{12}	femto(飞)	f	10^{-15}
giga(吉)	G	10^{9}	atto(阿)	a	10^{-18}
mega(兆)	M	10^{6}	zepto(仄)	z	10^{-21}
kilo(千)	k	10^{3}	yocto(幺)	y	10^{-24}
hecto(百)	h	10^{2}	xenno	x	10^{-27}
deca(十)	da	10^{1}	vendeco	v	10^{-30}

注：SI 词头表示的因数≥10^6 时，词头用大写字母；表示的因数≤10^3，词头用小写字母。

SI 词头不能单独使用，如 $l=20\mu$，应为 $l=20\mu$m。SI 词头也不能重叠使用，如 $l=20\mu\mu$m，应为 $l=20$nm。组合单位加 SI 词头时，一般只用 1 个词头，并用于组合单位中的第 1 单位之前，例如力矩 M 的单位可以写成 kN・m，不宜写成 N・km。再如摩尔热力学能 U_m 的单位可写成 kJ/mol，不宜写成 J/mmol 。又如某物质的浓度 c 的单位可写成 μmol/L，不宜写 nmol/mL（分子和分母都加词头了）。

13 Numeral Usage（数字的使用）

在科技英语的表述中，数字使用的频率相当高。常用的有基数词（cardinal numerals)，序数词（ordinal numerals）等，尤其是阿拉伯数字使用的场合更多。在科技论文中，数字除了用于论文章、节编号，图表、公式及结构式的编号，参考文献的编号外，还主要用于计量的表示。因此，数字的用法是否正确，是否符合国际统一规范，直接影响科技出版的质量。

13.1 阿拉伯数字使用的场合

(1) 用于物理量的量值

该量值与其后的单位符号之间留一空隙（常相当于一个小写字母所占空间)。如 5 h、25 mL、0.30 mol/L 等。

(2) 用于非物理量的计数数值

在句子的行文中凡是表示 10 以内的一般用英语基数词表示，10 以上（含 10）的用阿拉伯数字。如 four flasks，14flasks 等。

(3) 阿拉伯数字不用于句子的开头

举例如下。

① Twenty slides of each blood sample were prepared.

② Twenty-five milliliters of supernate was added to the reaction vessel.

③ Supernate（25 mL）was added to the reaction vessel.

④ Fifty-five samples were collected but only 22 were tested.

但化合物命名中取代基位置的阿拉伯数字编号不在此范畴之内。举例如下。

2-Aryl-1*H*-imidazo [4,5-*b*] porphyrins were prepared in good yields by the condensation of prophyrin-2，3-dione with the corresponding arylaldehyde in the presence of excess NH_4Ac.

（4）量和单位作为另一个名词的定语

量和单位之间应加连字号“-”，例如：

a 10-mg sample；a 20-mL aliquot；100-，200-and 300-mL aliquots 等。

当量和组合单位作定语时，则不加连字号，如 a 0.1 mol/L NaOH solution。

13.2 数值的范围表示

① 数值的范围符号采用“-”，其长度是连字号的 1 倍，也称为一字线。英文中常称为“en dash”，如 Figures 1-5。

② 书写百分数的范围时，每个百分数后面都要加“%”，如 28%-38%。

③ 书写具有相同幂次数的数值范围时，每个数值的幂指数都要写。例如 4×10^{-4}-5×10^{-4}，也可以简化成 (4-5)$\times10^{-4}$。

④ 单位相同的量值范围，单位符号只放在后一个量值之后，如 5-50 kg。

⑤ 单位不同的量值范围，每个量值的后面需写出各自的单位符号。例如，5 Hz-50 kHz。

⑥ 当带有计量单位的数值的范围作定语时，后一个量值与其单位之间需加连字号。如 a25-30-mL aliquot。

⑦ 具有相同计量单位的一系列数，单位符号置于最后一个数之后，如 10，20，30，40，and 50 mg。

13.3 数值公差的表示

① 参量及其公差的单位相同时，其单位可写成 15.5 nm±0.1 nm，也可写成 (15.5±0.1)nm，但不能写成 15.5±0.1 nm。表示带百分数公差的中心值时，“%”只写 1 次，且“%”之前的中心值和公差应放在圆括号之内。如 (25±1)%，不得写成 25±1%，也不能写成 25%±1%。

② 参量的上下公差不同时，公差分别写在参量的右上角和右下角。当参量与公差的单位相同时，单位只写 1 次，如，$10\ g^{+0.2}_{-0.1}\ g$ 可写成 $10^{+0.2}_{-0.1}\ g$。若参量与公差的单位不同时，其单位则需分别写出来，例如 $20\ cm^{+1}_{-0.1}\ mm$。

13.4 概数的表示

① 数字+more or less 表示左右，例如：

More or less than 20 reagents were used in this experiment.

② some+数字表示左右，例如：

Solvolysis rate of tosylate varies by some 8 powers of ten in acetic acid.（对甲苯磺酸酯在醋酸中溶解的速率为 10^8 左右。）

③ 数字+or so 表示左右，例如：

The number of the collected samples is 200 or so.

④ about [approximately，around，rounds，nearly，almost，ca (circa)，roughly，close to]+数词表示大约为。例如：

It was diluted with water until the total volume of solution was about 10mL.

Upon heating for a period of approximately 6 h, a yellow precipitate is formed.

All molecular weights were determined in ca. 0. 05 mol/L bezene solution by the vapor pressure as mometric method.

13.5 倍数的表示

① 数词（x）＋times＋as＋形容词或副词原级＋as 从句表示前者为后者的 x 倍或净增 $x-1$倍。例如：

The oxygen atom is almost 16 times as heavy as the hydrogen atom.（氧原子的质量几乎是氢原子的 16 倍或氧原子几乎比氢原子重 15 倍。）

数词（x）＋times＋形容词或副词的比较级＋than 从句亦表示前者为后者的 x 倍或净增 $x-1$ 倍。例如：

The oxygen atom is almost 16 times heavier than the hydrogen atom.

数词（x）＋times＋名词或代词（that）表示几倍于，前者为后者的 x 倍，或比后者净增 $x-1$ 倍。例如：

The mass of oxygen atom is almost 16 times that of hydrogen atom.

② 含有增加意义的动词（increase, rise, grow, raise, exceed 等）＋数词（x）＋times 表示增加到 x 倍，或净增 $x-1$ 倍。例如：

The rate of reaction (1) has been increased six times under this catalyst.

含有增加意义的动词＋数词（x）＋times＋as＋against…表示增加到 x 倍或净增 $x-1$ 倍。例如：

The rate of reaction (1) has been increased six times as against that of reaction (2).

含有增加意义的动词＋by a factor of＋数词（x）表示增加到 x 倍或净增 $x-1$ 倍。例如：

The rate of reaction was increased by a factor of six over this catalyst.

含有增加意义的动词＋n-fold 表示是……的 n 倍。例如：

The enzyme was purified 50-fold over crude extract.（经提纯该酶的纯度是粗萃物的 50 倍。）

③ 使用含有倍数意义的动词表示倍数，如 double（增加到 2 倍或 1 倍于），treble（增加到 3 倍或 2 倍于），quadruple（增加到 4 倍或 4 倍于）等，举例如下。

The reaction rate is approximately trebled for each 10 ℃ rise temperature.（温度每升高 10℃，反应速率提高到 3 倍。）

13.6 数增加或减少的表示

① 数字＋形容词比较级（more 或 less）than…表示比……多（或少）……

The yield of this reaction is expected to be 20% higher (or lower) under this condition than under that one. [在这种条件下该反应的收率预期要比那种条件下提高（降低）20%。]

② 形容词比较级＋than…＋by＋数字表示净增数，例如：

This chromatography column is longer (shorter) than that one by 20 cm. [这根色谱柱比那根长（短）20 cm。]

③ 百分数＋up on (over)＋比较对象表示净增数。例如：

The yield of estertification under such an improved condition was 20% up over that under the original condition.（在改进条件下酯化反应的收率比原条件下提高了20%。）

④ 含有增加（减少）意义的动词＋by＋数字表示净增（净减）数。举例如下。

The main cost is increased（decreased）by 70%.［主要成本提高（降低）了70%。］

⑤ 含有减少意义的动词＋数词（x）＋times 表示减少为 $1/x$［或减少了（$x-1$）$/x$］。举例如下。

The rate of this reaction is shortened 3 times.［该反应的速度缩短到1/3（缩短了2/3）。］

⑥ 含有减少意义的动词＋by a factor of＋数词（x）表示减少到 $1/x$ 或减少了（$x-1$）$/x$。举例如下。

The rate of this reaction is shortened by a factor of three.

The instrument being designed will reduce the error probability by a factor of 8.［正在设计的仪器将使误差概率降低到1/8（或降低了7/8）。］

13.7 不定数和近似数表述的常见短语

（1）表示“无数”的短语

hundreds of	数以百计	many thousands	成千上万
hundreds of thousands of	成千上万	millions of	数以百万计
thousands and thousands of	成千上万	thousands of millions of	亿万
thousands upon thousands of	成千上万	millions and millions of	千千万万
thousands of	数以千计	billions of	亿万

（2）表达“很多”，其后可接可数名词的短语

many of	许多	a quantity of	一些，大量
a good（great）many of	很多，大量	a large（great，considerable）quantity of	许多，大量
a lot of	许多，大量	a few	一些
lots of…	许多，大量	a good few	相当多，不少
a number of	若干，许多	quite a few	许多，相当不少
a great（large，considerable）number of	许多，大量	not a few	相当多，不少
numbers of	许多，大量	plenty of	很多
large numbers of	大批，许多	a multitude of	许多，大量

（3）表达“很多”，其后接不可数名词的短语

much of	许多，大量	a lot of	许多，大量
a great（good）deal of	许多，大量	lots of	许多，大量
quite a good deal of	相当多，许多	a quantity of	一些，大量
a large（great，considerable）amount of	许多，大量	a large（great，considerable）quantity of	许多，大量
not a little	不少，很多	plenty of	很多

（4）表达“很少”，其后接可数名词的短语

a few	几个	a small number of	少量
some few	少量	a small quantity of	少量
some of	几个	a handful of	少量

（5）表达“很少”，其后接不可数名词的短语

a little	少量，一些	a small quantity of	少量
some little	少量	a handful of	少量
a bit of	少量，一点	a particle of	少量
a small amount of	少量		

(6) 表达“多于”的短语

above＋数字	……以上	数字＋and more	……以上
over＋数字	……以上	数字＋and odd	……以上
beyond＋数字	……以上	数字＋and over	……以上
upwards of＋数字	……以上，超过	数字＋and upwards	……以上
more than＋数字	多于，……以上	not more than	不多于
数字＋odd	……以上	not less than	不少于
数字＋or odd	……以上		

(7) 表达“少于”的短语

less than＋数字	少于，……以下，不到	under＋数字	不到，……以下
below＋数字	不到，……以下		

13.8 罗马数字的构成

科技英语写作中有时也使用罗马数字，如系列论文、图、表及结构式的编序。

罗马数字有7个基本数字，即：Ⅰ(1)，Ⅴ(5)，Ⅹ(10)，L(50)，C(100)，D(500)，M(1000)。

这7个基本数字若重复几次，则表示该数增加到几倍，例如：ⅩⅩⅩ 表示30；LLL表示150。

各基本数字右边带有较小的数字，则表示各数之和，例如：ⅩⅢ 表示13；LⅡ表示52。

各基本数字左边带有较小的数字，则表示各数之差，例如：Ⅸ 表示9；ⅨC表示91。

罗马数字可用于编序号，例如：Asymmetric synthesis and Cram’s Rule（Ⅲ）（Ⅲ为系列论文序号）；TableⅠ（表序）；Scheme Ⅱ（图式序）；Structure Ⅳ（结构式序）；Appendix Ⅴ（附录序）。

罗马数字还可用于表示原子价态数，例如：Cu（Ⅱ），$Pb^{Ⅳ}O_2$，Hexacyanoferrate（Ⅱ)-(Ⅲ）。

13.9 数字词头

一个词中含有数的概念时，应使用数字词头（multiplying prefix）。这些词头中，一些来自拉丁文词头。常用的数字词头如下：deci-(十分之一)；centi-(百分之一)；milli-(千分之一)；hemi，semi（1/2）；mono-，uni-(一)；di-，bi，bis-(二)；tri-，tris-，ter-(三)；tetra-，tetrakis-，quadri-(四)；penta-，pentakis-，quinqui-，quinque-(五)；hexa-，hexakis-，sexi-(六)；hepta-，heptakis-，septi-(七)；oct-，octa-，octakis-，octo-(八)；nona-，nonakis-，ennea-(九)；deca-，decakis-(十)；undeca-，hendeca-(十一)；dodeca（十二）；trideca（十三）；tetradeca（十四）；pentadeca（十五）；hexadeca（十六）；heptadeca（十七）；octadeca（十八）；nondeca（十九）；eicosa（二十）；heneicosa（二十一）；docosa（二十二）；tricosa（二十三）；tetracosa（二十四）；pentacosa（二十五）；hexacosa（二十六）；heptacosa（二十七）；octacosa（二十八）；nonacosa（二十九）；triaconta（三十）；hentriaconta（三十一）；tetraconta（四十）；pentaconta（五十）；hexaconta（六十）；heptaconta（七十）；octaconta（八十）；enneaconta（九十）。

具体用法举例如下：monooxide（一氧化碳）；uniaxial（一重轴）；dichloride（二氯化物）；bimetallic（双金属的）；trichlorobenzene（三氯苯）；1,4-bis（3-bromo-1-oxopropyl）piperazine[1,4-双（3-溴-1-氧代丙基）哌嗪]；terbromide（三溴化物）；carbon tetrachloride（四氯化碳）；tris（ethylenediamine）cadmium dihydroxide [三（亚乙基二胺）氢氧化镉]；

tetrakishexahedron（四重六面体）；quadriphase system（四相系）；pentagon（五角形）；quinquevalence（五价）；sexavalence（六价）；hexachlorocyclohexane（六六六）；heptahedron（七面体）；septavalence（七价）；octagon（八角形）；octocyclic（八环的）；nonanol（壬醇）；enneahedron（九面体）；decahydronaphthalene（十氢化萘）；deciliter（分升）；centimeter（厘米）；milligram（毫克）。

14 Capitalization and Lower Case of English Letters（英文字母的大写和小写）

在英语科技论文的写作中，每个句子的第一个词的首字母均采用大写形式，在其他情况下均采用小写形式。因此，在一篇论文中，小写字母占绝大多数，而大写字母则相对较少。只要了解大写字母的使用规则，小写字母便不述自明了。

① 专有名词（proper nouns），如人名、地名、国家名和机构名等名词的首字母一律用大写，如Jerry March（人名）；New York（地名）；John Wiley & Sons，Inc.（出版公司名）；International Union of Pure and Applied Chemistry（机构名）；Amino Acid Analysis（书名）；American Journal of Human Genetics（杂志名）；School of Biochemistry and Molecular Genetics，University of New South Wales，Sydney，Australia（工作单位名）。

当专有名词作定语时，其首字母亦用大写形式。如Schiff base、Avogadro's number等。由专有名词派生出来的形容词，其首字母亦大写，如Coulombic、Hamiltonian等。

② 论文题目中的名词、动词、数词、形容词、副词和代词等实词的首字母一律用大写字母。介词、连词和冠词等虚词如果在题首，其首字母一律用大写字母；若处于其他位置，则用小写。举例如下。

Synthesis of Reactive γ-Lactams Related to Penicillins and Cephalosporins

A Dithiane-protected Benzoin Photolabile Safety Catch Linker for Solid-phase Synthesis

主副文题中主标题后面由冒号（:）引出副题，该副题的第一个词的首字母采用大写形式。例如：Acyl Isothiocyanate Resin：A Traceless Linker Approach to Substituted Guanidines

如果文题中含有动词不定式符号to，其首字母采用大写“T”。以上所述原则已被大多数科技期刊所采用。但有的科技期刊为了突出文题，则将文题中所有词中的每个字母都采用大写形式。还有些科技期刊的文题只要求第一个词的首字母采用大写形式，其后各词的所有字母都用小写形式。因此在为科技期刊撰写论文时，要先弄清楚该刊的要求。

③ 论文内各个部分的题目，如Abstract、Introduction、Experimental、Results、Discussion、Conclusion、Appendix、Acknowledgment、References、Bibliography等的首字母采用大写形式。

文内各层次标题第一个词的首字母采用大写形式。举例如下。

2.2 Influence of the substituents on the enantioselectivity and the reaction rate.

图序Fig. 1、表序Table 1、图式序Scheme 1等词的首字母均大写，其后的图题、表题和图式题的第一个词的首字母用大写形式。举例如下。

Fig. 2 Active site model for CAL-B.

Table 2 Relative reactivity of *R*-and *S*-β-substituted isopropylamines

④ 以小写缩写形式开头的词在任何时候均不能用大写，如pH。以大写缩写字母开头的词，在任何时候均不能用小写形式，如X-ray。

⑤ 化学元素符号无论在什么场合出现，其首字母一律用大写形式。如 H；Cl；$CuSO_4$；Hg^{2+}；*O*，*O*，*S*-trimethyl phosphorodithioate。

⑥ 化合物名称的大小写依其在文题和句首的位置不同而有差别。若在文题中，名词的首字母均需大写；若位于句首，只有第一个词的首字母大写，如 Benzyl Hydroperoxide（文题中），Benzyl hydroperoxide（句首）。

带有表示基团数目的数字词头的化合物名称，在文题中写成 Tetraphenyl Furan 形式，在句首时则写成 Tetraphenyl furan 形式。

带有基团位次数字编号化合物名称，在文题中写成 2-Naphthoyl Bromide 形式，在句首则写成 2-Naphthoyl bromide 形式。

带有希腊字母表示取代基团位置的化合物名称，在文题中写成 *β*-Hydroxy Butyrate Dehydrogenase，在句首时则写成 *β*-Hydroxy butyrate dehydrogenase。

带有表示不同异构体的相关词头，如 *n*-(正)、*iso*-(异)、*tert*-(叔)、*sec*-(仲)、*neo*-(新)、*p*-(对)、*m*-(间)、*o*-(邻)、*cis*-(顺式)、*trans*-(反式)、*R*-(构型)、*S*-(构型)、*E*-(构型)、*Z*-(构型)、*endo*-(内式)、*exo*-(外式) 等，在文题中则写成 *m*-Hydroxybenzyl Alcohol 形式，在句首则写成 *m*-Hydroxybenzyl alcohol 的形式。

高分子化合物的名称在文题或句首时，只第一个字母采用大写形式，如 Poly（ethylene terephthalate）。

⑦ 缩写词的大小写一般按缩略词词典中的规定处理。例如：UV-Vis（Ultraviolet-visible）；DNA（deoxyribonucleic acid）；dis（dissolve）；HPLC（High-pressure liquid chromatography）；ca.（circa）；cf.（compare）；m. p.（melting point）等。

⑧ 来源于人名的计量单位符号的首字母用大写，如 Pa（帕斯卡）、Hz（赫兹）、S（西门子）、A（安培）等。

15 Roman Type and Italic Type of English Letters（西文字母的正体和斜体）

科技英文论文的文字表述中，词汇或字母有正体或斜体两种印刷体例。在论文的总体框架中，国内外许多英文版科技期刊的英文文字大多采用正体形式排版印刷。为了引起读者重视或便于联系和查询，多数期刊将文章作者的工作单位，以及文后参考文献中的期刊名或专著名称排成斜体。举例如下 。

Michael A. Kwofie and William J. Lennarz

作者工作单位：*Department of Biochemistry and Cell Biology and the Institute for Cell and Developmental Biology*，*State University of New York at Stony Brook*，*Stony Brook*，*NY* 11794-5215，*USA*.

参考文献：Yang Z，Wensel T G. *J Biol Chem*，1992，267：24634-24640.

Butler L G. *Yeast and other Inorganic Pyrophosphatase in the Enzymes*. *Vol Ⅳ*. New York：Academic Press，1971：529-549.

在科技英语论文的表述中常涉及许多符号，或用 1 个或用几个字母表示的相关物理意义，这些符号或字母的正斜体的使用是相当严格的，且是不可滥用的。常见的规律如下。

15.1 正体使用的场合

① 物理量的单位符号用正体，如 kg、min、A、mL、mol 等。

② SI倍数词头（multiplying prefixes），如 M（10^6，兆）、μ（10^{-6}，微）、n（10^{-9}，纳）、p（10^{-12}，皮）等。

③ 有固定定义的函数符号。如 cos、sin、tan、cot（三角函数）等；exp（指数函数）等；log、lg、ln（对数函数）等。

④ 有特定意义的缩写字。如 max（最大）、min（最小）、mod（模量）、Re（实部）、Im（虚部）、T（转置符号）等。

⑤ 运算符号。如$\sum$（加和号）、$\prod$（连乘号）、d（微分号）、∂（偏微分号）、Δ（有限增量号）、lim（求极限号）等。

⑥ 值不变的数学常数符号。如自然对数底的符号 e（2.7182818…）；圆周率符号 π（3.1415926…）等。

⑦ 化学元素符号及化学分子式，如 O、S、Hg、Mg、H_3PO_4、C_6H_5OH、NaOH 等。

粒子的符号，如 p（质子）、n（中子）、e（电子）等。

射线的符号，如 X 射线、α射线、β射线、γ射线等。

⑧ 数字词头。如拉丁词头 di、tri、tetra、penta、hexa、hepta、octa、nona、deca 等；希腊词头 bis、tris、tetrakis、pentakis、hexakis、heptakis、octakis、nonakis、decakis 等。

⑨ 酸碱度符号（pH），洛氏硬度符号（HR），布氏硬度符号（HB）等。

⑩ 仪器元件型号或代号。JEOL JNM-EX90A NMR Spectrometer，JASCO FT/IR-200 Spectrophotometer 等。

⑪ 用非物理量意义符号所注的脚标。如 E_{a}（阳极电位）、I_{f}（法拉第电流）、P_{tot}（总压）、M_{r}（相对分子质量）、E_{B}（结合能）等。

⑫ 表示原子轨道和电子组态的符号。如 s、p、d、sp^3、$3d^84s^24p^0$、d_{xy}、d_{xz}、d_{yz}、d_{x2}、d_{y2}、d_{z2}等。

15.2 斜体使用的场合

① 物理量符号。如 I（电流）、c（浓度）、H（焓）、t（时间）、P（压力）、V（体积）等。

② 数学中使用的变量符号（如 x、y、z）；函数［如 $f(x)$］；常数或参数（如 a、b、c）；方程中的未知量［如 $\Delta S=nR\ln(V_f/V_i)$］；表示点、线、面、体的符号。如点 A，线段 BD，ΔABC，平面 ABC；坐标系的符号（如笛卡尔坐标 x，y，z；球坐标 γ，W，Φ）；矩阵符号 Δ，单位矩阵符号 E，矢量符号 a；用未知量构成的序数词（如 The ith，jth 等）。

③ 动植物、微生物种属名称，如 *Equus*（马属）、*Staphylococcus*（葡萄球菌属）等。

④ 用量名称符号所注的下角标（如 C_V，体积定容热容，V 为体积符号），用变动数字符号所注的下角标［如 $m_i(i=1, 2, 3\cdots)$］及用坐标符号所注的下角标（如 v_x，指速度 v 的 x 方向的分量，x 为坐标轴符号）等。

⑤ 表示化合物异构体名称的词头。如 *o*-dibromobenzene、*m*-hydroxybenzyl alcohol、*p*-benzene diacetic acid、*n*-butanol、*sec*-butanol、*tert*-butanol 等。

⑥ 表示化合物构型、构象异构体名称的词头。举例如下。

cis-1,2-dichloroethane，*trans*-1,2-dieuterio cyclobutane，(*E*)-diphenyldiazene，(*Z*)-5-chloro-4-pentenoic acid，(*R*)-2, 3-dihydroxy propanoic acid，(*S*)-1-bromo-3-chlorocyclo hexane，*meso*-tartatic acid，*d*-camphor，*l*-2-amino propanoic acid，*endo*-2-chlorobicyclo [2.2.1] heptane，*exo*-bicyclo [2.2.2] oct-5-en-2-ol，*syn*-7-methylbicyclo [2.2.1] heptane，*anti*-bicyclo [3.2.1] octan-8-amine，*erythro*-β-hydroxy aspartic acid，*threo*-2-amino-1-hydroxy-1-phenylpropane。

⑦ 表示化合物名称中取代基位置的元素符号。如 *N*,*N*′-dimethyl urea；*O*, *O*, *S*-trimethyl phosphorodithioate。

⑧ 表示杂环化合物名称中被氢饱和位置上的 H。如 4*H*-quinolizine。

⑨ 表示化合物命名中取代基位置的希腊字母。如 γ-hydroxy-β-aminobutyric acid。

⑩ 表示并合芳烃中环与环并合位置的符号。如 dibenzo［*c*，*g*］phenanthrene。

⑪ 在高分子化合物命名中表示单体聚合方式的字母或词头。如 poly（styrene-*co*-butadinene），poly（styrene-*g*-acrynitrile），poly［(methylmethacrylate)-*b*-（styrene-*co*-butadiene)］。

⑫ 表示对称群或空间群的符号。如 C_{2v}、Fd_{3m} 等。

16 Usage of Punctuation（科技英语论文中标点符号的用法）

标点符号是科技英语语言表述中的重要组成部分，它不但具有表示停顿、语气和语调的性质和作用的功能，而且还具有辅助修辞和增强语意表达效果的作用。因此科技英语论文的语言表达中应该正确地使用标点符号。英语写作中常用的标点符号见表 2.1。

表 2.1 科技英语表述中的标点符号

符号名称	符号英语名称	符号形式	符号名称	符号英语名称	符号形式
逗号	comma	,	连接号或范围号	en dash	-
分号	semi-colon	;	连字号	hyphen	-
冒号	colon	:	括号	bracket or parenthesis	( )
句号	period or full stop	.	方括号	square bracket	[ ]
问号	question mark	?	引号	quotation mark or inverted comma	" "
感叹号	exclamation mark	!	省略号	ellipsis point dot	…
破折号	dash or em mark dash	—	省字号	apostrophe	'

16.1 逗号的用法

逗号在句子中表示最短的停顿。

① 用于连接几个并列成分、词或短语等，最后两个并列成分之间常用 and 或 or 连接，有时 and 和 or 前也有逗号，举例如下。

In cancer research，the choice of target is often highlighted by the mutated gene underlying the cancer such as ras，P53，RB，P16，myc，and bcr-ab1 shown in Fig. 1.

Many molecular tools are available for target validation，including antisense oligonucleotide，ribozymes，dominant negative mutants，neutralizing antibodies，and mouse transgenics/knockouts.

② 用于并列句中各简单句的连接。例如：

The anticancer activity of nitrogen mustard is due to DNA alkylation，and many other cancer drugs were developed on the basis of this general concept of modification of DNA.

③ 用于作状语的不定式短语、分词短语、独立分词结构以及位于句首的某些状语之后。例如：

To determine how this small molecule promotes activation of caspase-3，we analyzed apoptosome formation by gel-filtration chromatography. Attempting to influence the enantioselectivity of the immobilized EH，the derivatized EH was imprinted with different molecules prior to cross-linking.

All other experimental conditions being equal，we first adjust the pressure meter to zero.

In the absence of antigen，the FV heterodimer of SPE7 crystallizad in two different conformations.

Alternatively，a normal gene product may be closely，correlated with cancer progression.

④ 用于非限制性定语之前。例如：

The PHP Ia and PPHP forms could be distinguished by PTH-stimulation testing，which was not possible here since the supply of synthetic PTH for human injection was discontinued.

Cell extracts from human cancer lines，including colon cancer，prostate cancer，promyelocytic leukemia，T cell leukemia，bone marrow leukemia，malignant melanoma，lymphoma，and glioblastoma celis，were responsive to PETCM.

⑤ 用于位于主句前的状语从句之后。例如：

Although doxorubicin and the taxanes were discovered on the basis of their toxicity to tumor cells in preclinical models，subsequent studies identified the biochemical mechanisms of action.

Since the nature of the solvent usually influences the enantioselectivity of enzymatic reactions，several solvents were tested in the lipase PS-catalyzed butyrylation of *N*-hydroxymethylated *β*-lactams.

⑥ 用于插入语的前后，例如：

However，the traditional model partially fails when applied to the comparison between amphetamine and 4-phenylbutan-2-amine.

In the United States，cancer is the second leading cause of death，in fact，behind cardiovascular disease.

⑦ 用于同位语的前后，例如：

The main curative therapies for cancer，surgery and radiation，are generally only successful if the cancer is found at an early localized stage.

⑧ 用于日期和地址的书写，例如：

On March 15，2006，an important event took place.

Genome sequencing center，School of Medicine Washington University，St Louis，Missouri 63108，USA.

⑨ 用于正文中不连续文献序号的分隔。例如：

Experimental investigation[10,13,16,18-25] concerned the relative importance of field and electronegativity effects.

⑩ 用于化合物名称中表示取代基位置编号间、元素符号间或希腊字母间的分隔。例如：

d-1，2，4-butanetriol；*N*，*N′*-dimethylurea，*ω*，*ω′*-dibromopolybutadiene；2，4-dichlorocyclohexane cyclohexane propionic acid 等。

⑪ 用于直接引语的引出。例如：

Richard Phillips Feynman said，“Everything is made of atoms That is the key hypothesis.”

16.2 分号的用法

分号主要用于比逗号相对较长的停顿。

① 用于两个意义上有联系且非独立的并列句子之间相对较长的停顿。例如：

All solvents were distilled from an appropriate drying agent；tetrahydrofuran and diethyl ether were also pretreated with activity alumina.

② 用于两个意义上有联系且通过 that is，however，so，then，hence，thus，therefore，nevertheless 等所连接的具有过渡性质的句子之间的相对较长的停顿。例如：

The proposed intermediate is nor easily accessible；therefore，the final product is observed initially.

③ 用于已有逗号分隔的，相互相对独立的有关情况的分离。例如：

Abbreviations：DTT，dithiothreitol；EDTA，ethylenediamine tetraacetate；BSA，bovine serum albumin；AOT，aerosol dioctyl sodium sulfosuccinate；TNBS，2，4，6-trinitrobenzene sulfonic acid.

The compounds studied were methyl ethyl ketone；sodium benzoate；and acetic，benzoic，and cinnamic acids.

16.3 冒号的用法

在句子中表示一个比分号更长的停顿。

（1）用于列举事物

Distillation temperatures and yields of the amines are as follows：45℃，80%；68℃，75%；72℃，72%；60℃，79% and 50℃ 74%.

（2）用于引起下文

The electron density was studied for the ground state of three groups of molecules：①methane-methanol-carbon dioxide；②water-hydrogen peroxide；and ③ferrous oxide-ferric oxide.

We now report a preliminary finding in the NMR spectrum：no chemical shifts changes were detected in the concentration range of 0.1-1.0 mol/L.

The following are our conclusions：The enantioselectivities of these processes allowed the isolation of remaining amines in very high yields and enantiomeric excesses. In all cases，amides were abtained with＞98% *e. e.* after a simple recrystallization of the enzymatic product. The enatioselectivity and reaction rates observed for some of these amines can be explained on the basis of the active site model proposed for CAL-B.

（3）在主副文题中用于副文题的引出

Adsorption of Polyampholyte Copolymers on the Solid/Liquid Interface：The influence of pH and Salt on the Adsorption Behaviour.

16.4 句号的用法

句号主要用于表明一个句子的结束。

① 用于陈述句或祈使句结束的停顿。例如：

These chromosomes are characterized by a heterochromatic short arm that contains essentially ribosomal RNA genes and a euchromatic long arm in which most，if not all of the protein-coding genes are located.

② 用于某些缩写字之后。如 ca.（circa），cf.（compare），e. g.（for example），i. e.（that is），i. d.（inside diameter），et al.（and others），etc.（and so forth），anal.（analysis），vs.（versus）等。

③ 用于桥环和螺环化合物的命名，如 bicyclo［3.2.1］heptane，spiro［4.5］decane。

16.5 问号的用法

用于疑问句的末尾，表示一个问句结束时的停顿。例如：What is the leading disease-related cause of death of the human population?

16.6 感叹号的用法

用于一个感叹句结束时的停顿。例如：How serious the problems of environmental pollution and decreasing resources are!

16.7 连接号或范围号的用法

科技英语论文中使用的连接号或范围号（en dash）是一条短线（—），其长度是连字号(-)的两倍。

① 用于构成两个等价词的复合词，表示"and"或"to"或"versus"的概念，如 structure—activity relationship，*cis*—*trans* isomerization，oxidation—reduction potential，carbon—oxygen bond，vapor—liquid equilibium，ethanol—ether mixture；temperature—time curve，producer—user communication。

② 用于连接作定语的两个人名，如 Fridel—Crafts reaction，John—Teler effect，Debye—Huckel theory，Beer Lambert law 等。

③ 用于混合溶剂中各组分的连接，如 The melting point was unchanged after two crys—tallizations from hexane-benzyne。

④ 用于表示数字或组成的起止范围，如 Figure1—5，25—50mg，Changchun—Beijing。

16.8 破折号的用法

破折号是一条长线（—），其长度是连接号的 2 倍，是连字号的 4 倍。主要用于引出同位语。例如：

All three exprimental parameters—temperature，time，and concentration，were strictly followed.

Linear relationships between the first deactivation rate constant and an inverse of temperature were found for all lipase—Fig. 3.

16.9 连字号的用法

连字号（hyphen）是最短的线，其长度是 en dash 的 1/2。

① 用于构成复合词，如 water-insoluble，free-radical，iron-exchange，three-phase，three-dimensional，six-membered，light-catalyzed，long-lived，fluorescence-quenching，ever-present，well-known，high frequency，second-order 等。

② 专有名词加词头（prefixes）或加词尾（suffixes）有些字要用连字号连接，如 non-Guassian，non-Newtonian，Kennedy-like 等。

③ 由词头加词根派生出来的新词，一般词头与词根之间不加连字号，在科技英语中常见的这样的词头如下：after，anti，auto，bi，bio，co，counter，de，di，down，electro，extra，hyper，hypo，infra，iso，metallo，mid，macro，micro，mini，mono，multi， non，over，photo，physico，poly，post，pre，pro，pseudo，re，semi，sub，stereo，super，supra，trans，tri，ultra，un，under，up，visco 等。但个别情况下，也有加连字号的，如 anti-infective，co-worker，un-ionize，co-ion 等。如果词头加到由两个字构成的复合词之前，需加连字号，如 non-radiation-caused effects，pseudo-first-order reaction，pre-steady-state condition 等。

④ 当量值和计量单位构成复合词作定语时，它们之间用连字号连接，例如：a 20-mL solution，a 12-min exposure，a 1－2-h sampling time，10-，20-，and 30-mL，4-mm-hick layer 等。但当计量单位为℃或复合单位时，则不加连字号，如 a 50 ℃ water bath，a 0.5 mol/L H_2SO_4 solution。

⑤ 用于化合物的名称中表示基团或重键位置的数字序号、希腊字母或元素符号，表示构型、构象或异构体的词头之后或前后，如 5-bromo-2-ethoxyacetanilide；2,3,5-triaziridinyl-*p*-bezoquinone；γ-hydroxy-β-aminobutyric acid；*N*,*N*-dimethyltryptamine；*cis*-1，3-butadiene；*p*-*ter*-butylphenol；sym-dibromoethane；*syn*-7-methylbicyclo ［2.2.1］ heptane；*anti*-bicyclo ［3.2.1］ octam-8-amine；（1*Z*，4*E*）-1,2,4,5-tetrachloro-1,4-pentadiene；（1*R*，3*S*）-bromo-3-chloro cyclohexane；*dl*-2-amino propanoic acid；*endo*-2-chlorobicyclo ［2.2.1］ hepatane；*exo*-5，6-dimethyl pentanoic acid。

16.10 圆括号和方括号的用法

（1）圆括号的用法

① 用于附加说明的部分。举例如下。

The total amount（20 mg）was recovered by modifying the procedure.

Curve a（Fig. 2）obeys the Beer-Lambert law.

② 用于枚举内容中各部分的序号。举例如下。

Three applications of this reaction are possible：（1） isomerization of sterically hindered aryl radicals，（2） enol-keto transformation，and （3） sigmatropic hydrogen shift.

③ 用于表示元素的氧化态数，如 iron（Ⅲ）或 Fe（Ⅲ）。

④ 用于表示试剂或仪器的商标型号和厂家及缩写字等。如（±）-1，2-epoxyoctane（Aldrich Chemical Co.），^{1}H NMR Spectrometer（JEOL JNMEX 90A），tetramethylsilane（TMS）。

⑤ 用于化合物的命名中，例如：2-(*O*-chlorophenyl)-1-naphthol，（1*R*，3*S*）-1-bromo-3-chlorocyclohexanol，3-bis（diethylamino）propane，poly（styrene-*co*-butadiene），Poly（sterene-*g*-acrylonitrile）。

（2）方括号的用法

① 用于表示并合多环芳烃名称中的并合位置，例如：dibenzo ［*c*，*g*］ phenanthrene，indeno ［1,2-*a*］ indene 等。用于表示桥环化合物名称中各桥上的原子数目，如 bicyclo ［3.2.0］ heptane。

② 用于表示螺环化合物名称中各环上的原子数目，如 Spiro ［4.5］ decane。

③ 用于表示化合物名称中同位素标记的原子，如 ［^{15}N］ alanine，［2-^{14}C］ leucine。

④ 用于表示化合物或离子的浓度，如 ［H_2SO_4］，［Ca^{2+}］。

⑤ 用于表示行文中参考文献引出的序号。举例如下。

Often multiple approaches must be evaluated，for example，the use of several of these tools has led to the recognition that telomerase enzyme [9,10,12] and the KDR receptor of VEGF [11,13] are good targets for drug development.

16.11 引号的用法

科技英文表述中的引号有双引号和单引号之分。

（1）双引号的使用

① 行文中一个新词或一个赋予新意的词第一次出现时，需用双引号。举例如下。

The so-called “electron-deficient” cations are well-established intermediates.

The integrated intensity of each diagnol in the spectrum is proportional to a “mixing coefficient”.

② 用于引出书名。举例如下。

Many immobilized lipases and their performances are introduced in a monograph called “Kinetics and mechanism of reactions catalyzed by immobilized lipases”.

③ 用于引证别人所说的话。举例如下。

“Hydrolytic enzymes such as lipases and esterases are frequently used because they accept a broad range of substrates and often exhibit a high enantioselectivity”, researchers can now choose from a large number of commercial lipases and esterases.

④ 用于引出直接引语，举例如下。

Robert Burgess from University of California said，“In the United States，Cancer is the second leading cause of death behind cardiovascular disease，and it is projected that cancer will become the leading cause of death within a few years”.

（2）单引号的使用

在双引号引用的内容中，若还有需要加引号引用的内容时，则加单引号。例如：
L. Ghosez said，“pyrazolidinone-containing compounds have been proven to bind to bacterial PBPs and to have been useful in vitro and in vivo antibacterial properties in ‘Recent Progress in the Chemical Synthesis of Antibiotics’，edited by G. Luka”.

16.12 省略号的用法

用于表示被省略的部分。举例如下。

Existing chemotherapeutic treatments are largely palliative in these advanced tumors, particularly in the case of the common epithelial cancers of lung，breast…

16.13 省字号的用法

① 用于表示名词的所有格或字母的省略，例如：

Van Deemster’s equation（范·德姆特方程）

It’s clear that new therapeutic options are necessary.

② 用于表示年份数字中前两个数字的省略。例如：

’99 International Symposium on the Separation of Proteins，Peptides and Polynucleotides（ISSPPP ’99）.

17 Submitting the Manuscripts and Publication（投稿与发表）

科研项目完成之后，总结的最好形式是撰写论文，交流的主要形式是在科技期刊上发表论文。因此作者必须了解科技期刊，了解如何投稿，了解论文的发表过程。

17.1 选择期刊

选择在哪种期刊上发表论文和如何投稿对于论文的公开发表是至关重要的。

通常，作者在撰写论文之前，首先应确定投哪一种刊物。投稿的原则是要发表的论文内容必须与刊物报道的学科范围相符合。如果一篇论文投到与其报道内容范围不相符的刊物，该论文就有可能未经同行专家审查就被退稿，即使呈送专家审查，也有可能由于该刊审稿专家不熟悉论文的专业内容而做出较差的甚至是不公正的评价，致使论文遭到拒绝；也有可能该论文通过专家的评审以致被刊物接受发表，但由于该刊物报道学科领域的读者对论文的内容很不熟悉，以致论文虽已公开发表，但却无人问津。科技期刊的专业性很强，甚至资格老的刊物、知名刊物的报道内容也随科学的发展在不断地变化，因此发表论文时要根据刊名、刊物的出版者、刊物的征稿简则以及刊物最近各期上报道的目录来选择。

当有几种刊物的报道内容与论文研究领域相符合时，一是要弄清该刊接受哪种类型的稿件，是论文还是快报，是简报还是综合评述，如果报道的文章类型不符，也会被退稿。二是要弄清该刊只是接受特约稿，还是接受自投稿，如果属于前者，后者很可能被退稿。三是要弄清该刊物的水平，即要投稿的论文水平应与刊物的级别相符。如果论文的水平低于刊物发表文章的整体水平，也会被退稿；如果论文的水平高于刊物的水平，即使被发表，也会影响交流的效果和产生的影响力。四是要弄清刊物的发表周期（即从投稿到出版之间的时间间隔），通常同一学科领域的刊物应首选双周刊（Biweekly）、半月刊（Semimonthly）、月刊（Monthly），因为这样的刊物出版周期短；其次为双月刊（Bimonthly）；最后为季刊（Quarterly）。如果出版周期太长，论文虽被接收发表，但却可能丧失领先权。五是要通过美国科技信息研究所（ISI）的期刊引证报告（JCR）来了解各个刊物的总被引频次（total citation）和影响因子（impact factor），通常引文频率相对较高、影响因子相对较大的刊物应作为优先的投稿刊物。六是要考虑刊物的知名度。通常学术团体如学会主办的刊物知名度较高，读者群体大，发表的文章易在同行中产生影响，故学术团体主办的刊物应是投稿优先的刊物。七是要考虑刊物的容量，通常刊物的容量大，发表文章的数量多，录用的概率亦相对较大。

17.2 阅读征稿简则

在选择刊物的过程中，应仔细阅读拟投稿刊物的征稿简则。因为征稿简则中不但介绍了刊物的特征，内容和文章的主要栏目，而且还详细提出了撰稿和投稿的具体要求。

每种英文科技刊物都附有自己的征稿简则，其英文表述常见的有：Instruction to Authors，Instruction to Contributors，Guide for Authors，Information for Contributors，Instruction for Authors，Notice to Authors of papers，Instructions for Preparation of Manuscripts for Direct Reproduction，Policies and Procedures 等。

17.3 打印稿的制作

论文打印稿打印质量也是其能否被投稿刊物接受发表的关键因素之一。绝大多数刊物编辑出版部门对投来的稿件都是先检查稿件打字排版质量。即使论文水平很高，创新性很强，但由于稿件打字排版质量太差，非但稿件不能被接受发表，而且编辑甚至可能直接将稿件退回。

在论文打字之前，应首先仔细阅读拟投稿刊物的征稿简则或作者须知和该刊物最新一期编辑排版的形式，尤其要注意该刊的编辑要求，如参考文献的著录体系（著者-出版年体系或顺序编码体系）、各级标题字号、黑体白体及位置、摘要的字号及位置、表格和插图的设计、字母的大小写、正斜体及上下角标等。总而言之，投稿时所准备的打字稿务必字迹清晰，无错误，编辑风格与所投刊物的要求是一致的。

论文打印稿需要编页号。通常，大多数刊物要求将论文文题、作者姓名和通信地址编为首页，摘要编为第二页，引言及后面各章节，如材料和方法、结果与讨论和结论等从第三页依次连续排版打印及编号。论文中所有的插图和表格也需要单独绘制，置于论文正文之后并依次编序号。图表说明（legend）（包括图题、表题、图注、表注）需要单独打印，置于插图和表格之前。论文打印完成之后，作者本人务必反复阅读校对，以纠正打字中出现的错误。如果可能的话，还应请业务水平较高和英语功底较深的同行审读，以最后避免专业、文字和语法方面的错误。

17.4 投 稿

向某一刊物投稿时，通常需要一式三份（其中包括图表及照片），一份为原件，两份为复印件。

投稿时，论文联系人应随稿件附一封致刊物主编或编辑部的信件，信中应介绍论文的题目、作者、联系人及其通讯地址，也可以扼要介绍一下文章的一般情况以及文章的总页数等。举例如下。

Dear Editor _____，

Enclosed are three complete copies of a manuscript written by Lu Haun，Liu Xiang and Zhao Dong-Fan titled：“Biosynthesis of Analogues of the Marine Antitumor Agent Curacin”，which is being submitted for possible publication in the Journal of Biological Chemistry.

This manuscript is new and not being considered elsewhere，correspondence regarding this manuscript should be sent to me at the address shown in the above letter head.

Sincerely

Zhao Dong-Fan

Dear Editor_____，

I am enclosing herewith three copies，one original and two xerox copies of the manuscript entitled “Determination of E3 in serum by EIA with Biotin-avidin Amplification system” by Chang Wei-Bao，Mei Xiao-Xiu，Wang Yong-Chen，and myself. The manuscript consists of 16 pages of text，three tables，two pages of legends to figures，and nine photocopies of figures.

We would be grateful if the manuscript could be reviewed and considered for publication in Journal of Biochemical Society.

Very sincerely yours

Li De-hua

Professor

Enclosed：Three manuscript copies

通常刊物的编辑出版部门收到稿件后会给作者一份收稿回执（acknowledgment of receipt）。回执中除告知收到了稿件外，还告诉稿件的编号，作者在以后与编辑联系有关稿件处理情况时，一定要使用此编号，以便于查询。举例如下。

Re：No. B2002382

Title：“Biosynthesis of Analogues of the Marine Antitumour Agent Curacin A”

Dear Prof. Zhao，

This is to acknowledgement receipt of the above manuscript. As soon as the refereeing process is completed, I will be in touch with you concerning its disposition. In the meantime, it will be helpful if you can provide your telephone number, a Fax number and E-mail address if available. Please refer to the above manuscript number when replying.

Sincerely

John F. Smith

Acting Editor

如果你在投出稿件2～3个星期后没接到编辑部门的收稿回执，作者则需要给编辑部门写信或打电话，确认稿件是否收到，若未收到，尚需补寄。

期刊的编辑部门在审阅稿件之后，首先要确认稿件内容是否符合该刊的办刊宗旨，其次要检查稿件的份数、稿件的制作是否符合要求，再次是要检查每份稿件页数、图表是否齐全，上述情况有一处不妥，通常都要被立即直接退稿。

对于满足上述先决条件的稿件，编辑部通常请两位同行专家评审，如果两位专家的审查意见大相径庭，尚需请第三位专家评审，编辑部根据评审结果，将对该稿件的取舍做出判断。通常，对那些内容无创新之处，或文内阐述的实验结果和得出的结论有重大错误的稿件予以退稿；对那些通过专家审查无需再作任何修改的稿件予以接受发表；对那些审稿专家提出若干问题的稿件，则需作者回答、解决和修改，并在编辑部规定的时间内将修改稿和修改说明寄回编辑部。稿件能否被录用，取决于修改后的稿件是否达到评审专家或编辑部的要求，有时尚需再次或多次修改后方能确定。

作为作者，要密切与编辑部配合。要认识到编辑部工作的宗旨与作者本身的工作宗旨是一致的，都想使水平高、创新性强、语言表述好的稿件公之于世。所以作者要把编辑转述的意见作为千金难买的东西，对于那些正确的并可接受的意见，要予以认真研究和逐一解决。如果作者认为评审意见是错误的，则可以有理有据地向编辑部提出申述。通常编辑部退稿有三种情况：一是完全拒绝（total rejection），二是稿件中存在某些错误，三是稿件尚有某些缺欠。若属于第一种情况，作者首先应考虑自己的稿件质量是否太差，如果太差，则需进一步深入工作，得到新的结果后再投稿；如果不是太差，可考虑改投其他刊物。若属于第二种情况，作者应该依据编辑的退稿理由和评审人提出的问题逐一解决，存在的问题解决后可再投同一刊物，也可以投往其他同类刊物。若属于第三种情况，作者应对原稿件作进一步补充和完善，然后再投至该刊编辑部。下面给出编辑和作者之间有关稿件修改方面的来往信函的实例。

（1）编辑致作者函

Dear Dr. ______,

On behalf of the Editorial Board, we thank you for submitting the paper entitled "Action of snake venom hemorrhagic principles on isolated glomerular basement membrane".

We are glad to be able to inform you that this paper should be accepted after condensation to no more than 60% of its present total length. In particular, the introduction should be written more concisely and the sections entitled Results and Discussion shortened and combined. You may also consider reducing the number of figures. In addition, one of the reviewers made the following comment: "It is important that the author characterize the material released by the hemorrhagic principles from basement membrane by means of amino acid and carbohydrate analyses. The nature of the peptides released would indicate the specificity of these principles". If this information is available, it should be included in the article.

The top copy will be returned under separate cover. If duplicate copies of the shortened version reach us by August 1 this year, the article will be published with the original receiving date.

Sincerely yours,

Editorial secretariat

（2）作者致编辑的答复函

Dear Editor,

Thank you for your kind letter of July 11, 2004. We were happy to hear that our paper entitled "Action snake venom hemorrhagic principles on isolated glomerular basemernt membrane" would be acceptable.

We tried to condense and revise the manuscript as much as possible in line with the suggestions made by the reviewers and yourself. I am herewith enclosing duplicate copies of the revised version.

According to your suggestion, we rewrote the INTRODUCTION more concisely, succeeding in condensing it to 60% of its original length. The two sections, RESULTS and DISCUSSION, were shortened and combined, as you requested, to form a new section, RESULTS AND DISCUSSION, where the general discussion was given the additional title of enzymatic nature of the action of venom hemorrhagic principles. Through these revisions we were able to shorten the manuscript by three pages in total.

No Figures were omitted from the manuscript since we think all of them are essential for this paper. We corrected some errors found in our original manuscript (page 18, Table 2→Table 1).

In the revised manuscript we cited one wore paper at the end of INTRODUCTION. This is the proceedings of the Darmstadt Meeting, where the present work was orally reported.

Concerning the characterization of the material released by the hemorrhagic principles form basement membrane, we realize it is important. At present, however, we don't have enough material to characterize it by means of amino acid and carbohydrate analyses. We appreciate the comment made by one of the reviewers on this point.

I hope all these corrections and revisions will be satisfactory. Although we have not succeeded in shortening the manuscript to 60% of original total length as you requested, I hope the revised version will be acceptable for publication in Biochemical Biophysical Acta. I also hope the revise manuscript will reach you before August 1, 2004.

Sincerely yours,

Prof. Wang Xingguang

March 10, 2004

（3）编辑和作者之间就一篇论文修改问题的四次往返信函

Dear Prof. Li,

Your manuscript has been reviewed by the Editorial Board and I am enclosing the comments of the referees.

Please consider these suggestions in revision of your contribution. Other instructions are enclosed herewith.

Sincerely yours,

Editor

May 8, 2004

Dear Editor,

Thank you very much for your letter of March 8, 2004, with regard to our manuscript together with the comments from the two reviewers. I am sending herewith two copies of our revised manuscript (pages 1-8 have been retyped. The revised manuscript includes another page).

Our incorporation of the reviewers, suggestions are as follows:

Reviewer A

① P2, lines 6-8. The sentence has been changed as suggested.

② P2, line 2 from bottom, "As described before (ref. 6)" has been added.

Reviewer B (major comments)

① P4, line 2, "At a mass concentration of 0.02 mg/mL" has been added.

② Table 2. A footnote "with derivative toxin—sensitized SRBC" has been added.

③ P6, line 6. "with SRbc sensitized with the derivative toxin" has been added.

④ P4, line 4. A sentence of explanation has been added.

⑤ "Fused" means partially identical and "coalesced" means antigenically identical.

⑥ P5, line 12. "Even after removal of DTT and urea by dialysis" has been inserted.

⑦ P8, line 6. from bottom. Two sentences have been added.

⑧ P4, line 14. Method for conventional disc electrophoresis has been inserted.

⑨ P8, line 5 from bottom. The sentence has been changed as suggested.

⑩ P5, line 7 from bottom "with anti-okra derivative toxin" has been added.

I believe that the manuscript has been improved satisfactorily and hope it will be accepted for publication in Infection and Immunity.

Sincerely yours,

Li Donghua

Professor

June 11, 2004

Dear Prof. Li,

Your revised manuscript was reviewed. Unfortunately, as you may note from the enclosed comments, one of the reviewers still has certain reservations. Please consider these reports. If they are valid, further modification will be required. Other instructions are enclosed.

Sincerely yours,

Editor

July 5, 2004

Dear Editor,

I have received your letter of June 11, our revised manuscript, and comments from one of the reviewers. We appreciate the additional comment and are sorry for causing your trouble.

I am enclosing herewith two copies of our further revised manuscript. Our alterations as a result of the reviewer's comments are as follows:

1. Page 9, Lines 1-7. These paragraphs were unclear and have been revised.

2. Fig. 6A. The photograph has been replaced with a new one. and the legend to Fig. 6A has been changed.

The line in front of wells 3 and 6 in the previous photograph was not a precipitin line but apparently an artifact caused by staining the precipitin plate.

3. Page 4, lines 1-2. Sensitization of SEBC with the three antigens at the same mass concentration 0.02 mg/mL was near the optimum, as half that concentration of the antigen gave less sensitivity and double the concentration gave no higher sensitivity with any of the antigens.

Table 2 is to show that anti-fragment Ⅰ has a higher neutralizing activity than anti-fragment Ⅱ and the same neutralizing activity as anti-derivative toxin on the same PHA titer basis. Even when all SRBC were not sensitized optimally, the relationship between the neutralizing activity and the PHA titer was the same in repeated experiments with the three antibodies.

The derivative toxin molecule seems to be made up of one molecule, each of fragments Ⅰ and Ⅱ, therefore, SRBC may be sensitized to the same degree with fragments Ⅰ and Ⅱ when the antigen is derivative toxin.

4. In the slide (5 cm×5 cm) Quchterlony test, a well usually holds about 0.03mL, but the volume of antigen or antibody is not always measured. So we can give only the concentration. We have often seen things expressed in this way in published reports.

5. Table 2 and Fig. 5, The IgG fraction was obtained from each antiserum by the method stated on page 4. The precipitated fraction was dissolved at an appropriate concentration, dialyzed, and the pHA titers of the serum and the IgG changed to be the same in all the three instances.

6. The purities of fragments Ⅰ and Ⅱ were examined by①SDS gel electrophoresis and ②agar gel diffusion tests.

The Semi-micro Quchterlony test on a plate of 5 cm×5 cm is better for detection of contaiminants than one might think.

Because even if one well holds only 0.03mL of antigen or antibody, the distance between every two wells is only 7mm, and you can add antigen or antibody to the wells twice, three times, or ever more of ten.

From the results of ① and ②, we think that neither fragment Ⅰ nor Ⅱ contained the counterpart fragment. The two tines between fragment Ⅰ and anti-fragment Ⅰ were formed only when QC derivative toxin was adjacent to fragment Ⅰ and, therefore, were interpreted as indicating that there were two antigens contained in it rather than a contaminant as stated on page 9.

I hope the 2nd revised manuscript will be acceptable for publication in Infection and Immunity.

Sincerely yours,
Li Donghua
Professor

August 1, 2004

Dear Prof. Li,

As you may note from the enclosed copy of the referees comments, further revision of your manuscript seems to be in order. Please consider these suggestions carefully in the further modification of your contribution. Other constructions are enclosed.

Sincerely yours,
Editor

August 25, 2004

Dear Editor,

I am sending you herewith two copies of the their revision of the manuscript of our paper. All the manuscript including legends to figures, except for Tables 1, 2 and 3, has been retyped and photos of Figs. 7A and 7B have been replaced with new ones. You will find in addition a new title given to the paper, "Development of antitoxin with each of two complementary fragments of clostridium botulinum type B derivative toxin". We want to report only the fact that either one of the two immunologically distinct nontoxic fragments of C. botulinum type B derivative and progenitor toxin.

I think the criticisms and question from one of the reviewers and our answers to these may have been diverting us a bit from the main subject of the paper.

We are responsible for these digressions, perhaps the title did not adequately describe our intention. We discussed too many of the implications that could be drawn from the results, and our English was not very functional. In the revised manuscript, we have deleted the points the reviewer questioned or disagreed with. Our reactions to each of the reviewer's comments are as follows:

① Item 3. 1st p. Further work seems necessary to conclude that several antigens at 0.02 mg/mL give optimal sensitization of sheep red blood cells.

We do not claim to have proved this point. We only was to state that this method gave the present results.

② We do not think we are entitled to debate the reviewer's theory, as we have no basis to think that one fragment has more free amine groups than the other, or one fragment is more accessible to the antibody than the other.

We titrated the immune sera by passive hemagglutination simply to check that each fragment had stimulated the rabbit to produce antibody, in case no antitoxic activity was detected.

③ P6 and P9 were modified according to the reviewer's suggestions.

④ Items 1 and 6. To avoid confusion the photos of Fig. 7 and 8 have been replaced with new ones.

We hope the third revised manuscript will be accepted for publication in Infection and Immunity. We apologize to you and the reviewer for causing so much trouble with our manuscript.

Best regards,

Sincerely yours,

Li Donghua

Professor

September 12, 2004

Dear prof. Li,

It affords me pleasure to inform you that your manuscript will be accepted for publication.

Before going to press, however, please consider the enclosed suggestions of the reviewer. Upon receipt of your contribution, the paper will go to press.

Sincerely yours,

Editor

September 25, 2004

Dear Editor,

Thank you very much for your encouraging letter of September 12, 2004, together with our 3rd revised manuscript and the comments from the reviewer. The 4th revised and, I hope, the final manuscript is enclosed.

You will find that we have corrected the manuscript taking in all the reviewer's suggestions.

We hope that the paper will go directly to press. We again apologize to you and the reviewers for all the trouble we have caused with this manuscript and our poor English.

Best regards,

Sincerely yours,

Li Donghua

Professor

17.5 校样与校对

通常科技期刊编辑部或出版社在论文定稿之后，需将作者文稿按照编辑出版的标准和规范重新录字和排版后，打印出版样。在正式上机印刷之前需要将版样中出现的所有错误全部找出来并予以纠正。这种版样称为校样。找出和纠正版样中出现的所有错误的过程叫做校对。

科技期刊是科学研究成果信息的重要载体，是将科学技术转化为生产力的主要传媒，是交流并启发人们不断创造新知识的源泉，是记录科技进步和发展的文献宝库。因此，科技论文在通过科技期刊正式出版发表之前，务需保证准确无误。而要做到消除版样中存在的所有的差错，一般是通过校对过程来实现的。

校对工作不但出版单位的编校人员要做，而且论文的作者也要做。尤其是论文作者对自己所撰写论文的内容很熟悉，经作者校对后的校样，即可付印。从某种意义上说，作者将自己的论文的校样校对完了之后，才算完成了此篇论文的全部工作。因此，当作者接到期刊编辑出版部门发来的校样时，一定要认真仔细地校对，任何微小的错误也不要放过。

作者校对大多采用对校的方法。所谓对校，就是将论文的原稿置于左侧，将校样置于右侧，先看原稿，再看校样，逐字逐句地校对下去。

校对过程中需要注意如下问题。

① 进一步核查论文的科学内容是否有误，如果尚有不妥之处，务必加以改正，但对校样不能进行改写和重写。因为印刷版样的版式已经确定，内容大幅度变动会导致页码和版式的变化，将给改版带来麻烦，甚至会带来新的差错。因此可改可不改之处，尽量不改。如果必须做较大改动时，需征得编辑部同意。

② 如果一篇论文有多个作者时，每个作者都应参加校对；若论文只有一个作者，可请同行帮助校对，这样可以避免一个人校对时可能会因习惯性而出现疏漏。

③ 要采用国际通用的校对符号纠正版样中的错误。除在错误处做出标记外，还要在距错误出现处较近的空白处做出标记或文字说明。

④ 校对时还需检查英语文法和修饰，错字、漏字和颠倒等差错。

⑤ 要特别注意表格中的数据和插图是否有误。

17.6 国际通用校对符号

国际通用校对符号（correction marks in common use）和页边标记或说明列入表 2.2 中。表 2.2 中所列符号是根据美国国家标准研究所（ANSI，1991）和英国标准研究所（BSI，1976）的规范整理的。如果两种标准所用符号不一致，则两种符号都给出，中间用 or 连接。

表 2.2 通用校对符号及其用法一览

American system		Alternative system		Explanation
Error with correction mark	Marginal correction	Error with correction mark	Marginal correction	
we have already Smith and Miller	tr. tr. [Miller and Smith]	we have already Smith and Miller	⊓ ⊓⊔⊓	*Transpose words*
results (1985) by Snyder	tr.	results (1985) by Snyder	○↷	*Rearrange words*
I have so far been only qualitatively evaluated	so far have been evaluated only qualitatively	…have so far been only qualitatively evaluated (3 1 2 4 6 7 5)	1–7	
Fisher (1950). The	insert from page X	Fisher (1950). The	((insert from page X))	*Insert phrase (or paragraph)*
sponed	×	sponed	○	*Reset damaged character(s)*
$W_ood^War_d$	=	$W_ood^War_d$	=	*Adjust line*
but never 70kg	# #	but never 70kg	Y ʅ	*Insert space (between words)*
proof reading	⁀‿	proof reading	⁀‿	*Delete space (between words)*

续表

American system		Altemative system		Explanation
Error with correction mark	Marginal correction	Error with correction mark	Marginal correction	
was ∨ crystallized	eq. #	was ↑ crystallized	↑	*Reduce space (between words)*
…the proton chemical shift < …correlations proved useful	ld.	…the proton chemical shift …correlations proved useful		*Insert space (between lines)*
…to lowering the …and solubility of	9 ld.	…to lowering the …and solubility of		*Reduce space (between lines)*
a) by automatic titration	fl. r.	a) by automatic titration		*Move to the right (indent)*
as that of glucose	fl. l.	as that of glueose		*Move to left (cancel indent)*
is a point of view of economy. The alkali sulfide	¶	is a point of view of economy. The alkalt sulfide		*Start new paragraph*
…increased. This result is…	no ¶	increased. This result is		*Run on (no new paragraph)*
disrotatory	stet.	disrotatory		*Leave as it stands*
the trans isomer	ital.	the trans isomer	((italics))	*Set in (or change to) italic type*

续表

American system		Alternative system		Explanation
Error with correction mark	Marginal correction	Error with correction mark	Marginal correction	
compound 13 a has	bf	compound 13 a has	((bold))	*Set in (or change to) boldface type*
dacron	Cap.	dacron	D	*Set in upper case (capitals). capitalize*
following Chapter	lc	following Chapter	c	*Set in lower case*
[( )] or			(( ))	*Follow instruction (as specified)*
				Re-set as indicated here

参考文献

[1] Janef S Dodd，Editor. The ACS Style Guide：A Manual for Authors and Editors. Washington，DC：ACS，1986.

[2] Robert A Day. How to Write and Publish a Scientific Paper. New York：Oryx press，1988.

[3] 谭炳煜，编著．怎样撰写科学论文．沈阳：辽宁人民出版社，1982.

[4] 任胜利，编著．英语科技论文撰写与投稿．北京：科学出版社，2004.

[5] 姚远，郑进保，张惠民，汪季贤．科技学术期刊撰稿指南．北京：光明日报出版社，1989.

[6] 陈浩元，主编．科技书刊标准化 18 讲．北京：北京师范大学出版社．

[7] 王立名，主编．科学技术期刊编辑教程．北京：人民军医出版社，1995.

[8] 高锦章，著．化学英语写作．兰州：甘肃教育出版社，1996.

[9] Marx V. Next-generation sequencing：The genome jigsaw. Nature. 2013；501（7466）：263-268.

[10] Cirillo E，Parnell LD，Evelo CT. A Review of Pathway-Based Analysis Tools That Visualize Genetic Variants. Front Genet. 2017；8：174.

[11] Feng Zhang，Yan Wen and Xiong Guo. Human Molecular Genetics，2014，23（R1）：R40-R46.

[12] Rivera RM，Bennett LB. Epigenetics in humans：an overview. Curr Opin Endocrinol Diabetes Obes. 2010；17（6）：493-499.

[13] Scott F. Gilbert and Michael J. F. Barresi. Developmental Biology，11th Edition，2016，Sinauer Associates，Inc.

[14] Serrano-Pozo A，Frosch MP，Masliah E，Hyman BT. Neuropathological alterations in Alzheimer disease. Cold Spring Harb Perspect Med. 2011；1（1）：a006189.